Mitsubishi FWD Models Automotive Repair Manual

by Larry Warren and John H Haynes
Member of the Guild of Motoring Writers

Models covered:
Mitsubishi Galant, Cordia, Tredia, Precis and Mirage
1983 through 1993
Does not cover V6 models or all-wheel drive

(7F14 - 68020)
(1669)

ABCDE
FGHI

Haynes Publishing Group
Sparkford Nr Yeovil
Somerset BA22 7JJ England

Haynes North America, Inc
861 Lawrence Drive
Newbury Park
California 91320 USA

Acknowledgements
Technical writers who contributed to this project include Robert Maddox and Brian Styve.

© **Haynes North America, Inc. 1991, 1993**
with permission from J.H. Haynes & Co. Ltd.

A book in the **Haynes Automotive Repair Manual Series**

Printed in the USA

All rights reserved. No part of this book may be reproduced or transmitted in any form or by any means, electronic or mechanical, including photocopying, recording or by any information storage or retrieval system, without permission in writing from the copyright holder.

ISBN 1 56392 091 3

Library of Congress Catalog Card Number 93-61245

While every attempt is made to ensure that the information in this manual is correct, no liability can be accepted by the authors or publishers for loss, damage or injury caused by any errors in, or omissions from, the information given.

Contents

Introductory pages
About this manual	0-6
Introduction to the Mitsubishi front-wheel drive models	0-6
Vehicle identification numbers	0-7
Buying parts	0-8
Maintenance techniques, tools and working facilities	0-8
Booster battery (jump) starting	0-15
Jacking and towing	0-16
Automotive chemicals and lubricants	0-17
Safety first!	0-18
Conversion factors	0-19
Troubleshooting	0-20

Chapter 1
Tune-up and routine maintenance — 1-1

Chapter 2 Part A
Engines — 2A-1

Chapter 2 Part B
General engine overhaul procedures — 2B-1

Chapter 3
Cooling, heating and air conditioning systems — 3-1

Chapter 4
Fuel and exhaust systems — 4-1

Chapter 5
Engine electrical systems — 5-1

Chapter 6
Emissions control systems — 6-1

Chapter 7 Part A
Manual transaxle — 7A-1

Chapter 7 Part B
Automatic transaxle — 7B-1

Chapter 8
Clutch and driveaxles — 8-1

Chapter 9
Brakes — 9-1

Chapter 10
Suspension and steering systems — 10-1

Chapter 11
Body — 11-1

Chapter 12
Chassis electrical system — 12-1

Wiring diagrams — 12-8

Index

1988 Mitsubishi Cordia Turbo

1988 Mitsubishi Tredia

1988 Mitsubishi Mirage

1988 Mitsubishi Precis

About this manual

Its purpose
The purpose of this manual is to help you get the best value from your vehicle. It can do so in several ways. It can help you decide what work must be done, even if you choose to have it done by a dealer service department or a repair shop; it provides information and procedures for routine maintenance and servicing; and it offers diagnostic and repair procedures to follow when trouble occurs.

We hope you use the manual to tackle the work yourself. For many simpler jobs, doing it yourself may be quicker than arranging an appointment to get the vehicle into a shop and making the trips to leave it and pick it up. More importantly, a lot of money can be saved by avoiding the expense the shop must pass on to you to cover its labor and overhead costs. An added benefit is the sense of satisfaction and accomplishment that you feel after doing the job yourself.

Using the manual
The manual is divided into Chapters. Each Chapter is divided into numbered Sections, which are headed in bold type between horizontal lines. Each Section consists of consecutively numbered paragraphs.

At the beginning of each numbered Section you will be referred to any illustrations which apply to the procedures in that Section. The reference numbers used in illustration captions pinpoint the pertinent Section and the Step within that Section. That is, illustration 3.2 means the illustration refers to Section 3 and Step (or paragraph) 2 within that Section.

Procedures, once described in the text, are not normally repeated. When it's necessary to refer to another Chapter, the reference will be given as Chapter and Section number. Cross references given without use of the word "Chapter" apply to Sections and/or paragraphs in the same Chapter. For example, "see Section 8" means in the same Chapter.

References to the left or right side of the vehicle assume you are sitting in the driver's seat, facing forward.

Even though we have prepared this manual with extreme care, neither the publisher nor the author can accept responsibility for any errors in, or omissions from, the information given.

NOTE

A **Note** provides information necessary to properly complete a procedure or information which will make the procedure easier to understand.

CAUTION

A **Caution** provides a special procedure or special steps which must be taken while completing the procedure where the **Caution** is found. Not heeding a **Caution** can result in damage to the assembly being worked on.

WARNING

A **Warning** provides a special procedure or special steps which must be taken while completing the procedure where the **Warning** is found. Not heeding a **Warning** can result in personal injury.

Introduction to the Mitsubishi front-wheel drive models

These models are available in two and four-door hatchback, coupe and sedan body styles.

The transversely-mounted inline four-cylinder engines used in these models are equipped with either a carburetor or fuel injection. Some models are turbocharged.

The engine drives the front wheels through a manual or automatic transaxle via independent driveaxles.

Independent suspension, featuring coil springs and struts, is used at the front wheels. Trailing arms or a torsion axle are used at the rear, depending on model. The rack and pinion steering unit is mounted behind the engine.

The brakes on most models are disc at the front and drums at the rear, with power assist standard. Some models are equipped with front and rear disc brakes.

Vehicle identification numbers

Modifications are a continuing and unpublicized process in vehicle manufacturing. Since spare parts lists are compiled on a numerical basis, the individual vehicle numbers are essential to correctly identify the component required.

Vehicle Identification Number (VIN)

This very important identification number is stamped on a plate attached to the left side of the dashboard, just inside the windshield on the driver's side of the vehicle (see illustration). The VIN also appears on the Vehicle Certificate of Title and Registration. It contains information such as the vehicle model, engine type and when it was manufactured.

Vehicle information code plate

This plate is attached to the engine compartment side of the firewall (see illustration). The plate contains the vehicle model code, engine model code and, on most models, the transaxle model code and body color or code.

Chassis number

The chassis number is also attached to the engine compartment side of the firewall (see illustration). It identifies the vehicle line, body type and engine displacement.

Vehicle safety certification label

The safety certification label is attached to the left front door or door pillar. The plate contains the name of the manufacturer, the month and year of production, the Gross Vehicle Weight Rating (GVWR) and the certification statement.

Engine identification number

The engine identification number is at the left corner of the top edge of the cylinder block on most models. It tells what type of engine it is, its displacement and when it was produced. This number is often required when ordering parts.

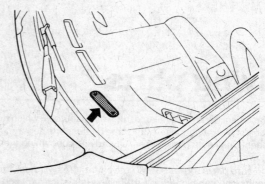

The Vehicle Identification Number (VIN) is visible from outside the vehicle through the driver's side of the windshield

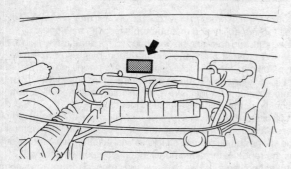

The Vehicle Information Code Plate is attached to the engine compartment side of the firewall

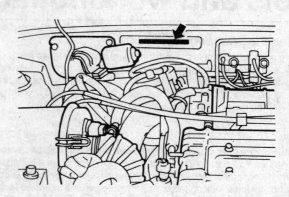

The chassis number is also attached to the firewall

Buying parts

Replacement parts are available from many sources, which generally fall into one of two categories – authorized dealer parts departments and independent retail auto parts stores. Our advice concerning these parts is as follows:

Retail auto parts stores: Good auto parts stores will stock frequently needed components which wear out relatively fast, such as clutch components, exhaust systems, brake parts, tune-up parts, etc. These stores often supply new or reconditioned parts on an exchange basis, which can save a considerable amount of money. Discount auto parts stores are often very good places to buy materials and parts needed for general vehicle maintenance such as oil, grease, filters, spark plugs, belts, touch-up paint, bulbs, etc. They also usually sell tools and general accessories, have convenient hours, charge lower prices and can often be found not far from home.

Authorized dealer parts department: This is the best source for parts which are unique to the vehicle and not generally available elsewhere (such as major engine parts, transmission parts, trim pieces, etc.).

Warranty information: If the vehicle is still covered under warranty, be sure that any replacement parts purchased – regardless of the source – do not invalidate the warranty!

To be sure of obtaining the correct parts, have engine and chassis numbers available and, if possible, take the old parts along for positive identification.

Maintenance techniques, tools and working facilities

Maintenance techniques

There are a number of techniques involved in maintenance and repair that will be referred to throughout this manual. Application of these techniques will enable the home mechanic to be more efficient, better organized and capable of performing the various tasks properly, which will ensure that the repair job is thorough and complete.

Fasteners

Fasteners are nuts, bolts, studs and screws used to hold two or more parts together. There are a few things to keep in mind when working with fasteners. Almost all of them use a locking device of some type, either a lockwasher, locknut, locking tab or thread adhesive. All threaded fasteners should be clean and straight, with undamaged threads and undamaged corners on the hex head where the wrench fits. Develop the habit of replacing all damaged nuts and bolts with new ones. Special locknuts with nylon or fiber inserts can only be used once. If they are removed, they lose their locking ability and must be replaced with new ones.

Rusted nuts and bolts should be treated with a penetrating fluid to ease removal and prevent breakage. Some mechanics use turpentine in a spout-type oil can, which works quite well. After applying the rust penetrant, let it work for a few minutes before trying to loosen the nut or bolt. Badly rusted fasteners may have to be chiseled or sawed off or removed with a special nut breaker, available at tool stores.

If a bolt or stud breaks off in an assembly, it can be drilled and removed with a special tool commonly available for this purpose. Most automotive machine shops can perform this task, as well as other repair procedures, such as the repair of threaded holes that have been stripped out.

Flat washers and lockwashers, when removed from an assembly, should always be replaced exactly as removed. Replace any damaged washers with new ones. Never use a lockwasher on any soft metal surface (such as aluminum), thin sheet metal or plastic.

Fastener sizes

For a number of reasons, automobile manufacturers are making wider and wider use of metric fasteners. Therefore, it is important to be able to tell the difference between standard (sometimes called U.S. or SAE) and metric hardware, since they cannot be interchanged.

All bolts, whether standard or metric, are sized according to diameter, thread pitch and length. For example, a standard 1/2 – 13 x 1 bolt is 1/2 inch in diameter, has 13 threads per inch and is 1 inch long. An M12 – 1.75 x 25 metric bolt is 12 mm in diameter, has a thread pitch of 1.75 mm (the distance between threads) and is 25 mm long. The two bolts are nearly identical, and easily confused, but they are not interchangeable.

In addition to the differences in diameter, thread pitch and length, metric and standard bolts can also be distinguished by examining the bolt heads. To begin with, the distance across the flats on a standard bolt head is measured in inches, while the same dimension on a metric bolt is sized in millimeters (the same is true for nuts). As a result, a standard wrench should not be used on a metric bolt and a metric wrench should not be used on a standard bolt. Also, most standard bolts have slashes radiating out from the center of the head to denote the grade or strength of the bolt, which is an indication of the amount of torque that can be applied to it. The greater the number of slashes, the greater the strength of the bolt. Grades 0 through 5 are commonly used on automobiles. Metric bolts have a property class (grade) number, rather than a slash, molded into their heads to indicate bolt strength. In this case, the higher the number, the stronger the bolt. Property class numbers 8.8, 9.8 and 10.9 are commonly used on automobiles.

Strength markings can also be used to distinguish standard hex nuts from metric hex nuts. Many standard nuts have dots stamped into one side, while metric nuts are marked with a number. The greater the number of dots, or the higher the number, the greater the strength of the nut.

Metric studs are also marked on their ends according to property class (grade). Larger studs are numbered (the same as metric bolts), while smaller studs carry a geometric code to denote grade.

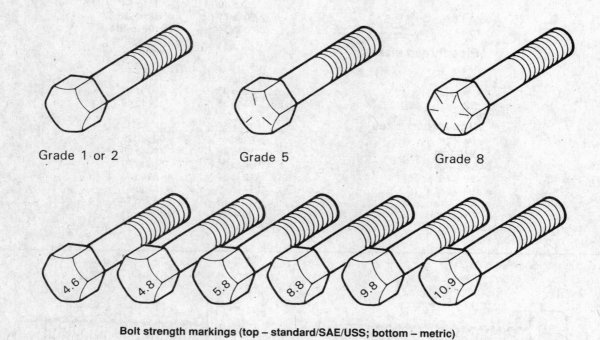

Bolt strength markings (top – standard/SAE/USS; bottom – metric)

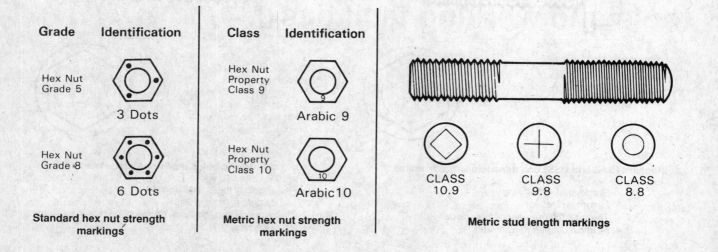

Standard hex nut strength markings

Metric hex nut strength markings

Metric stud length markings

0-10 Maintenance techniques, tools and working facilities

It should be noted that many fasteners, especially Grades 0 through 2, have no distinguishing marks on them. When such is the case, the only way to determine whether it is standard or metric is to measure the thread pitch or compare it to a known fastener of the same size.

Standard fasteners are often referred to as SAE, as opposed to metric. However, it should be noted that SAE technically refers to a non-metric *fine thread* fastener only. Coarse thread non-metric fasteners are referred to as USS sizes.

Since fasteners of the same size (both standard and metric) may have different strength ratings, be sure to reinstall any bolts, studs or nuts removed from your vehicle in their original locations. Also, when replacing a fastener with a new one, make sure that the new one has a strength rating equal to or greater than the original.

Tightening sequences and procedures

Most threaded fasteners should be tightened to a specific torque value (torque is the twisting force applied to a threaded component such as a nut or bolt). Overtightening the fastener can weaken it and cause it to break, while undertightening can cause it to eventually come loose. Bolts, screws and studs, depending on the material they are made of and their thread diameters, have specific torque values, many of which are noted in the Specifications at the beginning of each Chapter. Be sure to follow the torque recommendations closely. For fasteners not assigned a specific torque, a general torque value chart is presented here as a guide. These torque values are for dry (unlubricated) fasteners threaded into steel or cast iron (not aluminum). As was previously mentioned, the size and grade of a fastener determine the amount of torque that can safely be

Metric thread sizes	Ft-lbs	Nm
M-6	6 to 9	9 to 12
M-8	14 to 21	19 to 28
M-10	28 to 40	38 to 54
M-12	50 to 71	68 to 96
M-14	80 to 140	109 to 154
Pipe thread sizes		
1/8	5 to 8	7 to 10
1/4	12 to 18	17 to 24
3/8	22 to 33	30 to 44
1/2	25 to 35	34 to 47
U.S. thread sizes		
1/4 – 20	6 to 9	9 to 12
5/16 – 18	12 to 18	17 to 24
5/16 – 24	14 to 20	19 to 27
3/8 – 16	22 to 32	30 to 43
3/8 – 24	27 to 38	37 to 51
7/16 – 14	40 to 55	55 to 74
7/16 – 20	40 to 60	55 to 81
1/2 – 13	55 to 80	75 to 108

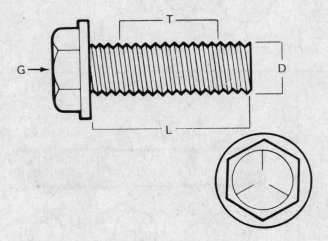

Standard (SAE and USS) bolt dimensions/grade marks

- G Grade marks (bolt length)
- L Length (in inches)
- T Thread pitch (number of threads per inch)
- D Nominal diameter (in inches)

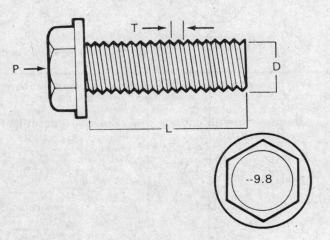

Metric bolt dimensions/grade marks

- P Property class (bolt strength)
- L Length (in millimeters)
- T Thread pitch (distance between threads in millimeters)
- D Diameter

Maintenance techniques, tools and working facilities

applied to it. The figures listed here are approximate for Grade 2 and Grade 3 fasteners. Higher grades can tolerate higher torque values.

Fasteners laid out in a pattern, such as cylinder head bolts, oil pan bolts, differential cover bolts, etc., must be loosened or tightened in sequence to avoid warping the component. This sequence will normally be shown in the appropriate Chapter. If a specific pattern is not given, the following procedures can be used to prevent warping.

Initially, the bolts or nuts should be assembled finger-tight only. Next, they should be tightened one full turn each, in a criss-cross or diagonal pattern. After each one has been tightened one full turn, return to the first one and tighten them all one-half turn, following the same pattern. Finally, tighten each of them one-quarter turn at a time until each fastener has been tightened to the proper torque. To loosen and remove the fasteners, the procedure would be reversed.

Component disassembly

Component disassembly should be done with care and purpose to help ensure that the parts go back together properly. Always keep track of the sequence in which parts are removed. Make note of special characteristics or marks on parts that can be installed more than one way, such as a grooved thrust washer on a shaft. It is a good idea to lay the disassembled parts out on a clean surface in the order that they were removed. It may also be helpful to make sketches or take instant photos of components before removal.

When removing fasteners from a component, keep track of their locations. Sometimes threading a bolt back in a part, or putting the washers and nut back on a stud, can prevent mix-ups later. If nuts and bolts cannot be returned to their original locations, they should be kept in a compartmented box or a series of small boxes. A cupcake or muffin tin is ideal for this purpose, since each cavity can hold the bolts and nuts from a particular area (i.e. oil pan bolts, valve cover bolts, engine mount bolts, etc.). A pan of this type is especially helpful when working on assemblies with very small parts, such as the carburetor, alternator, valve train or interior dash and trim pieces. The cavities can be marked with paint or tape to identify the contents.

Whenever wiring looms, harnesses or connectors are separated, it is a good idea to identify the two halves with numbered pieces of masking tape so they can be easily reconnected.

Gasket sealing surfaces

Throughout any vehicle, gaskets are used to seal the mating surfaces between two parts and keep lubricants, fluids, vacuum or pressure contained in an assembly.

Many times these gaskets are coated with a liquid or paste-type gasket sealing compound before assembly. Age, heat and pressure can sometimes cause the two parts to stick together so tightly that they are very difficult to separate. Often, the assembly can be loosened by striking it with a soft-face hammer near the mating surfaces. A regular hammer can be used if a block of wood is placed between the hammer and the part. Do not hammer on cast parts or parts that could be easily damaged. With any particularly stubborn part, always recheck to make sure that every fastener has been removed.

Avoid using a screwdriver or bar to pry apart an assembly, as they can easily mar the gasket sealing surfaces of the parts, which must remain smooth. If prying is absolutely necessary, use an old broom handle, but keep in mind that extra clean up will be necessary if the wood splinters.

After the parts are separated, the old gasket must be carefully scraped off and the gasket surfaces cleaned. Stubborn gasket material can be soaked with rust penetrant or treated with a special chemical to soften it so it can be easily scraped off. A scraper can be fashioned from a piece of copper tubing by flattening and sharpening one end. Copper is recommended because it is usually softer than the surfaces to be scraped, which reduces the chance of gouging the part. Some gaskets can be removed with a wire brush, but regardless of the method used, the mating surfaces must be left clean and smooth. If for some reason the gasket surface is gouged, then a gasket sealer thick enough to fill scratches will have to be used during reassembly of the components. For most applications, a non-drying (or semi-drying) gasket sealer should be used.

Hose removal tips

Warning: *If the vehicle is equipped with air conditioning, do not disconnect any of the A/C hoses without first having the system depressurized by a dealer service department or a service station.*

Hose removal precautions closely parallel gasket removal precautions. Avoid scratching or gouging the surface that the hose mates against or the connection may leak. This is especially true for radiator hoses. Because of various chemical reactions, the rubber in hoses can bond itself to the metal spigot that the hose fits over. To remove a hose, first loosen the hose clamps that secure it to the spigot. Then, with slip-joint pliers, grab the hose at the clamp and rotate it around the spigot. Work it back and forth until it is completely free, then pull it off. Silicone or other lubricants will ease removal if they can be applied between the hose and the outside of the spigot. Apply the same lubricant to the inside of the hose and the outside of the spigot to simplify installation.

As a last resort (and if the hose is to be replaced with a new one anyway), the rubber can be slit with a knife and the hose peeled from the spigot. If this must be done, be careful that the metal connection is not damaged.

If a hose clamp is broken or damaged, do not reuse it. Wire-type clamps usually weaken with age, so it is a good idea to replace them with screw-type clamps whenever a hose is removed.

Tools

A selection of good tools is a basic requirement for anyone who plans to maintain and repair his or her own vehicle. For the owner who has few tools, the initial investment might seem high, but when compared to the spiraling costs of professional auto maintenance and repair, it is a wise one.

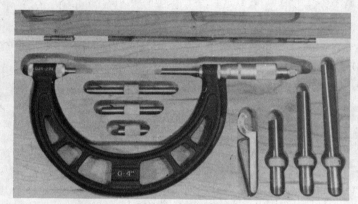

Micrometer set

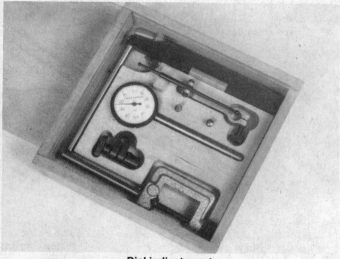

Dial indicator set

Maintenance techniques, tools and working facilities

Dial caliper

Hand-operated vacuum pump

Timing light

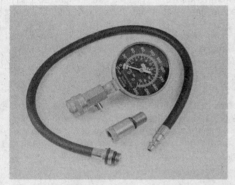
Compression gauge with spark plug hole adapter

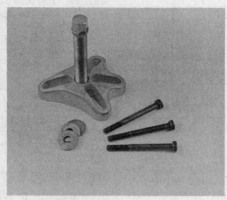
Damper/steering wheel puller

General purpose puller

Hydraulic lifter removal tool

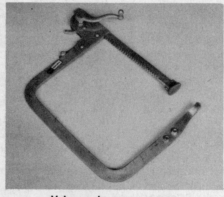
Valve spring compressor

Valve spring compressor

Ridge reamer

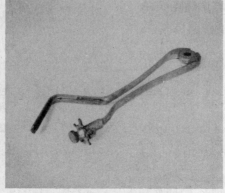
Piston ring groove cleaning tool

Ring removal/installation tool

Ring compressor

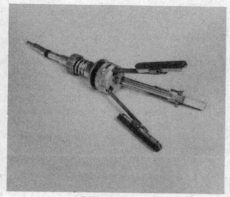
Cylinder hone

Brake hold-down spring tool

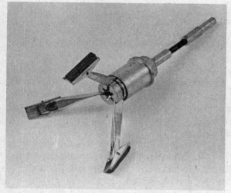
Brake cylinder hone

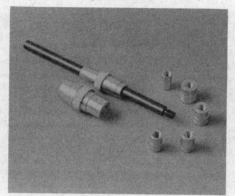
Clutch plate alignment tool

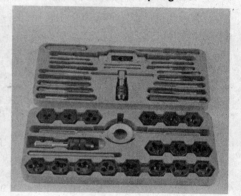
Tap and die set

To help the owner decide which tools are needed to perform the tasks detailed in this manual, the following tool lists are offered: *Maintenance and minor repair, Repair/overhaul* and *Special*.

The newcomer to practical mechanics should start off with the maintenance and minor repair tool kit, which is adequate for the simpler jobs performed on a vehicle. Then, as confidence and experience grow, the owner can tackle more difficult tasks, buying additional tools as they are needed. Eventually the basic kit will be expanded into the repair and overhaul tool set. Over a period of time, the experienced do-it-yourselfer will assemble a tool set complete enough for most repair and overhaul procedures and will add tools from the special category when it is felt that the expense is justified by the frequency of use.

Maintenance and minor repair tool kit

The tools in this list should be considered the minimum required for performance of routine maintenance, servicing and minor repair work. We recommend the purchase of combination wrenches (box-end and open-end combined in one wrench). While more expensive than open end wrenches, they offer the advantages of both types of wrench.

Combination wrench set (1/4-inch to 1 inch or 6 mm to 19 mm)
Adjustable wrench, 8 inch
Spark plug wrench with rubber insert
Spark plug gap adjusting tool
Feeler gauge set
Brake bleeder wrench
Standard screwdriver (5/16-inch x 6 inch)
Phillips screwdriver (No. 2 x 6 inch)
Combination pliers – 6 inch
Hacksaw and assortment of blades
Tire pressure gauge
Grease gun
Oil can
Fine emery cloth
Wire brush
Battery post and cable cleaning tool
Oil filter wrench
Funnel (medium size)
Safety goggles
Jackstands(2)
Drain pan

Note: *If basic tune-ups are going to be part of routine maintenance, it will be necessary to purchase a good quality stroboscopic timing light and combination tachometer/dwell meter. Although they are included in the list of special tools, it is mentioned here because they are absolutely necessary for tuning most vehicles properly.*

Repair and overhaul tool set

These tools are essential for anyone who plans to perform major repairs and are in addition to those in the maintenance and minor repair tool kit. Included is a comprehensive set of sockets which, though expensive, are invaluable because of their versatility, especially when various extensions and drives are available. We recommend the 1/2-inch drive over the 3/8-inch drive. Although the larger drive is bulky and more expensive, it has the capacity of accepting a very wide range of large sockets. Ideally, however, the mechanic should have a 3/8-inch drive set and a 1/2-inch drive set.

Socket set(s)
Reversible ratchet
Extension – 10 inch
Universal joint
Torque wrench (same size drive as sockets)
Ball peen hammer – 8 ounce
Soft-face hammer (plastic/rubber)
Standard screwdriver (1/4-inch x 6 inch)
Standard screwdriver (stubby – 5/16-inch)
Phillips screwdriver (No. 3 x 8 inch)
Phillips screwdriver (stubby – No. 2)

Pliers – vise grip
Pliers – lineman's
Pliers – needle nose
Pliers – snap-ring (internal and external)
Cold chisel – 1/2-inch
Scribe
Scraper (made from flattened copper tubing)
Centerpunch
Pin punches (1/16, 1/8, 3/16-inch)
Steel rule/straightedge – 12 inch
Allen wrench set (1/8 to 3/8-inch or 4 mm to 10 mm)
A selection of files
Wire brush (large)
Jackstands (second set)
Jack (scissor or hydraulic type)

Note: *Another tool which is often useful is an electric drill with a chuck capacity of 3/8-inch and a set of good quality drill bits.*

Special tools

The tools in this list include those which are not used regularly, are expensive to buy, or which need to be used in accordance with their manufacturer's instructions. Unless these tools will be used frequently, it is not very economical to purchase many of them. A consideration would be to split the cost and use between yourself and a friend or friends. In addition, most of these tools can be obtained from a tool rental shop on a temporary basis.

This list primarily contains only those tools and instruments widely available to the public, and not those special tools produced by the vehicle manufacturer for distribution to dealer service departments. Occasionally, references to the manufacturer's special tools are included in the text of this manual. Generally, an alternative method of doing the job without the special tool is offered. However, sometimes there is no alternative to their use. Where this is the case, and the tool cannot be purchased or borrowed, the work should be turned over to the dealer service department or an automotive repair shop.

Valve spring compressor
Piston ring groove cleaning tool
Piston ring compressor
Piston ring installation tool
Cylinder compression gauge
Cylinder ridge reamer
Cylinder surfacing hone
Cylinder bore gauge
Micrometers and/or dial calipers
Hydraulic lifter removal tool
Balljoint separator
Universal-type puller
Impact screwdriver
Dial indicator set
Stroboscopic timing light (inductive pick-up)
Hand operated vacuum/pressure pump
Tachometer/dwell meter
Universal electrical multimeter
Cable hoist
Brake spring removal and installation tools
Floor jack

Buying tools

For the do-it-yourselfer who is just starting to get involved in vehicle maintenance and repair, there are a number of options available when purchasing tools. If maintenance and minor repair is the extent of the work to be done, the purchase of individual tools is satisfactory. If, on the other hand, extensive work is planned, it would be a good idea to purchase a modest tool set from one of the large retail chain stores. A set can usually be bought at a substantial savings over the individual tool prices, and they often come with a tool box. As additional tools are needed, add-on sets, individual tools and a larger tool box can be purchased to expand the tool selection. Building a tool set gradually allows the cost of the tools to be spread over a longer period of time and gives the mechanic the freedom to choose only those tools that will actually be used.

Tool stores will often be the only source of some of the special tools that are needed, but regardless of where tools are bought, try to avoid cheap ones, especially when buying screwdrivers and sockets, because they won't last very long. The expense involved in replacing cheap tools will eventually be greater than the initial cost of quality tools.

Care and maintenance of tools

Good tools are expensive, so it makes sense to treat them with respect. Keep them clean and in usable condition and store them properly when not in use. Always wipe off any dirt, grease or metal chips before putting them away. Never leave tools lying around in the work area. Upon completion of a job, always check closely under the hood for tools that may have been left there so they won't get lost during a test drive.

Some tools, such as screwdrivers, pliers, wrenches and sockets, can be hung on a panel mounted on the garage or workshop wall, while others should be kept in a tool box or tray. Measuring instruments, gauges, meters, etc. must be carefully stored where they cannot be damaged by weather or impact from other tools.

When tools are used with care and stored properly, they will last a very long time. Even with the best of care, though, tools will wear out if used frequently. When a tool is damaged or worn out, replace it. Subsequent jobs will be safer and more enjoyable if you do.

Working facilities

Not to be overlooked when discussing tools is the workshop. If anything more than routine maintenance is to be carried out, some sort of suitable work area is essential.

It is understood, and appreciated, that many home mechanics do not have a good workshop or garage available, and end up removing an engine or doing major repairs outside. It is recommended, however, that the overhaul or repair be completed under the cover of a roof.

A clean, flat workbench or table of comfortable working height is an absolute necessity. The workbench should be equipped with a vise that has a jaw opening of at least four inches.

As mentioned previously, some clean, dry storage space is also required for tools, as well as the lubricants, fluids, cleaning solvents, etc. which soon become necessary.

Sometimes waste oil and fluids, drained from the engine or cooling system during normal maintenance or repairs, present a disposal problem. To avoid pouring them on the ground or into a sewage system, pour the used fluids into large containers, seal them with caps and take them to an authorized disposal site or recycling center. Plastic jugs, such as old antifreeze containers, are ideal for this purpose.

Always keep a supply of old newspapers and clean rags available. Old towels are excellent for mopping up spills. Many mechanics use rolls of paper towels for most work because they are readily available and disposable. To help keep the area under the vehicle clean, a large cardboard box can be cut open and flattened to protect the garage or shop floor.

Whenever working over a painted surface, such as when leaning over a fender to service something under the hood, always cover it with an old blanket or bedspread to protect the finish. Vinyl covered pads, made especially for this purpose, are available at auto parts stores.

Booster battery (jump) starting

Observe these precautions when using a booster battery to start a vehicle:
a) Before connecting the booster battery, make sure the ignition switch is in the Off position.
b) Turn off the lights, heater and other electrical loads.
c) Your eyes should be shielded. Safety goggles are a good idea.
d) Make sure the booster battery is the same voltage as the dead one in the vehicle.
e) The two vehicles MUST NOT TOUCH each other!
f) Make sure the transmission is in Neutral (manual) or Park (automatic).
g) If the booster battery is not a maintenance-free type, remove the vent caps and lay a cloth over the vent holes.

Connect the red jumper cable to the positive (+) terminals of each battery.

Connect one end of the black jumper cable to the negative (–) terminal of the booster battery. The other end of this cable should be connected to a good ground on the vehicle to be started, such as a bolt or bracket on the engine block **(see illustration)**. Make sure the cable will not come into contact with the fan, drivebelts or other moving parts of the engine.

Start the engine using the booster battery, then, with the engine running at idle speed, disconnect the jumper cables in the reverse order of connection.

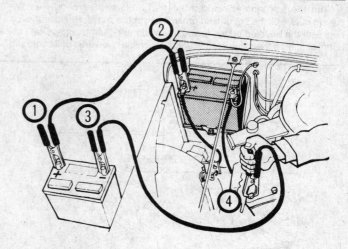

Make the booster battery cable connections in the numerical order shown (note that the negative cable of the booster battery is NOT attached to the negative terminal of the dead battery)

Jacking and towing

Jacking

Warning: *The jack supplied with the vehicle should only be used for raising the vehicle when changing a tire or placing jackstands under the frame. Never work under the vehicle or start the engine while the jack is being used as the only means of support.*

The vehicle must be on a level surface with the wheels blocked and the transaxle in Park (automatic) or Reverse (manual). Apply the parking brake if the front of the vehicle must be raised. Make sure no one is in the vehicle as it's being raised with the jack.

Remove the jack, lug nut wrench and spare tire (if needed) from the vehicle. If a tire is being replaced, use the lug wrench to remove the wheel cover or center cap (except aluminum wheel). The plastic wheel covers and center caps are easy to break, so pry carefully. **Warning:** *Wheel covers may have sharp edges – be very careful not to cut yourself.* Loosen the lug nuts one-half turn, but leave them in place until the tire is raised off the ground.

Position the jack under the vehicle at the indicated jacking point. There's a front and rear jacking point on each side of the vehicle **(see illustrations)**.

Turn the jack handle clockwise until the tire clears the ground. Remove the lug nuts, pull the tire off and replace it with the spare. Replace the lug nuts with the beveled edges facing in and tighten them snugly. Don't attempt to tighten them completely until the vehicle is lowered or it could slip off the jack.

Turn the jack handle counterclockwise to lower the vehicle. Remove the jack and tighten the lug nuts in a criss-cross pattern. If possible, tighten the nuts with a torque wrench (see Chapter 1 for the torque figures). If you don't have access to a torque wrench, have the nuts checked by a service station or repair shop as soon as possible.

Stow the tire, jack and wrench and unblock the wheels.

Towing

As a general rule, the vehicle can be towed with all four wheels on the ground. If one end must be lifted for towing, it should be the front (drive) wheel end. If the vehicle has an automatic transaxle and the front wheels will be on the ground, do not tow farther than 15 miles and do not exceed 30 mph. When towing any vehicle, be sure to release the parking brake and place the transaxle in Neutral. Also, the ignition key must be in the ACC position, since the steering lock mechanism isn't strong enough to hold the front wheels straight while towing.

Equipment specifically designed for towing should be used. It should be attached to the main structural members of the vehicle, not the bumpers or brackets.

Safety is a major consideration when towing and all applicable state and local laws must be obeyed. A safety chain must be used at all times. Remember that power steering and brakes won't work with the engine off.

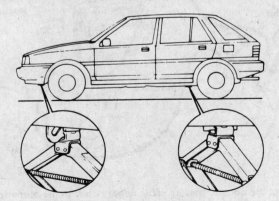

Jacking points for Cordia/Tredia and Precis models

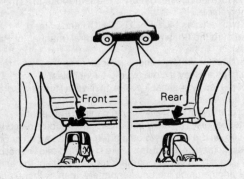

Jacking points for Mirage and Galant models

Automotive chemicals and lubricants

A number of automotive chemicals and lubricants are available for use during vehicle maintenance and repair. They include a wide variety of products ranging from cleaning solvents and degreasers to lubricants and protective sprays for rubber, plastic and vinyl.

Cleaners

Carburetor cleaner and choke cleaner is a strong solvent for gum, varnish and carbon. Most carburetor cleaners leave a dry-type lubricant film which will not harden or gum up. Because of this film it is not recommended for use on electrical components.

Brake system cleaner is used to remove grease and brake fluid from the brake system, where clean surfaces are absolutely necessary. It leaves no residue and often eliminates brake squeal caused by contaminants.

Electrical cleaner removes oxidation, corrosion and carbon deposits from electrical contacts, restoring full current flow. It can also be used to clean spark plugs, carburetor jets, voltage regulators and other parts where an oil-free surface is desired.

Demoisturants remove water and moisture from electrical components such as alternators, voltage regulators, electrical connectors and fuse blocks. They are non-conductive, non-corrosive and non-flammable.

Degreasers are heavy-duty solvents used to remove grease from the outside of the engine and from chassis components. They can be sprayed or brushed on and, depending on the type, are rinsed off either with water or solvent.

Lubricants

Motor oil is the lubricant formulated for use in engines. It normally contains a wide variety of additives to prevent corrosion and reduce foaming and wear. Motor oil comes in various weights (viscosity ratings) from 5 to 80. The recommended weight of the oil depends on the season, temperature and the demands on the engine. Light oil is used in cold climates and under light load conditions. Heavy oil is used in hot climates and where high loads are encountered. Multi-viscosity oils are designed to have characteristics of both light and heavy oils and are available in a number of weights from 5W-20 to 20W-50.

Gear oil is designed to be used in differentials, manual transmissions and other areas where high-temperature lubrication is required.

Chassis and wheel bearing grease is a heavy grease used where increased loads and friction are encountered, such as for wheel bearings, balljoints, tie-rod ends and universal joints.

High-temperature wheel bearing grease is designed to withstand the extreme temperatures encountered by wheel bearings in disc brake equipped vehicles. It usually contains molybdenum disulfide (moly), which is a dry-type lubricant.

White grease is a heavy grease for metal-to-metal applications where water is a problem. White grease stays soft under both low and high temperatures (usually from −100 to +190-degrees F), and will not wash off or dilute in the presence of water.

Assembly lube is a special extreme pressure lubricant, usually containing moly, used to lubricate high-load parts (such as main and rod bearings and cam lobes) for initial start-up of a new engine. The assembly lube lubricates the parts without being squeezed out or washed away until the engine oiling system begins to function.

Silicone lubricants are used to protect rubber, plastic, vinyl and nylon parts.

Graphite lubricants are used where oils cannot be used due to contamination problems, such as in locks. The dry graphite will lubricate metal parts while remaining uncontaminated by dirt, water, oil or acids. It is electrically conductive and will not foul electrical contacts in locks such as the ignition switch.

Moly penetrants loosen and lubricate frozen, rusted and corroded fasteners and prevent future rusting or freezing.

Heat-sink grease is a special electrically non-conductive grease that is used for mounting electronic ignition modules where it is essential that heat is transferred away from the module.

Sealants

RTV sealant is one of the most widely used gasket compounds. Made from silicone, RTV is air curing, it seals, bonds, waterproofs, fills surface irregularities, remains flexible, doesn't shrink, is relatively easy to remove, and is used as a supplementary sealer with almost all low and medium temperature gaskets.

Anaerobic sealant is much like RTV in that it can be used either to seal gaskets or to form gaskets by itself. It remains flexible, is solvent resistant and fills surface imperfections. The difference between an anaerobic sealant and an RTV-type sealant is in the curing. RTV cures when exposed to air, while an anaerobic sealant cures only in the absence of air. This means that an anaerobic sealant cures only after the assembly of parts, sealing them together.

Thread and pipe sealant is used for sealing hydraulic and pneumatic fittings and vacuum lines. It is usually made from a Teflon compound, and comes in a spray, a paint-on liquid and as a wrap-around tape.

Chemicals

Anti-seize compound prevents seizing, galling, cold welding, rust and corrosion in fasteners. High-temperature anti-seize, usually made with copper and graphite lubricants, is used for exhaust system and exhaust manifold bolts.

Anaerobic locking compounds are used to keep fasteners from vibrating or working loose and cure only after installation, in the absence of air. Medium strength locking compound is used for small nuts, bolts and screws that may be removed later. High-strength locking compound is for large nuts, bolts and studs which aren't removed on a regular basis.

Oil additives range from viscosity index improvers to chemical treatments that claim to reduce internal engine friction. It should be noted that most oil manufacturers caution against using additives with their oils.

Gas additives perform several functions, depending on their chemical makeup. They usually contain solvents that help dissolve gum and varnish that build up on carburetor, fuel injection and intake parts. They also serve to break down carbon deposits that form on the inside surfaces of the combustion chambers. Some additives contain upper cylinder lubricants for valves and piston rings, and others contain chemicals to remove condensation from the gas tank.

Miscellaneous

Brake fluid is specially formulated hydraulic fluid that can withstand the heat and pressure encountered in brake systems. Care must be taken so this fluid does not come in contact with painted surfaces or plastics. An opened container should always be resealed to prevent contamination by water or dirt.

Weatherstrip adhesive is used to bond weatherstripping around doors, windows and trunk lids. It is sometimes used to attach trim pieces.

Undercoating is a petroleum-based, tar-like substance that is designed to protect metal surfaces on the underside of the vehicle from corrosion. It also acts as a sound-deadening agent by insulating the bottom of the vehicle.

Waxes and polishes are used to help protect painted and plated surfaces from the weather. Different types of paint may require the use of different types of wax and polish. Some polishes utilize a chemical or abrasive cleaner to help remove the top layer of oxidized (dull) paint on older vehicles. In recent years many non-wax polishes that contain a wide variety of chemicals such as polymers and silicones have been introduced. These non-wax polishes are usually easier to apply and last longer than conventional waxes and polishes.

Safety first!

Regardless of how enthusiastic you may be about getting on with the job at hand, take the time to ensure that your safety is not jeopardized. A moment's lack of attention can result in an accident, as can failure to observe certain simple safety precautions. The possibility of an accident will always exist, and the following points should not be considered a comprehensive list of all dangers. Rather, they are intended to make you aware of the risks and to encourage a safety conscious approach to all work you carry out on your vehicle.

Essential DOs and DON'Ts

DON'T rely on a jack when working under the vehicle. Always use approved jackstands to support the weight of the vehicle and place them under the recommended lift or support points.
DON'T attempt to loosen extremely tight fasteners (i.e. wheel lug nuts) while the vehicle is on a jack – it may fall.
DON'T start the engine without first making sure that the transmission is in Neutral (or Park where applicable) and the parking brake is set.
DON'T remove the radiator cap from a hot cooling system – let it cool or cover it with a cloth and release the pressure gradually.
DON'T attempt to drain the engine oil until you are sure it has cooled to the point that it will not burn you.
DON'T touch any part of the engine or exhaust system until it has cooled sufficiently to avoid burns.
DON'T siphon toxic liquids such as gasoline, antifreeze and brake fluid by mouth, or allow them to remain on your skin.
DON'T inhale brake lining dust – it is potentially hazardous (see *Asbestos* below).
DON'T allow spilled oil or grease to remain on the floor – wipe it up before someone slips on it.
DON'T use loose fitting wrenches or other tools which may slip and cause injury.
DON'T push on wrenches when loosening or tightening nuts or bolts. Always try to pull the wrench toward you. If the situation calls for pushing the wrench away, push with an open hand to avoid scraped knuckles if the wrench should slip.
DON'T attempt to lift a heavy component alone – get someone to help you.
DON'T rush or take unsafe shortcuts to finish a job.
DON'T allow children or animals in or around the vehicle while you are working on it.
DO wear eye protection when using power tools such as a drill, sander, bench grinder, etc. and when working under a vehicle.
DO keep loose clothing and long hair well out of the way of moving parts.
DO make sure that any hoist used has a safe working load rating adequate for the job.
DO get someone to check on you periodically when working alone on a vehicle.
DO carry out work in a logical sequence and make sure that everything is correctly assembled and tightened.
DO keep chemicals and fluids tightly capped and out of the reach of children and pets.
DO remember that your vehicle's safety affects that of yourself and others. If in doubt on any point, get professional advice.

Asbestos

Certain friction, insulating, sealing, and other products – such as brake linings, brake bands, clutch linings, torque converters, gaskets, etc. – contain asbestos. *Extreme care must be taken to avoid inhalation of dust from such products, since it is hazardous to health.* If in doubt, assume that they *do* contain asbestos.

Fire

Remember at all times that gasoline is highly flammable. Never smoke or have any kind of open flame around when working on a vehicle. But the risk does not end there. A spark caused by an electrical short circuit, by two metal surfaces contacting each other, or even by static electricity built up in your body under certain conditions, can ignite gasoline vapors, which in a confined space are highly explosive. Do not, under any circumstances, use gasoline for cleaning parts. Use an approved safety solvent.

Always disconnect the battery ground (–) cable *at the battery* before working on any part of the fuel system or electrical system. Never risk spilling fuel on a hot engine or exhaust component.

It is strongly recommended that a fire extinguisher suitable for use on fuel and electrical fires be kept handy in the garage or workshop at all times. Never try to extinguish a fuel or electrical fire with water.

Fumes

Certain fumes are highly toxic and can quickly cause unconsciousness and even death if inhaled to any extent. Gasoline vapor falls into this category, as do the vapors from some cleaning solvents. Any draining or pouring of such volatile fluids should be done in a well ventilated area.

When using cleaning fluids and solvents, read the instructions on the container carefully. Never use materials from unmarked containers.

Never run the engine in an enclosed space, such as a garage. Exhaust fumes contain carbon monoxide, which is extremely poisonous. If you need to run the engine, always do so in the open air, or at least have the rear of the vehicle outside the work area.

If you are fortunate enough to have the use of an inspection pit, never drain or pour gasoline and never run the engine while the vehicle is over the pit. The fumes, being heavier than air, will concentrate in the pit with possibly lethal results.

The battery

Never create a spark or allow a bare light bulb near a battery. They normally give off a certain amount of hydrogen gas, which is highly explosive.

Always disconnect the battery ground (–) cable *at the battery* before working on the fuel or electrical systems.

If possible, loosen the filler caps or cover when charging the battery from an external source (this does not apply to sealed or maintenance-free batteries). Do not charge at an excessive rate or the battery may burst.

Take care when adding water to a non maintenance–free battery and when carrying a battery. The electrolyte, even when diluted, is very corrosive and should not be allowed to contact clothing or skin.

Always wear eye protection when cleaning the battery to prevent the caustic deposits from entering your eyes.

Household current

When using an electric power tool, inspection light, etc., which operates on household current, always make sure that the tool is correctly connected to its plug and that, where necessary, it is properly grounded. Do not use such items in damp conditions and, again, do not create a spark or apply excessive heat in the vicinity of fuel or fuel vapor.

Secondary ignition system voltage

A severe electric shock can result from touching certain parts of the ignition system (such as the spark plug wires) when the engine is running or being cranked, particularly if components are damp or the insulation is defective. In the case of an electronic ignition system, the secondary system voltage is much higher and could prove fatal.

Conversion factors

Length (distance)
Inches (in)	X 25.4	= Millimetres (mm)	X 0.0394	= Inches (in)
Feet (ft)	X 0.305	= Metres (m)	X 3.281	= Feet (ft)
Miles	X 1.609	= Kilometres (km)	X 0.621	= Miles

Volume (capacity)
Cubic inches (cu in; in^3)	X 16.387	= Cubic centimetres (cc; cm^3)	X 0.061	= Cubic inches (cu in; in^3)
Imperial pints (Imp pt)	X 0.568	= Litres (l)	X 1.76	= Imperial pints (Imp pt)
Imperial quarts (Imp qt)	X 1.137	= Litres (l)	X 0.88	= Imperial quarts (Imp qt)
Imperial quarts (Imp qt)	X 1.201	= US quarts (US qt)	X 0.833	= Imperial quarts (Imp qt)
US quarts (US qt)	X 0.946	= Litres (l)	X 1.057	= US quarts (US qt)
Imperial gallons (Imp gal)	X 4.546	= Litres (l)	X 0.22	= Imperial gallons (Imp gal)
Imperial gallons (Imp gal)	X 1.201	= US gallons (US gal)	X 0.833	= Imperial gallons (Imp gal)
US gallons (US gal)	X 3.785	= Litres (l)	X 0.264	= US gallons (US gal)

Mass (weight)
Ounces (oz)	X 28.35	= Grams (g)	X 0.035	= Ounces (oz)
Pounds (lb)	X 0.454	= Kilograms (kg)	X 2.205	= Pounds (lb)

Force
Ounces-force (ozf; oz)	X 0.278	= Newtons (N)	X 3.6	= Ounces-force (ozf; oz)
Pounds-force (lbf; lb)	X 4.448	= Newtons (N)	X 0.225	= Pounds-force (lbf; lb)
Newtons (N)	X 0.1	= Kilograms-force (kgf; kg)	X 9.81	= Newtons (N)

Pressure
Pounds-force per square inch (psi; lbf/in^2; lb/in^2)	X 0.070	= Kilograms-force per square centimetre (kgf/cm^2; kg/cm^2)	X 14.223	= Pounds-force per square inch (psi; lbf/in^2; lb/in^2)
Pounds-force per square inch (psi; lbf/in^2; lb/in^2)	X 0.068	= Atmospheres (atm)	X 14.696	= Pounds-force per square inch (psi; lbf/in^2; lb/in^2)
Pounds-force per square inch (psi; lbf/in^2; lb/in^2)	X 0.069	= Bars	X 14.5	= Pounds-force per square inch (psi; lbf/in^2; lb/in^2)
Pounds-force per square inch (psi; lbf/in^2; lb/in^2)	X 6.895	= Kilopascals (kPa)	X 0.145	= Pounds-force per square inch (psi; lbf/in^2; lb/in^2)
Kilopascals (kPa)	X 0.01	= Kilograms-force per square centimetre (kgf/cm^2; kg/cm^2)	X 98.1	= Kilopascals (kPa)

Torque (moment of force)
Pounds-force inches (lbf in; lb in)	X 1.152	= Kilograms-force centimetre (kgf cm; kg cm)	X 0.868	= Pounds-force inches (lbf in; lb in)
Pounds-force inches (lbf in; lb in)	X 0.113	= Newton metres (Nm)	X 8.85	= Pounds-force inches (lbf in; lb in)
Pounds-force inches (lbf in; lb in)	X 0.083	= Pounds-force feet (lbf ft; lb ft)	X 12	= Pounds-force inches (lbf in; lb in)
Pounds-force feet (lbf ft; lb ft)	X 0.138	= Kilograms-force metres (kgf m; kg m)	X 7.233	= Pounds-force feet (lbf ft; lb ft)
Pounds-force feet (lbf ft; lb ft)	X 1.356	= Newton metres (Nm)	X 0.738	= Pounds-force feet (lbf ft; lb ft)
Newton metres (Nm)	X 0.102	= Kilograms-force metres (kgf m; kg m)	X 9.804	= Newton metres (Nm)

Power
Horsepower (hp)	X 745.7	= Watts (W)	X 0.0013	= Horsepower (hp)

Velocity (speed)
Miles per hour (miles/hr; mph)	X 1.609	= Kilometres per hour (km/hr; kph)	X 0.621	= Miles per hour (miles/hr; mph)

*Fuel consumption**
Miles per gallon, Imperial (mpg)	X 0.354	= Kilometres per litre (km/l)	X 2.825	= Miles per gallon, Imperial (mpg)
Miles per gallon, US (mpg)	X 0.425	= Kilometres per litre (km/l)	X 2.352	= Miles per gallon, US (mpg)

Temperature
Degrees Fahrenheit = (°C x 1.8) + 32   Degrees Celsius (Degrees Centigrade; °C) = (°F − 32) x 0.56

*It is common practice to convert from miles per gallon (mpg) to litres/100 kilometres (l/100km), where mpg (Imperial) x l/100 km = 282 and mpg (US) x l/100 km = 235

Troubleshooting

Contents

Symptom	Section
Engine	
Engine backfires	13
Engine diesels (continues to run) after switching off	15
Engine hard to start or runs poorly when cold	4
Engine hard to start or runs poorly when hot	5
Engine lacks power	12
Engine lopes while idling or idles erratically	8
Engine misses at idle speed	9
Engine misses throughout driving speed range	10
Engine rotates but will not start	2
Engine stalls	11
Engine starts but stops immediately	7
Engine will not rotate when attempting to start	1
Pinging or knocking engine sounds during acceleration or uphill	14
Starter motor noisy or excessively rough in engagement	6
Starter motor operates without rotating engine	3
Engine electrical system	
Battery will not hold a charge	16
Ignition light fails to come on when key is turned on	18
Ignition light fails to go out	17
Fuel system	
Excessive fuel consumption	19
Fuel leakage and/or fuel odor	20
Cooling system	
Coolant loss	25
External coolant leakage	23
Internal coolant leakage	24
Overcooling	22
Overheating	21
Poor coolant circulation	26
Clutch	
Clutch slips (engine speed increases with no increase in vehicle speed)	28
Clutch pedal stays on floor when disengaged	31
Fails to release (pedal pressed to the floor – shift lever does not move freely in and out of Reverse)	27
Grabbing (chattering) as clutch is engaged	29
Squeal or rumble with clutch fully disengaged (pedal depressed)	30
Manual transaxle	
Difficulty in engaging gears	36
Noisy in all gears	33
Noisy in Neutral with engine running	32
Noisy in one particular gear	34
Oil leakage	37
Slips out of high gear	35
Automatic transaxle	
Fluid leakage	41
General shift mechanism problems	38
Transaxle slips, shifts rough, is noisy or has no drive in forward or reverse gears	40
Transaxle will not downshift with accelerator pedal pressed to the floor	39
Driveaxles	
Clicking noise in turns	42
Knock or clunk when accelerating after coasting	43
Shudder or vibration during acceleration	44
Rear axle	
Noise	45
Brakes	
Brake pedal feels spongy when depressed	49
Brake pedal pulsates during brake application	52
Excessive brake pedal travel	48
Excessive effort required to stop vehicle	50
Noise (high-pitched squeal with the brakes applied)	47
Pedal travels to the floor with little resistance	51
Vehicle pulls to one side during braking	46
Suspension and steering systems	
Excessive pitching and/or rolling around corners or during braking	55
Excessive play in steering	57
Excessive tire wear (not specific to one area)	59
Excessive tire wear on inside edge	61
Excessive tire wear on outside edge	60
Excessively stiff steering	56
Lack of power assistance	58
Shimmy, shake or vibration	54
Tire tread worn in one place	62
Vehicle pulls to one side	53

This section provides an easy reference guide to the more common problems which may occur during the operation of your vehicle. These problems and possible causes are grouped under various components or systems; i.e. Engine, Cooling System, etc., and also refer to the Chapter and/or Section which deals with the problem.

Remember that successful troubleshooting is not a mysterious black art practiced only by professional mechanics. It's simply the result of a bit of knowledge combined with an intelligent, systematic approach to the problem. Always work by a process of elimination, starting with the simplest solution and working through to the most complex – and never overlook the obvious. Anyone can forget to fill the gas tank or leave the lights on overnight, so don't assume that you are above such oversights.

Finally, always get clear in your mind why a problem has occurred and take steps to ensure that it doesn't happen again. If the electrical system fails because of a poor connection, check all other connections in the system to make sure that they don't fail as well. If a particular fuse continues to blow, find out why – don't just go on replacing fuses. Remember, failure of a small component can often be indicative of potential failure or incorrect functioning of a more important component or system.

Troubleshooting

Engine and performance

1 Engine will not rotate when attempting to start

1 Battery terminal connections loose or corroded. Check the cable terminals at the battery. Tighten the cable or remove corrosion as necessary.
2 Battery discharged or faulty. If the cable connections are clean and tight on the battery posts, turn the key to the On position and switch on the headlights and/or windshield wipers. If they fail to function, the battery is discharged.
3 Automatic transaxle not completely engaged in Park or Neutral or clutch pedal not completely depressed.
4 Broken, loose or disconnected wiring in the starting circuit. Inspect all wiring and connectors at the battery, starter solenoid, starter safety switch and ignition switch.
5 Starter motor pinion jammed in flywheel ring gear. If manual transaxle, place transaxle in gear and rock the vehicle to manually turn the engine. Remove starter and inspect pinion and flywheel at earliest convenience (see Chapter 5).
6 Starter solenoid faulty (see Chapter 5).
7 Starter motor faulty (see Chapter 5).
8 Ignition switch faulty (see Chapter 12).
9 Neutral start switch (automatic transaxle – see Chapter 7B) or clutch interlock switch (see Chapter 8) faulty.

2 Engine rotates but will not start

1 Fuel tank empty.
2 Fault in the carburetor or fuel injection system (see Chapter 4).
3 Battery discharged (engine rotates slowly). Check the operation of electrical components as described in the previous Section.
4 Battery terminal connections loose or corroded (see previous Section).
5 Fuel pump faulty (see Chapter 4).
6 Excessive moisture on, or damage to, ignition components (see Chapter 5).
7 Worn, faulty or incorrectly gapped spark plugs (see Chapter 1).
8 Broken, loose or disconnected wiring in the starting circuit (see previous Section).
9 Distributor loose, causing ignition timing to change. Turn the distributor as necessary to start the engine, then set the ignition timing as soon as possible (see Chapter 1).
10 Broken, loose or disconnected wires at the ignition coil or faulty coil (see Chapter 5).
11 Faulty igniter (see Chapter 5).

3 Starter motor operates without rotating engine

1 Starter pinion sticking. Remove the starter (see Chapter 5) and inspect.
2 Starter pinion or flywheel teeth worn or broken. Remove the flywheel/driveplate access cover and inspect.

4 Engine hard to start or runs poorly when cold

1 Battery discharged or low. Check as described in Section 1.
2 Fault in the fuel or electrical systems (see Chapters 4 and 5).
3 Carburetor in need of overhaul (see Chapter 4).
4 Distributor rotor carbon tracked and/or damaged (see Chapters 1 and 5).
5 Choke control stuck or inoperative (carbureted models) (see Chapters 1 and 4).
6 On early 1.5L Mirage models, engine hesitation during the first few miles of cold-engine driving may be caused by an incorrect carburetor base plate and sleeve. Ask your dealer service department about Service Bulletin STB-85-14-005. The kit to correct the problem is MD104628 (49-state) and MD104629 (California).
7 On early Cordia/Tredia 2.0L models, hesitation or surging when the engine is cold may be caused by an incorrect coolant temperature sensor or carburetor jets. Ask your dealer service department about Service Bulletins STB-84-14-004 and STB-85-14-006.
8 On early Mirage Turbo models, engine hesitation during the first few miles of cold-engine driving may be caused by an incorrect modulator. Ask your dealer service department about Service Bulletin STB-85-14-007.
9 On 1987 Galant models, an engine that dies out immediately after cold starting or being placed in gear may be caused by an incorrect modulator. Ask your dealer service department about Service Bulletin STB-88-14-002. The modulator kit is part number MD134375.

5 Engine hard to start or runs poorly when hot

1 Air filter clogged (see Chapter 1).
2 Fault in the fuel or electrical systems (see Chapters 4 and 5).
3 Fuel not reaching the carburetor (carburetor-equipped models) (see Section 2).
4 On early 1985 Mirage models, surging, stumbling or hesitation after the vehicle has been "hot soaked" (started 5 to 15 minutes after it's been shut down hot) may be caused by an incorrect float chamber cover. Ask your dealer service department about Service Bulletin STB-85-14-004.

6 Starter motor noisy or excessively rough in engagement

1 Pinion or flywheel gear teeth worn or broken. Remove the cover at the rear of the engine (if so equipped) and inspect.
2 Starter motor mounting bolts loose or missing.

7 Engine starts but stops immediately

1 Loose or faulty electrical connections at distributor, coil or alternator.
2 Fault in the fuel or electrical systems (see Chapters 4 and 5).
3 Insufficient fuel reaching the carburetor (carburetor-equipped models). Check the fuel pump (see Chapter 4).
4 Vacuum leak at the gasket surfaces of the intake manifold, or carburetor/throttle body. Make sure all mounting bolts/nuts are tightened securely and all vacuum hoses connected to the carburetor and manifold are positioned properly and in good condition.
5 On 1987 Galant models, an engine that dies out immediately after cold starting or being placed in gear may be caused by an incorrect modulator. Ask your dealer service department about Service Bulletin STB-88-14-002. The modulator kit is part number MD134375.
6 On 1985 through 1987 Galant models, incorrect air filter element type (see Chapter 1).

8 Engine lopes while idling or idles erratically

1 Vacuum leakage. Check the mounting bolts/nuts at the carburetor/throttle body and intake manifold for tightness. Make sure all vacuum hoses are connected and in good condition. Use a stethoscope or a length of fuel hose held against your ear to listen for vacuum leaks while the engine is running. A hissing sound will be heard. A soapy water solution will also detect leaks.
2 Fault in the fuel or electrical systems (see Chapters 4 and 5).
3 Leaking EGR valve or plugged PCV valve (see Chapters 1 and 6).
4 Air filter clogged or incorrect air filter type – particularly on 1985 through 1987 Galant models – (see Chapter 1).
5 Fuel pump not delivering sufficient fuel to the carburetor (carbureted models) (see Chapter 4).
6 Carburetor out of adjustment (see Chapter 4).
7 Leaking head gasket. Perform a compression check (see Chapter 2).

8 Camshaft lobes worn (see Chapter 2).
9 On early 1986 Galant models with MPI, a rough idle, miss or lack of power may be due to an improperly staked pintle in a fuel injector. The injector must be replaced. Ask your dealer service department about Service Bulletin STB-86-14-004.
10 A surging idle on 1984 feedback carburetor-equipped 2.0L Cordia/Tredia models may be caused by the Deceleration Control Valve. Ask your dealer service department about Service Bulletin STB-84-14-006.

9 Engine misses at idle speed

1 Spark plugs worn or not gapped properly (see Chapter 1).
2 Fault in the fuel or electrical systems (see Chapters 4 and 5).
3 Faulty spark plug wires or distributor cap (see Chapter 1).
4 On early 1986 Galant models with MPI, a rough idle, miss or lack of power may be due to an improperly staked pintle in a fuel injector. The injector must be replaced. Ask your dealer service department about Service Bulletin STB-86-14-004.
5 On 1985 through 1987 Galant models, incorrect air filter type (see Chapter 1).

10 Engine misses throughout driving speed range

1 Fuel filter clogged and/or impurities in the fuel system (see Chapter 1).
2 Faulty or incorrectly gapped spark plugs (see Chapter 1).
3 Fault in the fuel or electrical systems (see Chapters 4 and 5).
4 Incorrect ignition timing (see Chapter 1).
5 Check for cracked distributor cap, disconnected distributor wires and damaged distributor components (see Chapter 1).
6 Defective spark plug wires (see Chapter 1).
7 Faulty emissions system components (see Chapter 6).
8 Low or uneven cylinder compression pressures. Remove the spark plugs and test the compression with a gauge (see Chapter 2).
9 Weak or faulty ignition system (see Chapter 5).
10 Vacuum leaks at the carburetor/throttle body, intake manifold or vacuum hoses (see Section 8).
11 On early 1986 Galant models with MPI, a rough idle, miss or lack of power may be due to an improperly staked pintle in a fuel injector. The injector must be replaced. Ask your dealer service department about Service Bulletin STB-86-14-004.
12 On turbocharged models, a severe miss or engine harshness, particularly at intermediate engine speeds under boost conditions (such as ascending a freeway grade in fifth gear) may be "cross-fire shock." This is usually caused by incorrectly routed spark plug wires, allowing the spark to jump between them. Ask your dealer service department about Service Bulletin STB-85-08-011.
13 On 1987 and earlier models, except Galant, a light surging and engine speed fluctuation of 100 to 200 rpm when the vehicle is being driven at a steady speed may be caused by incomplete engagement of the torque converter damper clutch. Ask your dealer service department about Service Bulletin STB-87-21-003.

11 Engine stalls

1 Idle speed incorrect. Refer to the VECI label and Chapter 1.
2 Fuel filter clogged and/or water and impurities in the fuel system (see Chapter 1).
3 Distributor components damp or damaged (see Chapter 5).
4 Fault in the fuel system or sensors (see Chapters 4 and 6).
5 Faulty emissions system components (see Chapter 6).
6 Faulty or incorrectly gapped spark plugs (see Chapter 1). Also check the spark plug wires (see Chapter 1).
7 Vacuum leak at the carburetor/throttle body, intake manifold or vacuum hoses. Check as described in Section 8.
8 On 1985 through 1987 Galant models, incorrect air filter type (see Chapter 1).

12 Engine lacks power

1 Incorrect ignition timing (see Chapter 1).
2 Fault in the fuel or electrical systems (see Chapters 4 and 5).
3 Excessive play in the distributor shaft. At the same time, check for a damaged rotor, faulty distributor cap, wires, etc. (see Chapters 1 and 5).
4 Faulty or incorrectly gapped spark plugs (see Chapter 1).
5 Carburetor not adjusted properly or excessively worn (carbureted models) (see Chapter 4).
6 Faulty coil (see Chapter 5).
7 Brakes binding (see Chapter 1).
8 Automatic transaxle fluid level incorrect (see Chapter 1).
9 Clutch slipping (see Chapter 8).
10 Fuel filter clogged and/or impurities in the fuel system (see Chapter 1).
11 Emissions control system not functioning properly (see Chapter 6).
12 Use of substandard fuel. Fill the tank with the proper octane fuel.
13 Low or uneven cylinder compression pressures. Test with a compression tester, which will detect leaking valves and/or a blown head gasket (see Chapter 2).
14 On early 1986 Galant models with MPI, a rough idle, miss or lack of power may be due to an improperly staked pintle in a fuel injector. The injector must be replaced. Ask your dealer service department about Service Bulletin STB-86-14-004.
15 On 1985 through 1987 Galant models, incorrect air filter element type (see Chapter 1).

13 Engine backfires

1 Emissions system not functioning properly (see Chapter 6).
2 Fault in the fuel or electrical systems (see Chapters 4 and 5).
3 Ignition timing incorrect (see Chapter 1).
4 Faulty secondary ignition system (cracked spark plug insulator, faulty plug wires, distributor cap and/or rotor) (see Chapters 1 and 5).
5 Carburetor or fuel injection system in need of adjustment or worn excessively (see Chapter 4).
6 Vacuum leak at the carburetor/throttle body, intake manifold or vacuum hoses. Check as described in Section 8.
7 Valves sticking (see Chapter 2).

14 Pinging or knocking engine sounds during acceleration or uphill

1 Incorrect grade of fuel. Fill the tank with fuel of the proper octane rating.
2 Fault in the fuel or electrical systems (see Chapters 4 and 5).
3 Ignition timing incorrect (see Chapter 1).
4 Carburetor in need of adjustment (carbureted models) (see Chapter 4).
5 Improper spark plugs. Check the plug type against the VECI label located in the engine compartment. Also check the plugs and wires for damage (see Chapter 1).
6 Worn or damaged distributor components (see Chapter 5).
7 Faulty emissions system (see Chapter 6).
8 Vacuum leak. Check as described in Section 9.

Troubleshooting 0-23

15 Engine diesels (continues to run) after switching off

1 Idle speed too high. Refer to Chapter 1.
2 Fault in the fuel or electrical systems (see Chapters 4 and 5).
3 Ignition timing incorrectly adjusted (see Chapter 1).
4 Heated Air Intake (HAI) air cleaner system not operating properly (see Chapters 1 and 6).
5 Excessive engine operating temperature. Probable causes of this are a malfunctioning thermostat, clogged radiator, faulty water pump (see Chapter 3).

Engine electrical system

16 Battery will not hold a charge

1 Alternator drivebelt defective or not adjusted properly (see Chapter 1).
2 Electrolyte level low or battery discharged (see Chapter 1).
3 Battery terminals loose or corroded (see Chapter 1).
4 Alternator not charging properly (see Chapter 5).
5 Loose, broken or faulty wiring in the charging circuit (see Chapter 5).
6 Short in the vehicle wiring causing a continual drain on battery (refer to Chapter 12 and the Wiring Diagrams).
7 Battery defective internally.

17 Ignition light fails to go out

1 Fault in the alternator or charging circuit (see Chapter 5).
2 Alternator drivebelt defective or not properly adjusted (see Chapter 1).

18 Ignition light fails to come on when key is turned on

1 Instrument cluster warning light bulb defective (see Chapter 12).
2 Alternator faulty (see Chapter 5).
3 Fault in the instrument cluster printed circuit, dashboard wiring or bulb holder (see Chapter 12).

Fuel system

19 Excessive fuel consumption

1 Dirty or clogged air filter element (see Chapter 1).
2 Incorrectly set ignition timing (see Chapter 1).
3 Choke sticking or improperly adjusted (carbureted models) (see Chapter 1).
4 Emissions system not functioning properly (see Chapter 6).
5 Fault in the fuel or electrical systems (see Chapters 4 and 5).
6 Carburetor or fuel injection system internal parts excessively worn or damaged (see Chapter 4).
7 Low tire pressure or incorrect tire size (see Chapter 1).

20 Fuel leakage and/or fuel odor

1 Leak in a fuel feed or vent line (see Chapter 4).
2 Tank overfilled. Fill only to automatic shut-off.
3 Evaporative emissions system canister clogged (see Chapters 1 or 6).
4 Vapor leaks from system lines (see Chapter 4).
5 Carburetor or fuel injection system internal parts excessively worn or out of adjustment (see Chapter 4).

Cooling system

21 Overheating

1 Insufficient coolant in the system (see Chapter 1).
2 Water pump drivebelt defective or not adjusted properly (see Chapter 1).
3 Radiator core blocked or radiator grille dirty and restricted (see Chapter 3).
4 Thermostat faulty (see Chapter 3).
5 Fan blades broken or cracked (see Chapter 3).
6 Radiator cap not maintaining proper pressure. Have the cap pressure tested by gas station or repair shop.
7 Ignition timing incorrect (see Chapter 1).

22 Overcooling

Thermostat faulty (see Chapter 3).

23 External coolant leakage

1 Deteriorated or damaged hoses or loose clamps. Replace hoses and/or tighten the clamps at the hose connections (see Chapter 1).
2 Water pump seals defective. If this is the case, water will drip from the weep hole in the water pump body (see Chapter 3).
3 Leakage from radiator core or header tank. This will require the radiator to be professionally repaired (see Chapter 3 for removal procedures).
4 Engine drain plug leaking (see Chapter 1) or water jacket core plugs leaking (see Chapter 2).

24 Internal coolant leakage

Note: *Internal coolant leaks can usually be detected by examining the oil. Check the dipstick and inside of the cylinder head cover for water deposits and an oil consistency like that of a milkshake.*

1 Leaking cylinder head gasket. Have the cooling system pressure tested.
2 Cracked cylinder bore or cylinder head. Dismantle the engine and inspect (see Chapter 2).

25 Coolant loss

1 Too much coolant in the system (see Chapter 1).
2 Coolant boiling away due to overheating (see Section 15).
3 External or internal leakage (see Sections 23 and 24).
4 Faulty radiator cap. Have the cap pressure tested.

26 Poor coolant circulation

1 Inoperative water pump. A quick test is to pinch the top radiator hose closed with your hand while the engine is idling, then let it loose. You should feel the surge of coolant if the pump is working properly (see Chapter 1).
2 Restriction in the cooling system. Drain, flush and refill the system (see Chapter 1). If necessary, remove the radiator (see Chapter 3) and have it reverse flushed.

3 Water pump drivebelt defective or not adjusted properly (see Chapter 1).
4 Thermostat sticking (see Chapter 3).

Clutch

27 Fails to release (pedal pressed to the floor – shift lever does not move freely in and out of Reverse)

1 Leak in the clutch hydraulic system. Check the master cylinder, slave cylinder and lines (see Chapter 8).
2 Clutch plate warped or damaged (see Chapter 8).
3 Worn or dry clutch release shaft bushing (see Chapter 8).
4 On mechanical release systems, worn or improperly adjusted cable (see Chapters 1 and 8).

28 Clutch slips (engine speed increases with no increase in vehicle speed)

1 Linkage out of adjustment (see Chapter 8).
2 Clutch plate oil soaked or lining worn. Remove clutch (see Chapter 8) and inspect.
3 Clutch plate not seated. It may take 30 or 40 normal starts for a new one to seat.

29 Grabbing (chattering) as clutch is engaged

1 Oil on clutch plate lining. Remove (see Chapter 8) and inspect. Correct any leakage source.
2 Worn or loose engine or transaxle mounts. These units move slightly when the clutch is released. Inspect the mounts and bolts (see Chapter 2).
3 Worn splines on clutch plate hub. Remove the clutch components (see Chapter 8) and inspect.
4 Warped pressure plate or flywheel. Remove the clutch components and inspect.

30 Squeal or rumble with clutch fully disengaged (pedal depressed)

1 Worn, defective or broken release bearing (see Chapter 8).
2 Worn or broken pressure plate springs (or diaphragm fingers) (see Chapter 8).

31 Clutch pedal stays on floor when disengaged

Linkage or release bearing binding. Inspect the linkage or remove the clutch components as necessary.

Manual transaxle

32 Noisy in Neutral with engine running

1 Input shaft bearing worn.
2 Damaged main drive gear bearing.
3 Worn countershaft bearings.
4 Worn or damaged countershaft endplay shims.

33 Noisy in all gears

1 Any of the above causes, and/or:
2 Insufficient lubricant (see the checking procedures in Chapter 1).

34 Noisy in one particular gear

1 Worn, damaged or chipped gear teeth for that particular gear.
2 Worn or damaged synchronizer for that particular gear.

35 Slips out of high gear

1 Transaxle loose on clutch housing (see Chapter 7).
2 Shift rods interfering with the engine mounts or clutch lever (see Chapter 7).
3 Shift rods not working freely (see Chapter 7).
4 Dirt between the transaxle case and engine or misalignment of the transaxle (see Chapter 7).
5 Worn or improperly adjusted linkage (see Chapter 7).

36 Difficulty in engaging gears

1 Clutch not releasing completely (see clutch adjustment in Chapter 1).
2 Loose, damaged or out-of-adjustment shift linkage. Make a thorough inspection, replacing parts as necessary (see Chapter 7).

37 Oil leakage

1 Excessive amount of lubricant in the transaxle (see Chapter 1 for correct checking procedures). Drain lubricant as required.
2 Driveaxle oil seal or speedometer oil seal in need of replacement (see Chapter 7).

Automatic transaxle

Note: *Due to the complexity of the automatic transaxle, it's difficult for the home mechanic to properly diagnose and service this component. For problems other than the following, the vehicle should be taken to a dealer service department or a transmission shop.*

38 General shift mechanism problems

1 Chapter 7 deals with checking and adjusting the shift linkage on automatic transaxles. Common problems which may be attributed to poorly adjusted linkage are:
Engine starting in gears other than Park or Neutral.
Indicator on shifter pointing to a gear other than the one actually being selected.
Vehicle moves when in Park.
2 Refer to Chapter 7 to adjust the linkage.

39 Transaxle will not downshift with accelerator pedal pressed to the floor

Chapter 7 deals with adjusting the throttle cable to enable the transaxle to downshift properly.

Troubleshooting

40 Transaxle slips, shifts rough, is noisy or has no drive in forward or reverse gears

1 There are many probable causes for the above problems, but the home mechanic should be concerned with only one possibility – fluid level.
2 Before taking the vehicle to a repair shop, check the level and condition of the fluid as described in Chapter 1. Correct fluid level as necessary or change the fluid and filter if needed. If the problem persists, have a professional diagnose the probable cause.
3 On 1985 through 1987 Galant models, harsh or slow shifts may be caused by the kickdown servo piston not releasing properly. Ask your dealer service department about Service Bulletin STB-87-21-008.

41 Fluid leakage

1 Automatic transaxle fluid is a deep red color. Fluid leaks should not be confused with engine oil, which can easily be blown by air flow to the transaxle.
2 To pinpoint a leak, first remove all built-up dirt and grime from around the transaxle. Degreasing agents and/or steam cleaning will achieve this. With the underside clean, drive the vehicle at low speeds so air flow will not blow the leak far from its source. Raise the vehicle and determine where the leak is coming from. Common areas of leakage are:
  a) **Pan:** Tighten the mounting bolts and/or replace the pan gasket as necessary (see Chapter 7).
  b) **Filler pipe:** Replace the rubber seal where the pipe enters the transaxle case.
  c) **Transaxle oil lines:** Tighten the connectors where the lines enter the transaxle case and/or replace the lines.
  d) **Vent pipe:** Transaxle overfilled and/or water in fluid (see checking procedures, Chapter 1).
  e) **Speedometer connector:** Replace the O-ring where the speedometer cable enters the transaxle case (see Chapter 7).
3 On 1986 and later Galant models, a leak from the torque converter area may be caused by a leaking O-ring between the front pump cover and torque converter housing. Ask your dealer service department about Service Bulletin STB-86-21-007.

Driveaxles

42 Clicking noise in turns

Worn or damaged outer joint. Check for cut or damaged seals. Repair as necessary (see Chapter 8).

43 Knock or clunk when accelerating after coasting

Worn or damaged inner joint. Check for cut or damaged seals. Repair as necessary (see Chapter 8).

44 Shudder or vibration during acceleration

1 Excessive joint angle. Have checked and correct as necessary (see Chapter 8).
2 Worn or damaged CV joints. Repair or replace as necessary (see Chapter 8).
3 Sticking CV joint assembly. Correct or replace as necessary (see Chapter 8).

Rear axle

45 Noise

1 Road noise. No corrective procedures available.
2 Tire noise. Inspect tires and check tire pressures (see Chapter 1).
3 Rear wheel bearings loose, worn or damaged (see Chapter 1).

Brakes

Note: *Before assuming that a brake problem exists, make sure that the tires are in good condition and inflated properly (see Chapter 1), that the front end alignment is correct and that the vehicle is not loaded with weight in an unequal manner.*

46 Vehicle pulls to one side during braking

1 Defective, damaged or oil contaminated disc brake pads on one side. Inspect as described in Chapter 9.
2 Excessive wear of brake pad material or disc on one side. Inspect and correct as necessary.
3 Loose or disconnected front suspension components. Inspect and tighten all bolts to the specified torque (see Chapter 10).
4 Defective caliper assembly. Remove the caliper and inspect for a stuck piston or other damage (see Chapter 9).

47 Noise (high-pitched squeal with the brakes applied)

1 Disc brake pads worn out. The noise comes from the pad backing plate rubbing against the disc. Replace the pads with new ones immediately (see Chapter 9). If the pad material has worn completely away, the brake discs should be inspected for damage as described in Chapter 9.
2 On 1985 and earlier Galant models, the noise may be corrected using special anti-squeal shims (part number MB500400). Ask your dealer service department about Service Bulletin STB-85-05-001.

48 Excessive brake pedal travel

1 Partial brake system failure. Inspect the entire system (see Chapter 9) and correct as required.
2 Insufficient fluid in the master cylinder. Check (see Chapter 1), add fluid and bleed the system if necessary (see Chapter 9).
3 Rear brakes not adjusting properly. Make a series of starts and stops while the vehicle is in Reverse. If this does not correct the situation, inspect the self-adjusting mechanism (see Chapter 9).

49 Brake pedal feels spongy when depressed

1 Air in the hydraulic lines. Bleed the brake system (see Chapter 9).
2 Faulty flexible hoses. Inspect all system hoses and lines. Replace parts as necessary.
3 Master cylinder mounting bolts/nuts loose.
4 Master cylinder defective (see Chapter 9).

50 Excessive effort required to stop vehicle

1 Power brake booster not operating properly (see Chapter 9).
2 Excessively worn linings or pads. Inspect and replace if necessary (see Chapter 9).

Troubleshooting

3 One or more caliper pistons or wheel cylinders seized or sticking. Inspect and rebuild as required (see Chapter 9).
4 Brake linings or pads contaminated with oil or grease. Inspect and replace as required (see Chapter 9).
5 New pads or shoes installed and not yet seated. It will take a while for the new material to seat against the drum (or rotor).

51 Pedal travels to the floor with little resistance

Little or no fluid in the master cylinder reservoir caused by leaking wheel cylinder(s), leaking caliper piston(s), loose, damaged or disconnected brake lines. Inspect the entire system and correct as necessary.

52 Brake pedal pulsates during brake application

1 Caliper improperly installed. Remove and inspect (see Chapter 9).
2 Disc or drum defective. Remove (see Chapter 9) and check for excessive lateral runout and parallelism. Have the disc or drum resurfaced or replace it with a new one.

Suspension and steering systems

53 Vehicle pulls to one side

1 Tire pressures uneven (see Chapter 1).
2 Defective tire (see Chapter 1).
3 Excessive wear in suspension or steering components (see Chapter 10).
4 Front end in need of alignment.
5 Front brakes dragging. Inspect the brakes as described in Chapter 9.

54 Shimmy, shake or vibration

1 Tire or wheel out-of-balance or out-of-round. Have professionally balanced.
2 Loose, worn or out-of-adjustment rear wheel bearings (see Chapter 1).
3 Shock absorbers and/or suspension components worn or damaged (see Chapter 10).

55 Excessive pitching and/or rolling around corners or during braking

1 Defective shock absorbers. Replace as a set (see Chapter 10).
2 Broken or weak springs and/or suspension components. Inspect as described in Chapter 10.

56 Excessively stiff steering

1 Lack of fluid in power steering fluid reservoir (see Chapter 1).
2 Incorrect tire pressures (see Chapter 1).
3 Front end out of alignment.

57 Excessive play in steering

1 Excessive wear in suspension or steering components (see Chapter 10).
2 Steering gear damaged (see Chapter 10).

58 Lack of power assistance

1 Steering pump drivebelt faulty or not adjusted properly (see Chapter 1).
2 Fluid level low (see Chapter 1).
3 Hoses or lines restricted. Inspect and replace parts as necessary.
4 Air in power steering system. Bleed the system (see Chapter 10).

59 Excessive tire wear (not specific to one area)

1 Incorrect tire pressures (see Chapter 1).
2 Tires out-of-balance. Have professionally balanced.
3 Wheels damaged. Inspect and replace as necessary.
4 Suspension or steering components excessively worn (see Chapter 10).

60 Excessive tire wear on outside edge

1 Inflation pressures incorrect (see Chapter 1).
2 Excessive speed in turns.
3 Front end alignment incorrect (excessive toe-in). Have professionally aligned.
4 Suspension arm bent or twisted (see Chapter 10).

61 Excessive tire wear on inside edge

1 Inflation pressures incorrect (see Chapter 1).
2 Front end alignment incorrect. Have professionally aligned.
3 Loose or damaged steering components (see Chapter 10).

62 Tire tread worn in one place

1 Tires out-of-balance.
2 Damaged or buckled wheel. Inspect and replace if necessary.
3 Defective tire (see Chapter 1).

Chapter 1 Tune-up and routine maintenance

Contents

Section	No.
Air and PCV filter replacement	17
Automatic transaxle fluid and filter change	34
Automatic transaxle fluid level check	7
Battery check, maintenance and charging	11
Brake fluid replacement (Precis models only)	33
Brake check	16
Carburetor choke check and cleaning	29
Clutch pedal freeplay check and adjustment (cable-operated clutch)	10
Cooling system check	14
Cooling system servicing (draining, flushing and refilling)	31
Driveaxle boot check	23
Drivebelt check, adjustment and replacement	12
Engine idle speed check and adjustment (carbureted models only)	20
Engine oil and oil filter change	8
Evaporative emissions control system check and canister replacement	36
Exhaust system check	24
Fluid level checks	4
Fuel filter replacement	26
Fuel system check	21
Heated Air Intake (HAI) air cleaner check	30
Ignition timing check and adjustment	37
Introduction	1
Maintenance schedule	3
Manual transaxle lubricant change	35
Manual transaxle lubricant level check	25
Oxygen sensor replacement	39
Positive Crankcase Ventilation (PCV) valve check and replacement	38
Power steering fluid level check	6
Rear wheel bearing check, repack and adjustment (drum brakes only)	32
Spark plug replacement	27
Spark plug wire, distributor cap and rotor check and replacement	28
Steering and suspension check	22
Throttle position sensor check (feedback carburetor)	18
Tire and tire pressure checks	5
Tire rotation	15
Tune-up general information	2
Underhood hose check and replacement	13
Valve clearance check and adjustment (not all models)	19
Windshield wiper blade inspection and replacement	9

Specifications

Recommended lubricants and fluids

Note: *Listed here are manufacturers recommendations at the time this manual was written. Manufacturers occasionally upgrade their fluid and lubricant specifications, so check with your auto parts stores for current recommendations.*

Engine oil
- Type .... API grade SF or SF/CC multigrade and fuel efficient oil
- Viscosity .... See accompanying chart
- Capacity (approximate) .... 4.2 qts (4.0 liters)

Fuel .... Unleaded gasoline, 87 octane or higher

Automatic transaxle fluid
- Type .... ATF+3 type 7176 automatic transmission fluid
- Total capacity (approximate) .... 6.0 qts (5.71 liters)
- Refill capacity after fluid change (approximate) .... 4.2 qts (4.0 liters)

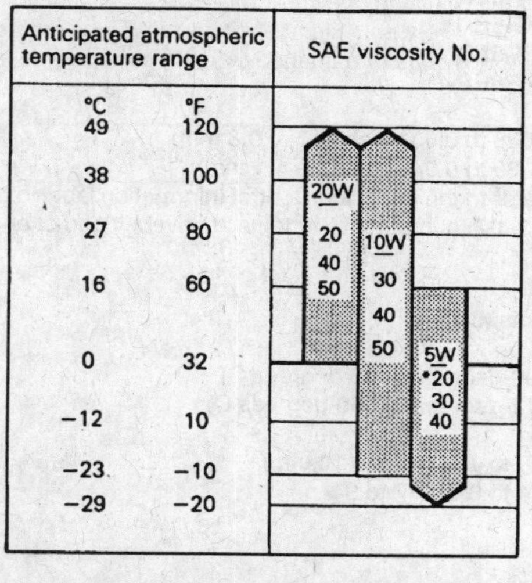

Engine oil viscosity chart – for best fuel economy and cold starting, select the lowest SAE viscosity number for the expected temperature range

Chapter 1 Tune-up and routine maintenance

 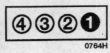

Front

1983 to 1988 Cordia, Tredia 1.8L
1985 to 1988 1.6L Turbo
1985 to 1990 Mirage, Precis 1.5L
1991 to 1993 Precis 1.5L
1988 to 1993 Galant 2.0L SOHC

1991 to 1993 Mirage 1.5L

Front

1989 to 1992 Mirage 1.6L
1989 to 1993 Galant 2.0L DOHC

Front

1984 to 1988 Tredia, Cordia 2.0L
1985 to 1987 Galant 2.4L

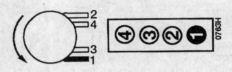

Front 1993 Galant 2.0L SOHC

The blackened terminal shown on the distributor cap indicates the Number One spark plug wire position

Cylinder location and distributor rotation

Recommended lubricants and fluids (continued)
Manual transaxle lubricant
  Type .... API GL-5 SAE 75/85W gear oil
  Capacity (approximate) .... 2.2 qts (2 liters)
Brake fluid type .... DOT 3 brake fluid
Power steering system fluid .... Dexron II automatic transmission fluid
Rear wheel bearing grease .... NLGI No. 2 EP wheel bearing grease

Ignition system
Spark plug
  Type
    G62B 1.8L and G32B 1.6L engines BPR7ES-11
    1990 and later Galant turbo BPR6ES
    1993 Galant SOHC 16-valve engine (non-turbo) .. BKR5E-11
    1993 Mirage 1.5L engine BPR5ES-11
    All others BPR6ES-11
  Gap
    1990 and later Galant turbo .... 0.028 to 0.031 inch (0.7 to 0.8 mm)
    All others .... 0.039 to 0.043 in (1.0 to 1.1 mm)
Ignition timing .... Refer to the Emission Control Information label in the engine compartment
Firing order .... 1-3-4-2 (number 1 cylinder is at drivebelt end of engine)

Cooling system
Coolant capacity (approximate) .... 6 qts (5 liters)
Thermostat rating
  Starts to open .... 190-degrees F (88-degrees C)
  Fully open .... 212-degrees F (100-degrees C)
Drivebelt deflection
  1990 and later Galant and 1.6L Mirage .... 3/8 to 1/2 in (10 to 14 mm)
  All others .... 1/4 to 3/8 in (7 to 9 mm)

Clutch (cable-operated)
Clutch pedal freeplay
  1990 and later Galant .... 1/4 to 1/2 in (6 to 12 mm)
  All others .... 1 in (25 mm)

Chapter 1 Tune-up and routine maintenance

Adjusting nut-to-insulator projection clearance
   Mirage and Precis .... 1/4 in (6.3 mm)
   Cordia/Tredia .... 1/8 in (3.175 mm)

Brakes
Disc brake pad lining thickness (minimum) .... 1/8 in (3.175 mm)
Drum brake shoe lining thickness (minimum) .... 1/8 in (3.175 mm)

Steering wheel freeplay limit .... 1 in (26 mm)

Valve clearances (engine hot)
Intake valve .... 0.006 in (0.15 mm)
Exhaust valve .... 0.010 in (0.25 mm)
Jet valve .... 0.010 in (0.25 mm)

Torque specifications
                                                                                 Ft-lbs
Automatic transaxle
   Pan bolts 8
   Filter bolts 7
   Drain plug 25
Manual transaxle drain and filler plugs 25
Engine oil pan drain plug 33
Spark plugs .... 15 to 21
Wheel lug nuts .... 65 to 80
Rear wheel bearing spindle nut
   castellated nut
      step 1 12
      step 2 .... loosen to 0
      step 3 .... 4 to 7
   self-locking nut .... 72 to 108

1 Introduction

This Chapter is designed to help the home mechanic maintain the Mitsubishi front wheel drive models with the goals of maximum performance, economy, safety and reliability in mind.

Included is a master maintenance schedule, followed by procedures dealing specifically with each item on the schedule. Visual checks, adjustments, component replacement and other helpful items are included. Refer to the accompanying illustrations of the engine compartment and the underside of the vehicle for the locations of various components.

Adhering to the mileage/time maintenance schedule and following the step-by-step procedures, which is simply a preventive maintenance program, will result in maximum reliability and vehicle service life. Keep in mind that it is a comprehensive program – maintaining some items but not others at the specified intervals will not produce the same results.

As you service the vehicle, you will discover that many of the procedures can – and should – be grouped together because of the nature of the particular procedure you're performing or because of the close proximity of two otherwise unrelated components to one another.

For example, if the vehicle is raised for chassis lubrication, you should inspect the exhaust, suspension, steering and fuel systems while you're under the vehicle. When you're rotating the tires, it makes good sense to check the brakes, since the wheels are already removed. Finally, let's suppose you have to borrow or rent a torque wrench. Even if you only need it to tighten the spark plugs, you might as well check the torque of as many critical fasteners as time allows.

The first step in this maintenance program is to prepare yourself before the actual work begins. Read through all the procedures you're planning to do, then gather up all the parts and tools needed. If it looks like you might run into problems during a particular job, seek advice from a mechanic or an experienced do-it-yourselfer.

2 Tune-up general information

The term tune-up is used in this manual to represent a combination of individual operations rather than one specific procedure.

If, from the time the vehicle is new, the routine maintenance schedule is followed closely and frequent checks are made of fluid levels and high wear items, as suggested throughout this manual, the engine will be kept in relatively good running condition and the need for additional work will be minimized.

More likely than not, however, there will be times when the engine is running poorly due to lack of regular maintenance. This is even more likely if a used vehicle, which has not received regular and frequent maintenance checks, is purchased. In such cases, an engine tune-up will be needed outside of the regular routine maintenance intervals.

The first step in any tune-up or diagnostic procedure to help correct a poor running engine is a cylinder compression check (see Chapter 2). This check will help determine the condition of internal engine components and should be used as a guide for tune-up and repair procedures. For instance, if a compression check indicates serious internal engine wear, a conventional tune-up will not improve the performance of the engine and would be a waste of time and money. Because of its importance, the compression check should be done by someone with the right equipment and the knowledge to use it properly.

The following procedures are those most often needed to bring a generally poor running engine back into a proper state of tune.

Minor tune-up
   Clean, inspect and test the battery (Section 11)
   Check all engine related fluids (Section 4)
   Check and adjust the drivebelts (Section 12)
   Replace the spark plugs (Section 27)
   Inspect the distributor cap and rotor (Section 28)
   Inspect the spark plug and coil wires (Section 28)
   Check and adjust the idle speed (Section 20)
   Check the PCV valve (Section 38)
   Check the air filter (Section 17)
   Check the cooling system (Section 14)
   Check all underhood hoses (Section 13)

Major tune-up
   All items listed under Minor tune-up, plus . . .
   Check the EGR system (see Chapter 6)
   Check the ignition system (see Chapter 5)
   Check the charging system (see Chapter 5)
   Check the fuel system (see Chapter 4)
   Replace the air filter (Section 17)
   Replace the distributor cap and rotor (Section 28)
   Replace the spark plug wires (Section 28)

1-4 Chapter 1 Tune-up and routine maintenance

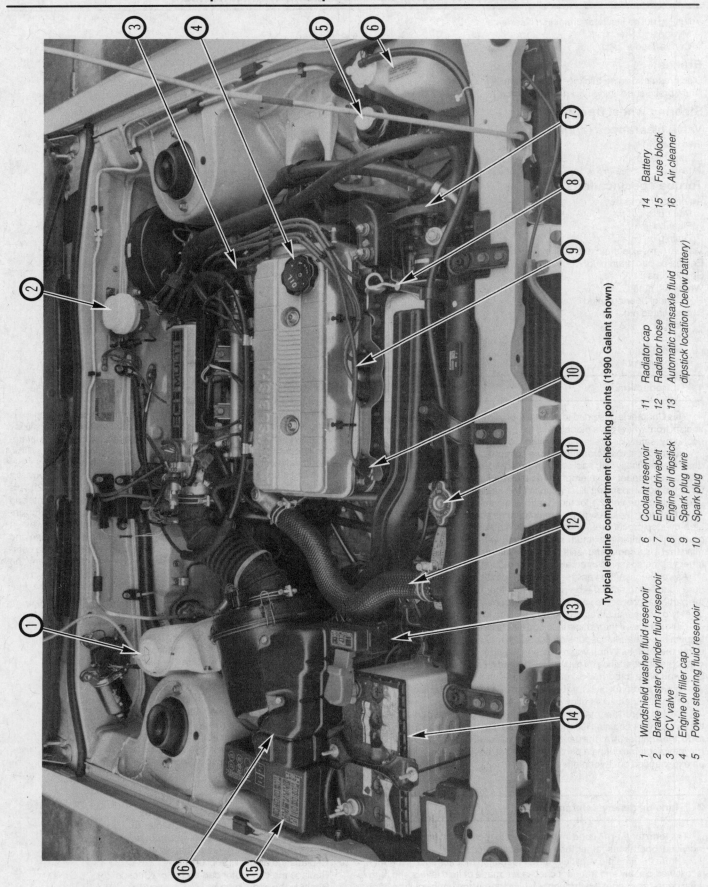

Typical engine compartment checking points (1990 Galant shown)

1. Windshield washer fluid reservoir
2. Brake master cylinder fluid reservoir
3. PCV valve
4. Engine oil filler cap
5. Power steering fluid reservoir
6. Coolant reservoir
7. Engine drivebelt
8. Engine oil dipstick
9. Spark plug wire
10. Spark plug
11. Radiator cap
12. Radiator hose
13. Automatic transaxle fluid dipstick location (below battery)
14. Battery
15. Fuse block
16. Air cleaner

Chapter 1 Tune-up and routine maintenance

Typical engine compartment underside checking points

1. Automatic transaxle fluid dipstick
2. Lower radiator hose
3. Engine oil filter
4. Radiator drain
5. Engine drivebelt
6. Disc brake caliper
7. Driveaxle CV joint boot
8. Engine oil pan drain plug
9. Exhaust system
10. Automatic transaxle pan

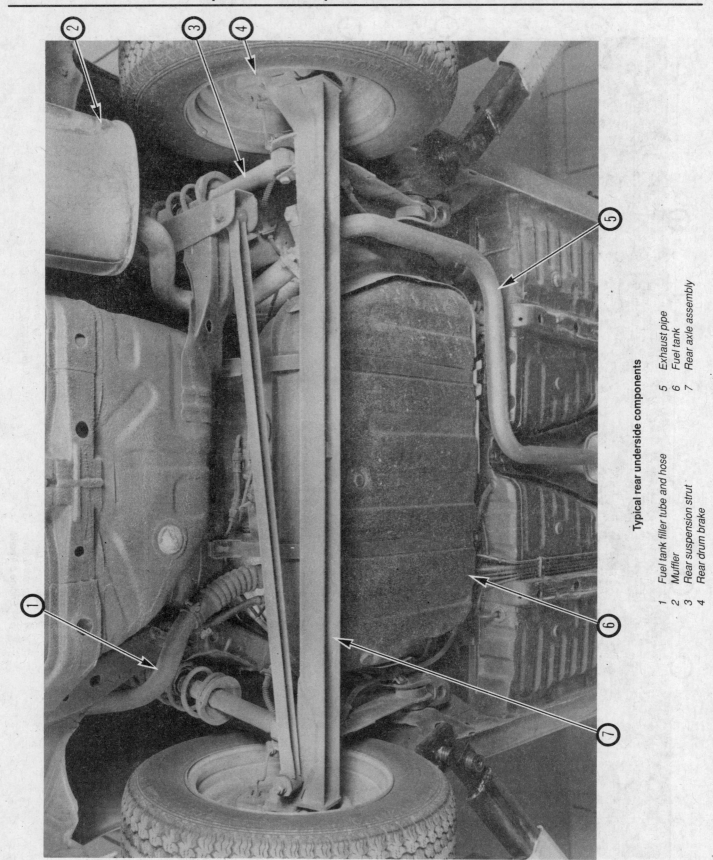

Typical rear underside components

1 Fuel tank filler tube and hose
2 Muffler
3 Rear suspension strut
4 Rear drum brake
5 Exhaust pipe
6 Fuel tank
7 Rear axle assembly

3 Mitsubishi Maintenance schedule

The following maintenance intervals are based on the assumption that the vehicle owner will be doing the maintenance or service work, as opposed to having a dealer service department do the work. Although the time/mileage intervals are loosely based on factory recommendations, most have been shortened to ensure, for example, that such items as filters, lubricants and fluids are checked/changed at intervals that promote maximum engine/driveline service life. Also, subject to the preference of the individual owner interested in keeping his or her vehicle in peak condition at all times, and with the vehicle's ultimate resale in mind, many of the maintenance procedures may be performed more often than recommended in the following schedule. We encourage such owner initiative.

When the vehicle is new it should be serviced initially by a factory authorized dealer service department to protect the factory warranty. In many cases the initial maintenance check is done at no cost to the owner (check with your dealer service department for more information).

Every 250 miles or weekly, whichever comes first

Check the engine oil level (Section 4)
Check the engine coolant level (Section 4)
Check the windshield washer fluid level (Section 4)
Check the water (electrolyte) in the battery (Section 4)
Check the brake fluid level (Section 4)
Check the tires and tire pressures (Section 5)

Every 3500 miles or 3 months, whichever comes first

All items listed above plus . . .
Check the power steering fluid level (Section 6)
Check the automatic transaxle fluid level (Section 7)
Change the engine oil and oil filter (Section 8)

Every 7500 miles or 6 months, whichever comes first

Inspect/replace the windshield wiper blades (Section 9)
Check/adjust the clutch pedal freeplay (cable-operated models) (Section 10)
Check and service the battery (Section 11)
Check/adjust the engine drivebelts (Section 12)
Inspect/replace all underhood hoses (Section 13)
Check the cooling system (Section 14)
Rotate the tires (Section 15)

Every 15,000 miles or 12 months, whichever comes first

All items listed above plus . . .
Inspect the brakes (Section 16)
Check/replace the air and PCV filters (Section 17)
Check the throttle positioner sensor system (1.5L carbureted, 1986 and earlier Mirage and 1989 and earlier Precis models) (Section 18)
Check/adjust the valve clearances (some models) (Section 19)
Check/adjust the engine idle speed – carbureted models only – (Section 20)
Inspect the fuel system (Section 21)
Inspect the steering and suspension components (Section 22)*
Check the driveaxle boots (Section 23)*
Inspect the exhaust system (Section 24)
Check the manual transaxle lubricant level (Section 25)

Every 30,000 miles or 24 months, whichever comes first

Replace the fuel filter (Section 26)
Check/replace the spark plugs (Section 27)
Inspect/replace the spark plug wires, distributor cap and rotor (Section 28)*
Check the carburetor choke and clean the linkage (Section 29)
Check the Heated Air Intake (HAI) air cleaner (Section 30)
Replace the drivebelts for the water pump and alternator (Section 12)
Drain, flush and refill the cooling system (Section 31)
Check/repack the rear wheel bearings (Section 32)
Drain the brake system and refill it with new fluid (Section 33)
If the vehicle is equipped with an automatic transaxle, change the fluid and filter (Section 34)**
If the vehicle is equipped with a manual transaxle, drain and refill it with new lubricant (Section 35)

Every 50,000 miles or 40 months, whichever comes first

Inspect the evaporative emissions control system and replace the canister (Section 36)
Check/adjust the ignition timing (Section 37)
Check/replace the PCV valve (Section 38)
Replace the oxygen sensor (Section 39)

Every 60,000 miles or 48 months, whichever comes first

Install a new timing belt (see Chapter 2)
Replace the turbocharger air intake and oil hoses (See Chapter 4 – should be done by a dealer service department)

* This item is affected by "severe" operating conditions as described below. If the vehicle in question is operated under "severe" conditions, perform all maintenance indicated with an asterisk (*) at 7500 mile/6 month intervals. Consider the conditions "severe" if most driving is done . . .
In dusty areas
When towing a trailer
At low speeds or with extended periods of engine idling
When outside temperatures remain below freezing and most trips are less than four miles

** If most driving is done under one or more of the following conditions, change the automatic transaxle fluid every 15,000 miles:
In heavy city traffic where the outside temperature regularly reaches 90-degrees F (32-degrees C) or higher
In hilly or mountainous terrain
Frequent trailer pulling

Chapter 1 Tune-up and routine maintenance

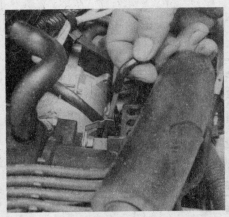

4.2a The engine oil dipstick is located by a clip on some models

4.2b Typical oil dipstick location (Galant shown)

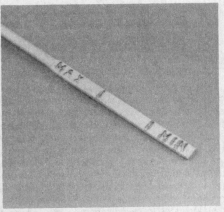

4.4 The oil level should be between the MIN and MAX marks – if it isn't, add enough oil to bring the level to or near the MAX mark (it takes one full quart to raise the level from the MIN to the MAX mark)

4.6 Rotate the oil filler cap half a turn to remove it – always make sure the area around the opening is clean before removing the cap; this prevents dirt from contaminating the engine

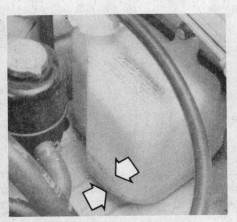

4.8 Make sure the coolant level in the reservoir is between the upper and lower marks – if it's below the lower mark, add the prescribed mixture of antifreeze and water until the level is where it should be (see text)

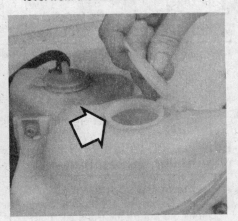

4.14a On some models the windshield washer fluid reservoir is located next to the coolant reservoir – be careful not to confuse these two reservoirs!

4 Fluid level checks

Note: *The following fluid level checks should be done on a 250 mile or weekly basis. Additional fluid level checks can be found in specific maintenance procedures which follow. Regardless of intervals, be alert for fluid leaks under the vehicle which would indicate a leak to be fixed immediately.*

Warning: *The electric cooling fan can activate at any time, even when the ignition is in the Off position. Disconnect the fan motor or negative battery cable when working in the vicinity of the fan.*

1 Fluids are an essential part of the lubrication, cooling, brake and windshield washer systems. Because the fluids gradually become depleted and/or contaminated during normal operation of the vehicle, they must be periodically replenished. See *Recommended lubricants and fluids* at the beginning of this Chapter before adding fluid to any of the following components. **Note:** *The vehicle must be on level ground when fluid levels are checked.*

Engine oil

Refer to illustrations 4.2a, 4.2b, 4.4 and 4.6

2 The engine oil level is checked with a dipstick located on the side of the engine at the front (drivebelt) end **(see illustrations)**. It extends through a tube and into the oil pan at the bottom of the engine.

3 The oil level should be checked before the vehicle has been driven, or about 15 minutes after the engine has been shut off. If the oil is checked immediately after driving the vehicle, some of the oil will remain in the upper engine components, resulting in an inaccurate reading on the dipstick.

4 Pull the dipstick out and wipe all the oil off the end with a clean rag or paper towel. Insert the clean dipstick all the way back into the tube, then pull it out again. Note the oil at the end of the dipstick. Add oil as necessary to keep the level between the MIN mark and the MAX mark on the dipstick **(see illustration)**.

5 Don't overfill the engine by adding too much oil, since it may result in oil fouled spark plugs, oil leaks or oil seal failures.

6 Oil is added to the engine after removing the cap from the valve cover **(see illustration)**. An oil can spout or funnel may help to reduce spills.

7 Checking the oil level is an important preventive maintenance step. A consistently low oil level indicates oil leakage through damaged seals, defective gaskets or past worn rings or valve guides. If the oil looks milky in color or has water droplets in it, the cylinder head gasket may be blown or the head or block may be cracked. The engine should be checked immediately. The condition of the oil should also be noted. Whenever you check the oil level, slide your thumb and index finger up the dipstick before wiping off the oil. If you see small dirt or metal particles clinging to the dipstick, the oil should be changed (Section 8).

Chapter 1 Tune-up and routine maintenance

1 – 9

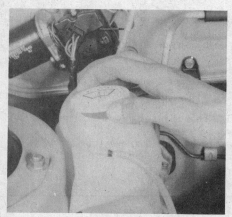

4.14b On other models the washer fluid reservoir is located separately from the coolant reservoir – flip up the cap to add fluid

4.17 Remove the cell caps to check the electrolyte (water) level in the battery – if the level is low, add distilled water only

4.19a The brake fluid level should be between the MIN and MAX marks (don't let it drop below MIN) – unscrew the cap to add fluid

Engine coolant

Refer to illustration 4.8

Warning: *Do not allow antifreeze to come in contact with your skin or painted surfaces of the vehicle. Flush contaminated areas immediately with plenty of water. Don't store new coolant or leave old coolant lying around where it's accessible to children or pets – they're attracted by its sweet taste. Ingestion of even a small amount of coolant can be fatal! Wipe up garage floor and drip pan coolant spills immediately. Keep antifreeze containers covered and repair leaks in your cooling system as soon as they are noted.*

8 All vehicles covered by this manual are equipped with a pressurized coolant recovery system. A white plastic coolant reservoir located in the engine compartment is connected by a hose to the radiator filler neck **(see illustration)**. If the engine overheats, coolant escapes through a valve in the radiator cap and travels through the hose into the reservoir. As the engine cools, the coolant is automatically drawn back into the cooling system to maintain the correct level.

9 The coolant level in the reservoir should be checked regularly. **Warning:** *Do not remove the radiator cap to check the coolant level when the engine is warm.* The level in the reservoir varies with the temperature of the engine. When the engine is cold, the coolant level should be at or slightly above the LOW mark on the reservoir **(see illustration 4.8)**. Once the engine has warmed up, the level should be at or near the FULL mark. If it isn't, allow the engine to cool, then remove the cap from the reservoir and add a 50/50 mixture of ethylene glycol-based antifreeze and water.

10 Drive the vehicle and recheck the coolant level. If only a small amount of coolant is required to bring the system up to the proper level, water can be used. However, repeated additions of water will dilute the antifreeze and water solution. In order to maintain the proper ratio of antifreeze and water, always top up the coolant level with the correct mixture. An empty plastic milk jug or bleach bottle makes an excellent container for mixing coolant. Do not use rust inhibitors or additives.

11 If the coolant level drops consistently, there may be a leak in the system. Inspect the radiator, hoses, filler cap, drain plugs and water pump (see Section 14). If no leaks are noted, have the radiator cap pressure tested by a service station.

12 If you have to remove the radiator cap, wait until the engine has cooled, then wrap a thick cloth around the cap and turn it to the first stop. If coolant or steam escapes, let the engine cool down longer, then remove the cap.

13 Check the condition of the coolant as well. It should be relatively clear. If it's brown or rust colored, the system should be drained, flushed and refilled. Even if the coolant appears to be normal, the corrosion inhibitors wear out, so it must be replaced at the specified intervals.

Windshield washer fluid

Refer to illustration 4.14a and 4.14b

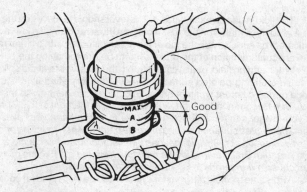

4.19b On some models the level should be kept between the MAX and A marks (it shouldn't drop below the B mark)

14 Fluid for the windshield washer system is located in a plastic reservoir in the engine compartment **(see illustrations)**.

15 In milder climates, plain water can be used in the reservoir, but it should be kept no more than 2/3 full to allow for expansion if the water freezes. In colder climates, use windshield washer system antifreeze, available at any auto parts store, to lower the freezing point of the fluid. Mix the antifreeze with water in accordance with the manufacturer's directions on the container. **Caution:** *Don't use cooling system antifreeze – it will damage the vehicle's paint.*

16 To help prevent icing in cold weather, warm the windshield with the defroster before using the washer.

Battery electrolyte

Refer to illustration 4.17

17 All vehicles with which this manual is concerned are equipped with a battery which is permanently sealed (except for vent holes) and has no filler caps. Water doesn't have to be added to these batteries at any time. If an aftermarket maintenance-type battery is installed, the caps on top of the battery should be removed periodically to check for a low water level **(see illustration)**. This check is most critical during the warm summer months.

Brake and clutch fluid

Refer to illustrations 4.19a and 4.19b

18 The brake master cylinder is mounted on the front of the power booster unit in the engine compartment. The clutch master cylinder (used on later manual transaxle-equipped models) is located next to the brake master cylinder. Clutch fluid level is checked in the same manner as brake fluid.

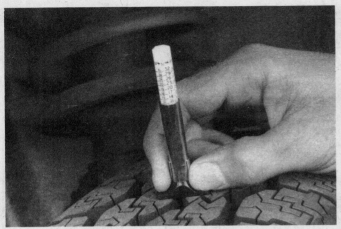

5.2 Use a tire tread depth indicator to monitor tire wear — they are available at auto parts stores and service stations and cost very little

19 The fluid inside is readily visible. The level should be between the MIN (or A) and MAX marks on the reservoir **(see illustrations)**. If a low level is indicated, be sure to wipe the top of the reservoir cover with a clean rag to prevent contamination of the brake system before removing the cover.

20 When adding fluid, pour it carefully into the reservoir to avoid spilling it onto surrounding painted surfaces. Be sure the specified fluid is used, since mixing different types of brake fluid can cause damage to the system. See Recommended lubricants and fluids at the front of this Chapter or your owner's manual. **Warning:** *Brake fluid can harm your eyes and damage painted surfaces, so use extreme caution when handling or pouring it. Do not use brake fluid that has been standing open or is more than one year old. Brake fluid absorbs moisture from the air. Excess moisture can cause a dangerous loss of braking effectiveness.*

21 At this time the fluid and master cylinder can be inspected for contamination. The system should be drained and refilled if deposits, dirt particles or water droplets are seen in the fluid (see Section 33).

22 After filling the reservoir to the proper level, make sure the cover is on tight to prevent fluid leakage.

23 The brake fluid level in the master cylinder will drop slightly as the pads and the brake shoes at each wheel wear down during normal operation. If the master cylinder requires repeated additions to keep it at the proper level, it's an indication of leakage in the brake system, which should be corrected immediately. Check all brake lines and connections (see Section 16 for more information).

24 If, upon checking the master cylinder fluid level, you discover the reservoir empty or nearly empty, the brake system should be bled (see Chapter 9).

5 Tire and tire pressure checks

Refer to illustrations 5.2, 5.3, 5.4a, 5.4b and 5.8

1 Periodic inspection of the tires may spare you the inconvenience of being stranded with a flat tire. It can also provide you with vital information regarding possible problems in the steering and suspension systems before major damage occurs.

2 The original tires on this vehicle are equipped with 1/2-inch side bands that will appear when tread depth reaches 1/16-inch, but they don't appear until the tires are worn out. Tread wear can be monitored with a simple, inexpensive device known as a tread depth indicator **(see illustration)**.

3 Note any abnormal tread wear **(see illustration)**. Tread pattern irregularities such as cupping, flat spots and more wear on one side than the other are indications of front end alignment and/or balance problems. If any of these conditions are noted, take the vehicle to a tire shop or service station to correct the problem.

4 Look closely for cuts, punctures and embedded nails or tacks. Sometimes a tire will hold air pressure for a short time or leak down very slowly after a nail has embedded itself in the tread. If a slow leak persists, check the valve stem core to make sure it's tight **(see illustration)**. Examine the

Condition	Probable cause	Corrective action	Condition	Probable cause	Corrective action
Shoulder wear	• Underinflation (both sides wear) • Incorrect wheel camber (one side wear) • Hard cornering • Lack of rotation	• Measure and adjust pressure. • Repair or replace axle and suspension parts. • Reduce speed. • Rotate tires.	Toe wear (Feathered edge)	• Incorrect toe	• Adjust toe-in.
Center wear	• Overinflation • Lack of rotation	• Measure and adjust pressure. • Rotate tires.	Uneven wear	• Incorrect camber or caster • Malfunctioning suspension • Unbalanced wheel • Out-of-round brake drum • Lack of rotation	• Repair or replace axle and suspension parts. • Repair or replace suspension parts. • Balance or replace. • Turn or replace. • Rotate tires.

5.3 This chart will help you determine the condition of the tires, the probable cause(s) of abnormal wear and the corrective action necessary

Chapter 1 Tune-up and routine maintenance

1–11

5.4a If a tire loses air on a steady basis, check the valve core first to make sure it's snug (special inexpensive wrenches are commonly available at auto parts stores)

5.4b If the valve core is tight, raise the corner of the vehicle with the low tire and spray a soapy water solution onto the tread as the tire is turned slowly – leaks will cause small bubbles to appear

5.8 To extend the life of the tires, check the air pressure at least once a week with an accurate gauge (don't forget the spare!)

6.2 The power steering fluid filler cap/dipstick is located on the left side of the engine compartment – unscrew the cap to check the dipstick and/or add fluid

6.6 The power steering fluid level should be kept between the MIN and MAX lines on the dipstick

tread for an object that may have embedded itself in the tire or for a "plug" that may have begun to leak (radial tire punctures are repaired with a plug that's installed in a puncture). If a puncture is suspected, it can be easily verified by spraying a solution of soapy water onto the suspected area **(see illustration)**. The soapy solution will bubble if there's a leak. Unless the puncture is unusually large, a tire shop or service station can usually repair the tire.

5 Carefully inspect the inner sidewall of each tire for evidence of brake fluid. If you see any, inspect the brakes immediately.

6 Correct air pressure adds miles to the lifespan of the tires, improves mileage and enhances overall ride quality. Tire pressure cannot be accurately estimated by looking at a tire, especially if it's a radial. A tire pressure gauge is essential. Keep an accurate gauge in the vehicle. The pressure gauges attached to the nozzles of air hoses at gas stations are often inaccurate.

7 Always check tire pressure when the tires are cold. Cold, in this case, means the vehicle has not been driven over a mile in the three hours preceding a tire pressure check. A pressure rise of four to eight pounds is not uncommon once the tires are warm.

8 Unscrew the valve cap protruding from the wheel or hubcap and push the gauge firmly onto the valve stem **(see illustration)**. Note the reading on the gauge and compare the figure to the recommended tire pressure shown on the label attached to the inside of the glove compartment door. Be sure to reinstall the valve cap to keep dirt and moisture out of the valve stem mechanism. Check all four tires and, if necessary, add enough air to bring them up to the recommended pressure.

9 Don't forget to keep the spare tire inflated to the specified pressure (refer to your owner's manual or the tire sidewall).

6 Power steering fluid level check

Refer to illustrations 6.2 and 6.6

Warning: *The electric cooling fan can activate at any time, even when the ignition is in the Off position. Disconnect the fan motor or negative battery cable when working in the vicinity of the fan.*

1 Unlike manual steering, the power steering system relies on fluid which may, over a period of time, require replenishing.

2 The fluid reservoir for the power steering pump is located on the inner fender panel near the front of the engine **(see illustration)**.

3 For the check, the front wheels should be pointed straight ahead and the engine should be off.

4 Use a clean rag to wipe off the reservoir cap and the area around the cap. This will help prevent any foreign matter from entering the reservoir during the check.

5 Remove the cap and note the dipstick attached to it.

6 Wipe off the fluid with a clean rag, reinsert the dipstick, then withdraw it and read the fluid level. The level should be between the MIN and MAX marks **(see illustration)**. Never allow the fluid level to drop below the MIN mark.

7 If additional fluid is required, pour the specified type directly into the

Chapter 1 Tune-up and routine maintenance

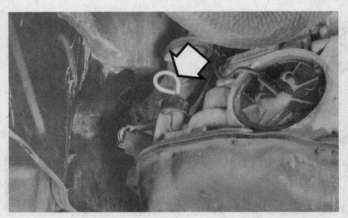

7.4 The automatic transaxle fluid dipstick (arrow) is located near the battery (view is from underneath the vehicle)

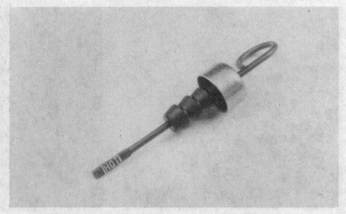

7.6 Check the fluid with the transaxle at normal operating temperature – the level should be kept in the HOT range (between the two lines)

reservoir, using a funnel to prevent spills.

8 If the reservoir requires frequent fluid additions, all power steering hoses, hose connections and the power steering pump should be carefully checked for leaks.

7 Automatic transaxle fluid level check

Refer to illustrations 7.4 and 7.6

Warning: *The electric cooling fan can activate at any time, even when the ignition is in the Off position. Disconnect the fan motor or negative battery cable when working in the vicinity of the fan.*

1 The level of the automatic transaxle fluid should be carefully maintained. Low fluid level can lead to slipping or loss of drive, while overfilling can cause foaming, loss of fluid and transaxle damage.

2 The transaxle fluid level should only be checked when the engine is at normal operating temperature. **Caution:** *If the vehicle has just been driven for a long time at high speed or in city traffic in hot weather, or if it has been pulling a trailer, an accurate fluid level reading cannot be obtained. Allow the fluid to cool down for about 30 minutes.*

3 Park on level ground, apply the parking brake and start the engine. While the engine is idling, depress the brake pedal and move the selector lever through all the gear ranges, beginning and ending in Park.

4 With the engine still idling, remove the dipstick **(see illustration)**.

5 Wipe the fluid off the dipstick with a clean rag and reinsert it until the cap seats.

6 Pull the dipstick out again. The fluid level should be in the HOT range **(see illustration)**. If the level is at the low side of the range, add the specified automatic transmission fluid through the dipstick tube with a funnel.

7 Add the fluid a little at a time and keep checking the level until it's correct.

8 The condition of the fluid should also be checked along with the level. If the fluid at the end of the dipstick is black or a dark reddishbrown color, or if it smells burned, the fluid should be changed (Section 33). If you're in doubt about the condition of the fluid, purchase some new fluid and compare the two for color and odor.

8 Engine oil and filter change

Refer to illustrations 8.3, 8.9, 8.14 and 8.18

1 Frequent oil changes are the most important preventive maintenance procedures that can be done by the home mechanic. As engine oil ages, it becomes diluted and contaminated, which leads to premature engine wear.

2 Although some sources recommend oil filter changes every other oil change, a new filter should be installed every time the oil is changed.

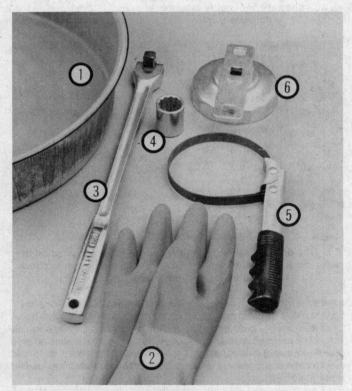

8.3 These tools are required when changing the engine oil and filter

1 **Drain pan** – *It should be fairly shallow in depth, but wide to prevent spills*
2 **Rubber gloves** – *When removing the drain plug and filter, you will get oil on your hands (the gloves will prevent burns)*
3 **Breaker bar** – *Sometimes the oil drain plug is tight and a long breaker bar is needed to loosen it*
4 **Socket** – *To be used with the breaker bar or a ratchet (must be the correct size to fit the drain plug – six-point preferred)*
5 **Filter wrench** – *This is a metal band-type wrench, which requires clearance around the filter to be effective*
6 **Filter wrench** – *This type fits on the bottom of the filter and can be turned with a ratchet or breaker bar (different size wrenches are available for different types of filters)*

3 Gather together all necessary tools and materials before beginning this procedure **(see illustration)**.

Chapter 1 Tune-up and routine maintenance

8.9 Use the proper size box-end wrench or six-point socket to remove the oil drain plug to avoid rounding it off

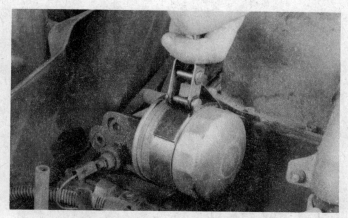

8.14 The oil filter is usually on very tight and will require a special wrench for removal – DO NOT use the wrench to tighten the new filter

4   You should have plenty of clean rags and newspapers handy to mop up any spills. Access to the underside of the vehicle is greatly improved if the vehicle can be lifted on a hoist, driven onto ramps or supported by jackstands. **Warning:** *Do not work under a vehicle which is supported only by a bumper, hydraulic or scissors-type jack.*
5   If this is your first oil change, get under the vehicle and familiarize yourself with the locations of the oil drain plug and the oil filter. The engine and exhaust components will be warm during the actual work, so note how they are situated to avoid touching them when working under the vehicle.
6   Warm the engine to normal operating temperature. If the new oil or any tools are needed, use this warm-up time to gather everything necessary for the job. Refer to *Recommended lubricants and fluids* at the beginning of this Chapter for the type of oil required.
7   With the engine oil warm (warm engine oil will drain better and more built-up sludge will be removed with it), raise and support the vehicle. Make sure it's safely supported!
8   Move all necessary tools, rags and newspapers under the vehicle. Set the drain pan under the drain plug. Keep in mind that the oil will initially flow from the pan with some force; position the pan accordingly.
9   Being careful not to touch any of the hot exhaust components, use a wrench to remove the drain plug near the bottom of the oil pan **(see illustration)**. Depending on how hot the oil is, you may want to wear gloves while unscrewing the plug the final few turns.
10 Allow the old oil to drain into the pan. It may be necessary to move the pan as the oil flow slows to a trickle.
11 After all the oil has drained, wipe off the drain plug with a clean rag. Small metal particles may cling to the plug and would immediately contaminate the new oil.
12 Clean the area around the drain plug opening and reinstall the plug. Tighten it securely with the wrench. If a torque wrench is available, use it to tighten the plug.
13 Move the drain pan into position under the oil filter.
14 Use the filter wrench to loosen the oil filter **(see illustration)**. Chain or metal band filter wrenches may distort the filter canister, but it doesn't matter since the filter will be discarded anyway.
15 Completely unscrew the old filter. Be careful; it's full of oil. Empty the oil inside the filter into the drain pan.
16 Compare the old filter with the new one to make sure they're the same type.
17 Use a clean rag to remove all oil, dirt and sludge from the area where the oil filter mounts to the engine.
18 Apply a light coat of clean oil to the rubber gasket on the new oil filter **(see illustration)**.
19 Attach the new filter to the engine, following the tightening directions printed on the filter canister or packing box. Most filter manufacturers recommend against using a wrench due to the possibility of overtightening the filter and damaging the seal.
20 Remove all tools, rags, etc. from under the vehicle, being careful not to spill the oil in the drain pan, then lower the vehicle.

8.18 Lubricate the oil filter gasket with clean engine oil before installing the filter on the engine

21 Move to the engine compartment and locate the oil filler cap.
22 Pour the fresh oil through the filler opening into the engine. A funnel should be used to prevent spills.
23 Pour four quarts of fresh oil into the engine. Wait a few minutes to allow the oil to drain into the pan, then check the level on the oil dipstick (see Section 4 if necessary). If the oil level is above the MIN mark, start the engine and allow the new oil to circulate.
24 Run the engine for only about a minute and then shut it off. Immediately look under the vehicle and check for leaks at the oil pan drain plug and around the oil filter. If either is leaking, tighten with a bit more force.
25 With the new oil circulated and the filter now completely full, recheck the level on the dipstick and add more oil as necessary.
26 During the first few trips after an oil change, make it a point to check frequently for leaks and proper oil level.
27 The old oil drained from the engine cannot be reused in its present state and should be disposed of. Oil reclamation centers, auto repair shops and gas stations will normally accept the oil, which can be refined and used again. After the oil has cooled it can be drained into a container (capped plastic jugs, topped bottles, milk cartons, etc.) for transport to a disposal site.

9 Windshield wiper blade inspection and replacement

1   The windshield wiper and blade assembly should be inspected periodically for damage, loose components and cracked or worn blade elements.
2   Road film can build up on the wiper blades and affect their efficiency,

Chapter 1 Tune-up and routine maintenance

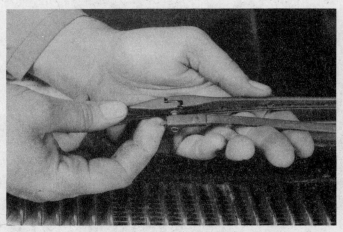

9.5 Lift up on the release lever and slide the blade assembly pin out of the arm

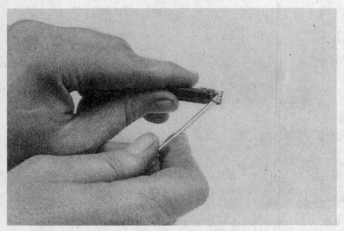

9.6 Use a small screwdriver to pry the lock up and over the end of the blade insert, then slide the insert out of the wiper arm, away from the lock

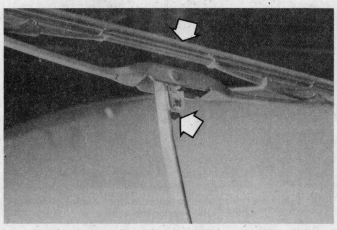

9.8 Press the retaining tab in, then slide the wiper assembly down and out of the hook in the end of the wiper arm

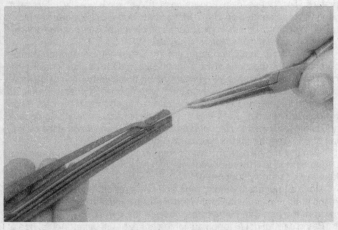

9.9a Use needle-nose pliers to pull the two support rods out of the blade element

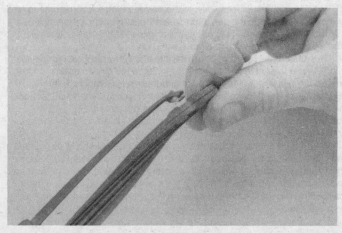

9.9b With the rods out, it's an easy job to remove the element from the frame

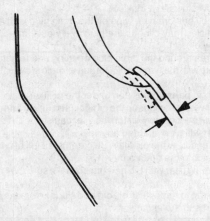

10.1 To check the clutch pedal freeplay, measure the distance between the natural resting place of the pedal and the point at which you encounter resistance

so they should be washed regularly with a mild detergent solution.

3 The action of the wiping mechanism can loosen bolts, nuts and fasteners, so they should be checked and tightened, as necessary, at the same time the wiper blades are checked.

4 If the wiper blade elements are cracked, worn or warped, or no longer clean adequately, they should be replaced with new ones.

Chapter 1 Tune-up and routine maintenance

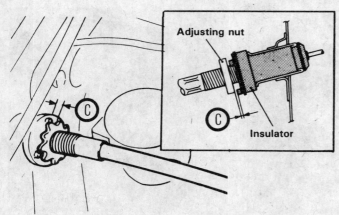

10.2 Adjust the clutch pedal freeplay by turning the cable adjusting nut until the specified clearance (C) is achieved

Early models

Refer to illustrations 9.5 and 9.6

5 Lift the arm assembly away from the glass for clearance, lift up on the release lever and detach the blade assembly from the arm **(see illustration)**.

6 Pry the metal lock on the end of the wiper arm up with a small screwdriver until it clears the metal tab on the end of the insert, then slide the insert away from the lock, out of the arm **(see illustration)**.

7 Slide the new insert into place until the hole in the end snaps over the tab.

Later models

Refer to illustrations 9.8, 9.9a and 9.9b

8 Pull the wiper blade/arm away from the glass, depress the retaining tab and slide the assembly out of the hook in the arm **(see illustration)**.

9 Bend the end of the element out of the way and use needle-nose pliers to pull the two metal support rods out of the rubber element **(see illustration)**. With the rods removed, slide the element out of the wiper frame **(see illustration)**.

10 Slide the new element into place in the frame, then insert the rods to lock it in place.

10 Clutch pedal freeplay check and adjustment (cable-operated clutch)

Refer to illustrations 10.1 and 10.2

1 Push down on the clutch pedal and use a small steel ruler to measure the distance that it moves freely before the clutch resistance is felt **(see illustration)**. The freeplay should be within the limits listed in this Chapter's Specifications. If it isn't, it must be adjusted.

2 Working in the engine compartment, turn the outer cable adjusting nut (located at the point where the cable enters the bulkhead) until the clearance between the nut and insulator projection listed in the Specifications is achieved **(see illustration)**.

3 Recheck the freeplay. Repeat the adjustment as necessary.

11 Battery check, maintenance and charging

Refer to illustrations 11.1, 11.6, 11.7a, 11.7b and 11.7c

Warning: Certain precautions must be followed when checking and servicing the battery. Hydrogen gas, which is highly flammable, is always present in the battery cells, so keep lighted tobacco and all other open flames and sparks away from the battery. The electrolyte inside the battery is ac-

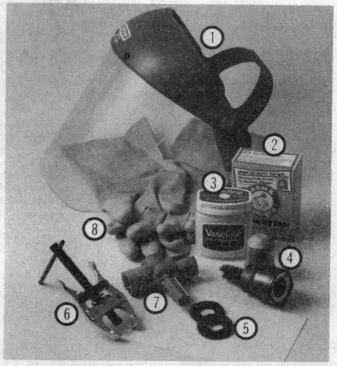

11.1 Tools and materials required for battery maintenance

1 **Face shield/safety goggles** – When removing corrosion with a brush, the acidic particles can easily fly up into your eyes
2 **Baking soda** – A solution of baking soda and water can be used to neutralize corrosion
3 **Petroleum jelly** – A layer of this on the battery posts will help prevent corrosion
4 **Battery post/cable cleaner** – This wire brush cleaning tool will remove all traces of corrosion from the battery posts and cable clamps
5 **Treated felt washers** – Placing one of these on each post, directly under the cable clamps, will help prevent corrosion
6 **Puller** – Sometimes the cable clamps are very difficult to pull off the posts, even after the nut/bolt has been completely loosened. This tool pulls the clamp straight up and off the post without damage.
7 **Battery post/cable cleaner** – Here is another cleaning tool which is a slightly different version of number 4 above, but it does the same thing
8 **Rubber gloves** – Another safety item to consider when servicing the battery; remember that's acid inside the battery!

tually dilute sulfuric acid, which will cause injury if splashed on your skin or in your eyes. It will also ruin clothes and painted surfaces. When removing the battery cables, always detach the negative cable first and hook it up last!

Check and maintenance

1 Battery maintenance is an important procedure which will help ensure that you aren't stranded because of a dead battery. Several tools are required for this procedure **(see illustration)**.

2 When checking/servicing the battery, always turn the engine and all accessories off.

3 A sealed (sometimes called maintenance-free), battery is standard equipment on these vehicles. The cell caps cannot be removed, no electrolyte checks are required and water cannot be added to the cells. However, if a standard aftermarket battery has been installed, the following maintenance procedure can be used.

11.6 Use a wrench to check the tightness of the battery cable bolts; when removing corroded bolts, it may be necessary to use special battery pliers

11.7a Battery terminal corrosion

11.7b When cleaning the cable clamps, all corrosion must be removed (the inside of the clamp is tapered to match the taper on the post, so don't remove too much material)

11.7c Regardless of the type of tool used on the battery posts, a clean, shiny surface should be the result

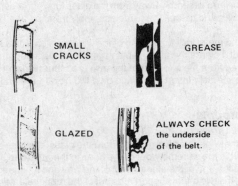
12.3a Here are some of the more common problems associated with drivebelts (check the belts very carefully to prevent an untimely breakdown)

4 Remove the caps and check the electrolyte (water) level in each of the battery cells (see Section 4). It must be above the plates. There's usually a split-ring indicator in each cell to indicate the correct level. If the level is low, add distilled water only, then reinstall the cell caps. **Caution:** *Overfilling the cells may cause electrolyte to spill over during periods of heavy charging, causing corrosion and damage to nearby components.*
5 The external condition of the battery should be checked periodically. Look for damage such as a cracked case.
6 Check the tightness of the battery cable bolts **(see illustration)** to ensure good electrical connections. Inspect the entire length of each cable, looking for cracked or abraded insulation and frayed conductors.
7 If corrosion (visible as white, fluffy deposits) **(see illustration)** is evident, remove the cables from the terminals, clean them with a battery brush and reinstall them **(see illustrations)**. Corrosion can be kept to a minimum by applying a layer of petroleum jelly or grease to the terminals.
8 Make sure the battery carrier is in good condition and the hold-down clamp is tight. If the battery is removed (see Chapter 5 for the removal and installation procedure), make sure that no parts remain in the bottom of the carrier when it's reinstalled. When reinstalling the hold-down clamp, don't overtighten the nuts.
9 Corrosion on the carrier, battery case and surrounding areas can be removed with a solution of water and baking soda. Apply the mixture with a small brush, let it work, then rinse it off with plenty of clean water.
10 Any metal parts of the vehicle damaged by corrosion should be coated with a zinc-based primer, then painted.

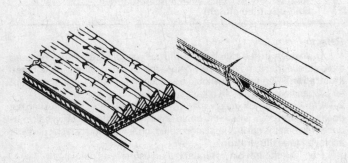
12.3b Check V-ribbed belts for signs of wear like these – if the belt looks worn, replace it

Charging

11 Remove all of the cell caps (if equipped) and cover the holes with a clean cloth to prevent spattering electrolyte. Disconnect the negative battery cable and hook the battery charger leads to the battery posts (positive

Chapter 1 Tune-up and routine maintenance

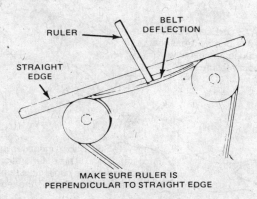

12.4 Measuring drivebelt deflection with a straightedge and ruler

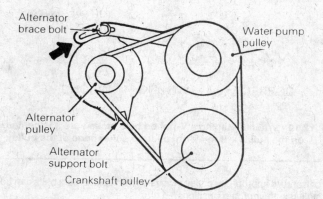

12.6 Move the alternator to adjust the drivebelt tension

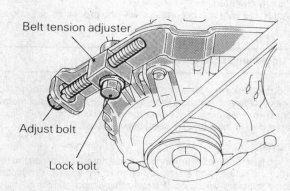

12.7 Loosen the lock bolt and turn the adjusting bolt to tighten or loosen the drivebelt

to positive, negative to negative), then plug in the charger. Make sure it is set at 12 volts if it has a selector switch.

12 If you're using a charger with a rate higher than two amps, check the battery regularly during charging to make sure it doesn't overheat. If you're using a trickle charger, you can safely let the battery charge overnight after you've checked it regularly for the first couple of hours.

13 If the battery has removeable cell caps, measure the specific gravity with a hydrometer every hour during the last few hours of the charging cycle. Hydrometers are available inexpensively from auto parts stores — follow the instructions that come with the hydrometer. Consider the battery charged when there's no change in the specific gravity reading for two hours and the electrolyte in the cells is gassing (bubbling) freely. The specific gravity reading from each cell should be very close to the others. If not, the battery probably has a bad cell(s).

14 Some batteries with sealed tops have built-in hydrometers on the top that indicate the state of charge by the color displayed in the hydrometer window. Normally, a bright-colored hydrometer indicates a full charge and a dark hydrometer indicates the battery still needs charging. Check the battery manufacturer's instructions to be sure you know what the colors mean.

15 If the battery has a sealed top and no built-in hydrometer, you can hook up a digital voltmeter across the battery terminals to check the charge. A fully charged battery should read 12.6 volts or higher.

12 Drivebelt check, adjustment and replacement

Refer to illustrations 12.3a, 12.3b, 12.4, 12.6, 12.7 and 12.10

Check

1 The drivebelts are either V-belts or V-ribbed belts. Sometimes referred to as "fan" belts, the drivebelts are located at the left end of the engine. The good condition and proper adjustment of the belts is critical to the operation of the engine. Because of their composition and the high stresses to which they are subjected, drivebelts stretch and deteriorate as they get older. They must therefore be periodically inspected.

2 The number of belts used on a particular vehicle depends on the accessories installed. One belt transmits power from the crankshaft to the alternator and water pump. The air conditioning compressor and power steering pumps are driven by other belts.

3 With the engine off, open the hood and locate the drivebelts at the left end of the engine. With a flashlight, check each belt: On V-belts, check for cracks and separation of the belt plies (see illustration). On V-ribbed belts, check for separation of the adhesive rubber on both sides of the core, core separation from the belt side and a severed core. Also on V-ribbed belts, check for separation of the ribs from the adhesive rubber, cracking or separation of the ribs, and torn or worn ribs or cracks in the inner ridges of the ribs (see illustration). On both belt types, check for fraying and glazing, which gives the belt a shiny appearance. Both sides of the belt should be inspected, which means you will have to twist the belt to check the underside. Use your fingers to feel the belt where you can't see it. If any of the above conditions are evident, replace the belt (go to Step 8).

4 The tightness of each belt is checked by pushing on it at a distance halfway between the pulleys (see illustration). Apply about 10 pounds of force with your thumb and see how much the belt moves down (deflects). Refer to the Specifications listed in this Chapter for the amount of deflection allowed in each belt.

Adjustment

5 If adjustment is necessary, it is done by moving the belt-driven accessory on the bracket.

6 On some components, if a belt must be adjusted, loosen the adjustment (brace) bolt that secures the component to the slotted bracket and pivot the component (away from the engine block to tighten the belt, toward the block to loosen the belt) (see illustration). It's helpful to lever the component or power steering pump with a large prybar when adjusting the belt because the prybar enables you to precisely position the component until the adjuster bolt is tightened. Be very careful not to damage the housing of the component, particularly the aluminum housing of the alternator. Recheck the belt tension using the above method.

7 On some components, loosen the lock bolt. Turn the adjusting bolt to tension the belt, then retighten the lock bolt (see illustration).

Replacement

8 To replace a belt, follow the above procedures for drivebelt adjustment, but slip the belt off the crankshaft pulley and remove it. To replace some belts, you'll have to remove other belts to get at the one you're replacing because of the way the belts are arranged on the crankshaft pulley. Because of this and because belts tend to wear out more or less together, it is a good idea to replace all belts at the same time. Mark each

Chapter 1 Tune-up and routine maintenance

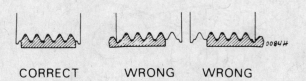

12.10 When installing a V-ribbed belt, make sure it is centered on the pulley – it must not overlap either edge of the pulley

belt and its appropriate pulley groove so the replacement belts can be installed in their proper positions.
9 Take the old belts to the parts store in order to make a direct comparison for length, width and design.
10 After replacing a V-ribbed drivebelt, make sure it fits properly in the ribbed grooves in the pulleys **(see illustration)**. It is essential that the belt be properly centered.
11 Adjust the belt(s) in accordance with the procedure outlined above.

13 Underhood hose check and replacement

Caution: *Replacement of air conditioning hoses must be left to a dealer service department or air conditioning shop that has the equipment to depressurize the system safely. Never remove air conditioning components or hoses until the system has been depressurized.*

General

1 High temperatures in the engine compartment can cause the deterioration of the rubber and plastic hoses used for engine, accessory and emission systems operation. Periodic inspection should be made for cracks, loose clamps, material hardening and leaks.
2 Information specific to the cooling system hoses can be found in Section 14.
3 Some, but not all, hoses are secured to the fittings with clamps. Where clamps are used, check to be sure they haven't lost their tension, allowing the hose to leak. If clamps aren't used, make sure the hose has not expanded and/or hardened where it slips over the fitting, allowing it to leak.

Vacuum hoses

4 It's quite common for vacuum hoses, especially those in the emissions system, to be color coded or identified by colored stripes molded into them. Various systems require hoses with different wall thicknesses, collapse resistance and temperature resistance. When replacing hoses, be sure the new ones are made of the same material.
5 Often the only effective way to check a hose is to remove it completely from the vehicle. If more than one hose is removed, be sure to label the hoses and fittings to ensure correct installation.
6 When checking vacuum hoses, be sure to include any plastic T-fittings in the check. Inspect the fittings for cracks and the hose where it fits over the fitting for distortion, which could cause leakage.
7 A small piece of vacuum hose (1/4-inch inside diameter) can be used as a stethoscope to detect vacuum leaks. Hold one end of the hose to your ear and probe around vacuum hoses and fittings, listening for the "hissing" sound characteristic of a vacuum leak. **Warning:** *When probing with the vacuum hose stethoscope, be very careful not to come into contact with moving engine components such as the drivebelts, cooling fan, etc.*

Fuel hose

Warning: *There are certain precautions which must be taken when inspecting or servicing fuel system components. Work in a well ventilated area and do not allow open flames (cigarettes, appliance pilot lights, etc.) or bare light bulbs near the work area. Mop up any spills immediately and do not store fuel soaked rags where they could ignite.*

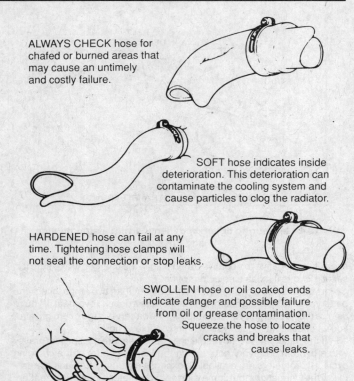

14.4 Hoses, like drivebelts, have a habit of failing at the worst possible time – to prevent the inconvenience of a blown radiator or heater hose, inspect them carefully as shown here

Caution: *On fuel-injected models, relieve the fuel system pressure before disconnecting any fuel lines (see Chapter 4)*

8 Check all rubber fuel lines for deterioration and chafing. Check especially for cracks in areas where the hose bends and just before fittings, such as where a hose attaches to the fuel filter.
9 High quality fuel line, usually identified by the word Fluroelastomer printed on the hose, should be used for fuel line replacement. Never, under any circumstances, use unreinforced vacuum line, clear plastic tubing or water hose for fuel lines. On fuel-injected models, special high-pressure fuel hoses are used. Be sure to obtain an exact replacement.
10 Spring-type clamps are commonly used on fuel lines. These clamps often lose their tension over a period of time, and can be "sprung" during removal. Replace all spring-type clamps with screw clamps whenever a hose is replaced.

Turbocharger hoses

11 On turbocharged models, the air intake and oil supply hoses must be periodically inspected.
12 Check the air intake hoses for leaks or cracks. Replace the hose with a new one if it is damaged.
13 Inspect the oil supply hoses for leaks, replacing as necessary.

Metal lines

14 Sections of metal line are often used for fuel line between the fuel pump and fuel injection system or carburetor. Check carefully to be sure the line has not been bent or crimped and that cracks have not started in the line.
15 If a section of metal fuel line must be replaced, only seamless steel tubing should be used, since copper and aluminum tubing don't have the strength necessary to withstand normal engine vibration.
16 Check the metal brake lines where they enter the master cylinder and brake proportioning unit (if used) for cracks in the lines or loose fittings. Any sign of brake fluid leakage calls for an immediate thorough inspection of the brake system.

Chapter 1 Tune-up and routine maintenance

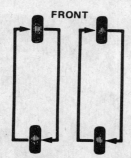

15.2 The recommended tire rotation pattern for these vehicles

14 Cooling system check

Refer to illustration 14.4

1 Many major engine failures can be attributed to a faulty cooling system. If the vehicle is equipped with an automatic transmission, the cooling system also cools the transmission fluid and thus plays an important role in prolonging transmission life.
2 The cooling system should be checked with the engine cold. Do this before the vehicle is driven for the day or after the engine has been shut off for at least three hours.
3 Remove the radiator cap by turning it to the left until it reaches a stop. If you hear a hissing sound (indicating there is still pressure in the system), wait until it stops. Now press down on the cap with the palm of your hand and continue turning to the left until the cap can be removed. Thoroughly clean the cap, inside and out, with clean water. Also clean the filler neck on the radiator. All traces of corrosion should be removed. The coolant inside the radiator should be relatively transparent. If it's rust colored, the system should be drained and refilled (Section 31). If the coolant level isn't up to the top, add additional antifreeze/coolant mixture (see Section 4).
4 Carefully check the large upper and lower radiator hoses along with the smaller diameter heater hoses which run from the engine to the firewall. Inspect each hose along its entire length, replacing any hose which is cracked, swollen or shows signs of deterioration. Cracks may become more apparent if the hose is squeezed **(see illustration)**. Regardless of condition, it's a good idea to replace hoses with new ones every two years.
5 Make sure that all hose connections are tight. A leak in the cooling system will usually show up as white or rust colored deposits on the areas adjoining the leak. If wire-type clamps are used at the ends of the hoses, it may be a good idea to replace them with more secure screw-type clamps.
6 Use compressed air or a soft brush to remove bugs, leaves, etc. from the front of the radiator or air conditioning condenser. Be careful not to damage the delicate cooling fins or cut yourself on them.
7 Every other inspection, or at the first indication of cooling system problems, have the cap and system pressure tested. If you don't have a pressure tester, most gas stations and repair shops will do this for a minimal charge.

15 Tire rotation

Refer to illustration 15.2

1 The tires should be rotated at the specified intervals and whenever uneven wear is noticed. Since the vehicle will be raised and the tires removed anyway, check the brakes (Section 16) at this time.
2 Radial tires must be rotated in a specific pattern **(see illustration)**.
3 Refer to the information in Jacking and towing at the front of this manual for the proper procedures to follow when raising the vehicle and changing a tire. If the brakes are to be checked, do not apply the parking brake as stated. Make sure the tires are blocked to prevent the vehicle from rolling.
4 Preferably, the entire vehicle should be raised at the same time. This can be done on a hoist or by jacking up each corner and then lowering the vehicle onto jackstands placed under the frame rails. Always use four jackstands and make sure the vehicle is firmly supported.

16.5 You will find an inspection hole like this in each caliper – placing a steel ruler across the hole should enable you to determine the thickness of the remaining pad material for both inner and outer pads (arrow)

5 After rotation, check and adjust the tire pressures as necessary and be sure to check the lug nut tightness.
6 For further information on the wheels and tires, refer to Chapter 10.

16 Brake check

Note: *For detailed photographs of the brake system, refer to Chapter 9.*
1 In addition to the specified intervals, the brakes should be inspected every time the wheels are removed or whenever a defect is suspected. Any of the following symptoms could indicate a potential brake system defect: The vehicle pulls to one side when the brake pedal is depressed; the brakes make squealing or dragging noises when applied; brake travel is excessive; the pedal pulsates; brake fluid leaks, usually onto the inside of the tire or wheel.
2 Loosen the wheel lug nuts.
3 Raise the vehicle and place it securely on jackstands.
4 Remove the wheels (see *Jacking and towing* at the front of this book, or your owner's manual, if necessary).

Disc brakes

Refer to illustration 16.5

5 There are two pads – an outer and an inner – in each caliper. The pads are visible through small inspection holes in each caliper **(see illustration)**.
6 Check the pad thickness by looking at each end of the caliper and through the inspection hole in the caliper body. If the lining material is less than the specified thickness, replace the pads. **Note:** *Keep in mind that the lining material is riveted or bonded to a metal backing plate and the metal portion is not included in this measurement.*
7 If it is difficult to determine the exact thickness of the remaining pad material by the above method, or if you are at all concerned about the condition of the pads, remove the caliper(s), then remove the pads from the calipers for further inspection (refer to Chapter 9).
8 Once the pads are removed from the calipers, clean them with brake cleaner and remeasure them with a small steel pocket ruler or a vernier caliper.
9 Check the disc. Look for score marks, deep scratches and burned spots. If these conditions exist, the hub/disc assembly will have to be removed (see Chapter 9).
10 Before installing the wheels, check all brake lines and hoses for damage, wear, deformation, cracks, corrosion, leakage, bends and twists, particularly in the vicinity of the rubber hoses at the calipers. Check the clamps for tightness and the connections for leakage. Make sure that all hoses and lines are clear of sharp edges, moving parts and the exhaust system. If any of the above conditions are noted, repair, reroute or replace the lines and/or fittings as necessary (refer to Chapter 9).

Chapter 1 Tune-up and routine maintenance

16.12 The rear brake shoe lining thickness is measured from the outer surface of the lining to the metal shoe (arrows)

16.14 Peel the wheel cylinder boot back carefully and check for leaking fluid, indicating the cylinder must be replaced or rebuilt

17.2 The top of the filter housing is removed by releasing the clips on the sides and unscrewing the wing nut on the top

17.4 Lift out the air filter element and wipe out the inside of the air cleaner housing with a clean rag

Rear drum brakes

Refer to illustrations 16.12 and 16.14

11 Refer to Section 32 and remove the rear brake drums. **Warning:** *Brake dust produced by lining wear and deposited on brake components contains asbestos, which is hazardous to your health. DO NOT blow it out with compressed air and DO NOT inhale it! DO NOT use gasoline or solvents to remove the dust. Brake system cleaner should be used to flush the dust into a drain pan. After the brake components are wiped clean with a damp rag, dispose of the contaminated rag(s) and solvent in a covered and labelled container. Try to use non-asbestos replacement parts whenever possible.*

12 Note the thickness of the lining material on the rear brake shoes **(see illustration)** and look for signs of contamination by brake fluid and grease. If the lining material is within 1/16-inch of the recessed rivets or metal shoes, replace the brake shoes with new ones. The shoes should also be replaced if they are cracked, glazed (shiny lining surfaces) or contaminated with brake fluid or grease. See Chapter 9 for the replacement procedure.

13 Check the shoe return and hold-down springs and the adjusting mechanism to make sure they're installed correctly and in good condition.

Deteriorated or distorted springs, if not replaced, could allow the linings to drag and wear prematurely.

14 Check the wheel cylinders for leakage by carefully peeling back the rubber boots **(see illustration)**. If brake fluid is noted behind the boots, the wheel cylinders must be replaced (see Chapter 9).

15 Check the drums for cracks, score marks, deep scratches and hard spots, which will appear as small discolored areas. If imperfections cannot be removed with emery cloth, the drums must be resurfaced by an automotive machine shop (see Chapter 9 for more detailed information).

16 Refer to Chapter 9 and install the brake drums.

17 Install the wheels and tighten the wheel lug nuts finger tight.

18 Remove the jackstands and lower the vehicle.

19 Tighten the wheel lug nuts to the torque listed at the beginning of this Chapter.

Parking brake

20 A simple method of checking the parking brake is to park the vehicle on a steep hill with the parking brake set and the transmission in Neutral. If the parking brake cannot prevent the vehicle from rolling, it is in need of adjustment (see Chapter 9).

Chapter 1 Tune-up and routine maintenance

17.7 Pull the PCV filter out of the housing

17.11a Detach the clips

17.11b Be careful when pulling the air flow sensor out of the air cleaner

17.12 Support the air flow sensor and lift the filter element out

18.3 The cam-shaped actuator located on the left side of the carburetor (arrow) must move smoothly through its arc and press progressively on the throttle position sensor, located directly to the rear of the actuator

17 Air and PCV filter replacement

Warning: *The electric cooling fan can activate at any time, even when the ignition is in the Off position. Disconnect the fan motor or negative battery cable when working in the vicinity of the fan.*

1 At the specified intervals, the air filter and PCV filter (if equipped) should be replaced with new ones. The engine air cleaner also supplies filtered air to the PCV system.

Carbureted models

Refer to illustrations 17.2, 17.4 and 17.7

2 The filter is located on top of the carburetor and is replaced by unscrewing the wing nut, detaching the clips from the top of the filter housing and lifting off the cover **(see illustration)**.
3 While the top plate is off, be careful not to drop anything down into the carburetor or air cleaner assembly.
4 Lift the air filter element out of the housing **(see illustration)** and wipe out the inside of the air cleaner housing with a clean rag.
5 Place the new filter in the air cleaner housing. Make sure it seats properly in the bottom of the housing.
6 The PCV filter is also located inside the air cleaner housing. Remove the top plate and air filter as previously described, then locate the PCV filter on the inside of the housing.
7 Remove the old filter **(see illustration)**.
8 Install the new PCV filter and the new air filter.

9 Install the top plate and any hoses which were disconnected. Don't overtighten the wing nut.

Fuel-injected models

Refer to illustrations 17.11a, 17.11b and 17.12

10 If equipped, unplug the airflow sensor connector. Remove the air intake hose.
11 Detach the clips and pull the air flow sensor out of the air cleaner housing **(see illustrations)**.
12 Remove the old air filter element **(see illustration)**.
13 Insert the new filter element in the air cleaner housing and align the tabs, then install the air flow sensor body and secure it with the clips.
14 Connect the air intake hose securely, making sure the are no air leaks.

18 Throttle position sensor check (feedback carburetor)

Refer to illustrations 18.3 and 18.4

1 Some models are equipped with a feedback carburetor and the throttle position sensor and linkage must work properly or vehicle driveability and exhaust emissions will be affected.
2 Have an assistant open and close the throttle while you watch the sensor and linkage.
3 On some models, a cam-shaped actuator is used which must move smoothly throughout its arc, progressively depressing and releasing the throttle position sensor **(see illustration)**.

1–22 Chapter 1 Tune-up and routine maintenance

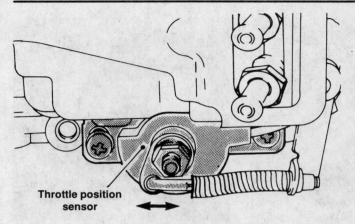

18.4 On some models, check the throttle position sensor lever for smooth operation and make sure the two retaining screws are tight

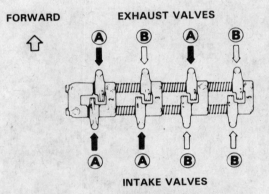

19.5 With the number one piston at Top Dead Center (TDC), adjust the valves marked A – with the number 4 piston at TDC, adjust the valves marked B

19.1 The rocker arms on engines that don't require valve adjustment look like these (arrow) – they don't have adjustment screws (if the rocker arms do have adjustment screws for the intake and exhaust or smaller jet valves, follow the procedure in the text)

19.7 There should be a slight drag as the feeler gauge is pulled between the jet valve adjustment screw and the valve stem

4   On other models, check at the lever itself for smooth operation and make sure the two retaining screws are tight (see illustration). Caution: Be sure you don't loosen the screws or disturb the sensor. Setting the throttle position sensor is an exacting process that often requires special tools.
5   If the linkage or the sensor bind, refer to Chapter 4 for more information on the throttle position sensor.

19   Valve clearance check and adjustment (not all models)

Refer to illustrations 19.1, 19.5, 19.7 and 19.12
Warning: The electric cooling fan can activate at any time, even when the ignition is in the Off position. Disconnect the fan motor or negative battery cable when working in the vicinity of the fan.
Note: This procedure applies to single overhead camshaft (SOHC) engines only.
1   The valve clearances must be checked and adjusted at the specified intervals with the engine at normal operating temperature. Not all models require valve adjustment, while others require only adjustment of the jet valves. Consult your owner's manual or dealer to determine if your vehicle requires valve adjustment. If this information is not readily available, the only sure way to tell that adjustment is required is to remove the engine valve cover. Valves that do not require adjustment have no adjusting screws (see illustration).
2   Remove the air cleaner assembly (see Chapter 4).
3   Remove the valve cover (see Chapter 2). On some models it will be necessary to support the engine and remove the front engine mount for access to the front valve cover bolts. Note: The engine can be supported with a hoist or support fixture from above or with a floor jack under the oil pan. If using a floor jack, be sure to place a piece of wood between the jack head and the oil pan to spread the load.
4   Place the number one piston at Top Dead Center (TDC) on the compression stroke (see Chapter 2). The number one cylinder rocker arms (closest to the timing belt end of the engine) should be loose (able to move up and down slightly) and the camshaft lobes should be facing away from the rocker arms.
5   With the crankshaft in this position the valves labeled A (plus the jet valves adjacent to the intake valves) can be checked and adjusted (see illustration). Always check and adjust the jet valve clearance first.
6   The intake valve and jet valve adjusting screws are located on a common rocker arm. Make sure the intake valve adjusting screw has been backed off two full turns, then loosen the locknut on the jet valve adjusting screw.
7   Turn the jet valve adjusting screw counterclockwise and insert the appropriate size feeler gauge (see this Chapter's Specifications) between the valve stem and the adjusting screw. Carefully tighten the adjusting screw until you can feel a slight drag on the feeler gauge as you withdraw it from between the stem and adjusting screw (see illustration).
8   Since the jet valve spring is relatively weak, use special care not to force the jet valve open. Be particularly careful if the adjusting screw is hard to turn. Hold the adjusting screw with a screwdriver (to keep it from turning) and tighten the locknut. Recheck the clearance to make sure it hasn't changed.

Chapter 1 Tune-up and routine maintenance

19.12 To make sure the adjusting screw doesn't move when the locknut is tightened, use a box-end wrench and have a good grip on the screwdriver

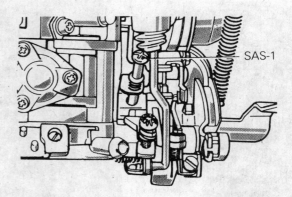

20.8 When adjusting the engine idle speed, make sure to turn only the SAS 1 screw – the other screws are preset at the factory and require special equipment for proper adjustment

9 Next, check and adjust the intake valve clearance. Insert the appropriate size feeler gauge between the intake valve stem and the adjusting screw. Carefully tighten the adjusting screw until you can feel a slight drag on the feeler gauge as you withdraw it from between the stem and adjusting screw.
10 Hold the adjusting screw with a screwdriver (to keep it from turning) and tighten the locknut. Recheck the clearance to make sure it hasn't changed.
11 Loosen the locknut on the exhaust valve adjusting screw. Turn the adjusting screw counterclockwise and insert the appropriate size feeler gauge between the valve stem and the adjusting screw. Carefully tighten the adjusting screw until you can feel a slight drag on the feeler gauge as you withdraw it from between the stem and adjusting screw.
12 Hold the adjusting screw with a screwdriver (to keep it from turning) and tighten the locknut **(see illustration)**. Recheck the clearance to make sure it hasn't changed.
13 Rotate the crankshaft until the number four piston is at TDC on the compression stroke. The number four cylinder rocker arms (closest to the transaxle end of the engine) are loose with the camshaft lobes facing away from the rocker arms.
14 Adjust the valves labelled B as described in Steps 4 through 8 **(see illustration 19.5)**.
15 Install the rocker arm cover and the air cleaner assembly.

20 Engine idle speed check and adjustment (carbureted models only)

Refer to illustration 20.8

1 Engine idle speed is the speed at which the engine operates when no accelerator pedal pressure is applied, as when stopped at a traffic light. This speed is critical to the performance of the engine itself, as well as many engine subsystems.
2 Set the parking brake firmly set and block the wheels to prevent the vehicle from rolling. Put the transaxle in Neutral.
3 Hook up a hand-held tachometer.
4 Start the engine and allow it to reach normal operating temperature.
5 Check, and adjust if necessary, the ignition timing (Section 37).
6 Run the engine up to 2000 to 3000 rpm for more than five seconds and then allow the engine to idle for two minutes.
7 Check the engine idle speed on the tachometer and compare it to the Emission Control Information label in the engine compartment.
8 If the idle speed is too low or too high, turn the speed adjusting screw (SAS 1) **(see illustration)** until the specified idle speed is obtained. Turn only the Speed Adjusting Screw (SAS 1) as the other adjustment screws are preset at the factory and require special equipment for proper adjustment.

21 Fuel system check

Warning: *Gasoline is extremely flammable, so take extra precautions when you work on any part of the fuel system. Don't smoke or allow open flames or bare light bulbs near the work area, and don't work in a garage where a natural gas-type appliance (such as a water heater or clothes dryer) with a pilot light is present. If you spill any fuel on your skin, rinse it off immediately with soap and water. When you perform any kind of work on the fuel tank, wear safety glasses and have a Class B type fire extinguisher on hand.*
Caution: *On fuel-injected models, do not disconnect any fuel lines until you relieve the fuel system pressure (see Chapter 4).*

1 If you smell gasoline while driving or after the vehicle has been sitting in the sun, inspect the fuel system immediately.
2 Remove the gas filler cap and inspect if for damage and corrosion. The gasket should have an unbroken sealing imprint. If the gasket is damaged or corroded, replace the cap.
3 Inspect the fuel feed and return lines for cracks. Check the metal fuel line connections to make sure they are tight.
4 Since some components of the fuel system – the fuel tank and part of the fuel feed and return lines, for example – are underneath the vehicle, they can be inspected more easily with the vehicle raised on a hoist. If that's not possible, raise the vehicle and secure it on jackstands.
5 With the vehicle raised and safely supported, inspect the gas tank and filler neck for punctures, cracks and other damage. The connection between the filler neck and the tank is particularly critical. Sometimes a rubber filler neck will leak because of loose clamps or deteriorated rubber. These are problems a home mechanic can usually rectify. **Warning:** *Do not, under any circumstances, try to repair a fuel tank (except rubber components). A welding torch or any open flame can easily cause fuel vapors inside the tank to explode.*
6 Carefully check all rubber hoses and metal lines leading away from the fuel tank. Check for loose connections, deteriorated hoses, crimped lines and other damage. Carefully inspect the lines from the tank to the carburetor. Repair or replace damaged sections as necessary.

22 Steering and suspension check

Refer to illustration 22.6
Note: *For detailed illustrations of the steering and suspension components, refer to Chapter 10.*

With the wheels on the ground

1 With the vehicle stopped and the front wheels pointed straight ahead, rock the steering wheel gently back and forth. If freeplay is excessive, a

1–24 Chapter 1 Tune-up and routine maintenance

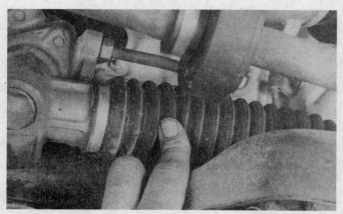

22.6 Push on the steering gear boot to check for cracks and lubricant leaks

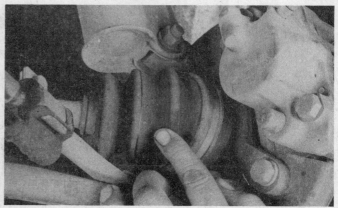

23.2 Flex the driveaxle boots by hand to check for cracks or leaking grease

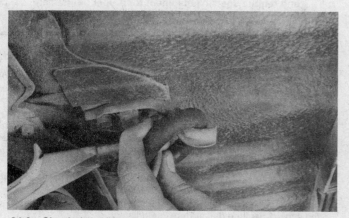

24.2 Check the rubber exhaust system hangers for cracks and deterioration – replace any that are in poor condition

front wheel bearing, main shaft yoke, intermediate shaft yoke, lower arm balljoint or steering system joint is worn or the steering gear is out of adjustment or broken. Refer to Chapter 10 for the appropriate repair procedure.

2   Other symptoms, such as excessive vehicle body movement over rough roads, swaying (leaning) around corners and binding as the steering wheel is turned, may indicate faulty steering and/or suspension components.

3   Check the shock absorbers by pushing down and releasing the vehicle several times at each corner. If the vehicle does not come back to a level position within one or two bounces, the shocks/struts are worn and must be replaced. When bouncing the vehicle up and down, listen for squeaks and noises from the suspension components. Additional information on suspension components can be found in Chapter 10.

With the vehicle raised

4   Raise the vehicle and support it securely on jackstands. See Jacking and towing at the front of this book for the proper jacking points.

5   Check the tires for irregular wear patterns (see Section 5) and proper inflation. See Section 5 in this chapter for information regarding tire wear and Section 32 or Chapter 10 for the wheel bearing replacement procedures.

6   Inspect the universal joint between the steering shaft and the steering gear housing. Check the steering gear housing for grease leakage or oozing. Make sure that the dust seals and boots are not damaged and that the boot clamps are not loose (see illustration). Check the steering linkage for looseness or damage. Check the tie-rod ends for excessive play. Look for loose bolts, broken or disconnected parts and deteriorated rubber bushings on all suspension and steering components. While an assistant turns the steering wheel from side to side, check the steering components

for free movement, chafing and binding. If the steering components do not seem to be reacting with the movement of the steering wheel, try to determine where the slack is located.

7   Inspect the balljoint boots for damage and leaking grease. Replace the boots with new ones if they are damaged (see Chapter 10).

23 Driveaxle boot check

Refer to illustration 23.2

1   The driveaxle boots are very important because they prevent dirt, water and foreign material from entering and damaging the constant velocity (CV) joints. Oil and grease can cause the boot material to deteriorate prematurely, so it's a good idea to wash the boots with soap and water.

2   Inspect the boots for tears and cracks as well as loose clamps (see illustration). If there is any evidence of cracks or leaking lubricant, they must be replaced as described in Chapter 8.

24 Exhaust system check

Refer to illustration 24.2

1   With the engine cold (at least three hours after the vehicle has been driven), check the complete exhaust system from its starting point at the engine to the end of the tailpipe. This should be done on a hoist where unrestricted access is available.

2   Check the pipes and connections for evidence of leaks, severe corrosion or damage. Make sure that all brackets and hangers are in good condition and tight (see illustration).

3   At the same time, inspect the underside of the body for holes, corrosion, open seams, etc. which may allow exhaust gases to enter the passenger compartment. Seal all body openings with silicone or body putty.

4   Rattles and other noises can often be traced to the exhaust system, especially the mounts and hangers. Try to move the pipes, muffler and catalytic converter. If the components can come in contact with the body or suspension parts, secure the exhaust system with new mounts.

5   Check the running condition of the engine by inspecting inside the end of the tailpipe. The exhaust deposits here are an indication of engine state-of-tune. If the pipe is black and sooty or coated with white deposits, the engine is in need of a tune-up, including a thorough fuel system inspection and adjustment.

25 Manual transaxle lubricant level check

Refer to illustration 25.1

1   The manual transaxle does not have a dipstick. To check the lubricant level, raise the vehicle and support it securely on jackstands. On the lower rear side of the transaxle housing, you will see a plug (see illustration).

Chapter 1 Tune-up and routine maintenance

1–25

25.1 Remove the manual transaxle plug from the location shown – the lubricant level must be at the lower edge of the hole

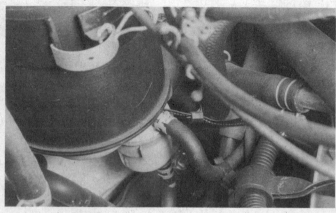

26.1 On most carbureted models, the fuel filter is mounted on the firewall, below the evaporative emissions system canister

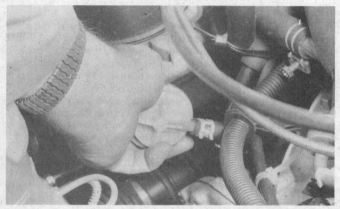

26.2 Unsnap the fuel filter from the clip and pull it out for better access to the hoses

26.10 Disconnect the fuel lines (arrows), then detach the filter from the bracket – use a flare-nut wrench on the lower fitting

Remove it. If the lubricant level is correct, it should be up to the lower edge of the hole.
2 If the transaxle needs more lubricant (if the level is not up to the hole), use a syringe to add more. Stop filling the transaxle when the lubricant begins to run out the hole.
3 Install the plug and tighten it securely. Drive the vehicle a short distance, then check for leaks.

26 Fuel filter replacement

Warning: *Gasoline is extremely flammable, so take extra precautions when you work on any part of the fuel system. Don't smoke or allow open flames or bare light bulbs near the work area, and don't work in a garage where a natural gas-type appliance (such as a water heater or clothes dryer) with a pilot light is present. If you spill any fuel on your skin, rinse it off immediately with soap and water. When you perform any kind of work on the fuel tank, wear safety glasses and have a Class B type fire extinguisher on hand.*

Carbureted models
Refer to illustrations 26.1 and 26.2

1 The fuel filter is normally located on the firewall, below the emissions canister **(see illustration)**.
2 For easier access, pull the filter out of the spring clip so both fittings can be reached **(see illustration)**.
3 Release the hose clamps at the filter fittings and slide them back up the hoses.

4 Disconnect the hoses and remove the filter. Now would be a good time to replace the hoses if they're deteriorated.
5 Push the hoses onto the new filter and position the clamps approximately 1/4-inch back from the ends.
6 Push the filter back into the spring clip. Check to make sure it is held securely and the hoses are not kinked.
7 Start the engine and check for fuel leaks at the filter.

Fuel-injected models
Refer to illustration 26.10

8 Depressurize the fuel system (see Chapter 4).
9 The fuel filter is normally located on the firewall, next to the emissions canister.
10 Disconnect the fuel lines, detach the filter and lift it out of the engine compartment **(see illustration)**.
11 Installation is the reverse of removal. Be sure to use new sealing washers on each side of the banjo fitting.

27 Spark plug replacement

Refer to illustrations 27.1, 27.4a, 27.4b, 27.6, 27.8 and 27.10

1 Spark plug replacement requires a spark plug socket which fits onto a ratchet wrench. This socket is lined with a rubber grommet to protect the porcelain insulator of the spark plug and to hold the plug while you insert it into the spark plug hole. You will also need a wire-type feeler gauge to check and adjust the spark plug gap and a torque wrench to tighten the

Chapter 1 Tune-up and routine maintenance

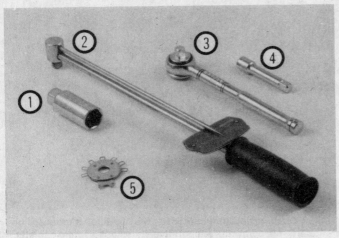

27.1 Tools required for changing spark plugs

1. **Spark plug socket** – This will have special padding inside to protect the spark plug's porcelain insulator
2. **Torque wrench** – Although not mandatory, using this tool is the best way to ensure the plugs are tightened properly
3. **Ratchet** – Standard hand tool to fit the spark plug socket
4. **Extension** – Depending on model and accessories, you may need special extensions and universal joints to reach one or more of the plugs
5. **Spark plug gap gauge** – This gauge for checking the gap comes in a variety of styles. Make sure the gap for your engine is included.

new plugs to the torque listed in this Chapter's specifications **(see illustration)**.

2 When replacing the plugs, purchase the new plugs in advance, adjust them to the proper gap and then replace each plug one at a time. **Note:** *When buying new spark plugs, it's essential that you obtain the correct plugs for your specific vehicle. This information can be found on the Vehicle Emissions Control Information (VECI) label located on the underside of the hood or in the owner's manual. If these two sources specify different plugs, purchase the spark plug type specified on the VECI label because that information is provided specifically for your engine.*

3 Inspect each of the new plugs for defects. If there are any signs of cracks in the porcelain insulator of a plug, don't use it.

4 Check the electrode gaps of the new plugs. Check the gap by inserting the wire gauge of the proper thickness between the electrodes at the tip of the plug **(see illustration)**. The gap between the electrodes should be identical to that listed in this chapter's Specifications Section or as specified on the VECI label. If the gap is incorrect, use the notched adjuster on the feeler gauge body to bend the curved side electrode slightly **(see illustration)**.

5 If the side electrode is not exactly over the center electrode, use the notched adjuster to align them. **Caution:** *If the gap of a new plug must be adjusted, bend only the base of the ground electrode; do not touch the tip.*

Removal

6 To prevent the possibility of mixing up spark plug wires, work on one spark plug at a time. Remove the wire and boot from one spark plug. Grasp the boot – not the cable – as shown, give it a half twisting motion and pull it off **(see illustration)**.

7 If compressed air is available, blow any dirt or foreign material away from the spark plug area before proceeding (a common bicycle pump will also work).

8 Remove the spark plug **(see illustration)**.

9 Compare each old spark plug with those shown in the accompanying photos to determine the overall running condition of the engine.

Installation

10 It's often difficult to insert spark plugs into their holes without cross-threading them. To avoid this possibility, fit a short piece of 3/16-inch ID rubber hose over the end of the spark plug **(see illustration)**. The flexible hose acts as a universal joint to help align the plug with the plug hole. Should the plug begin to cross-thread, the hose will slip on the spark plug, preventing thread damage. Tighten the plug securely.

11 Attach the plug wire to the new spark plug, again using a twisting motion on the boot until it is firmly seated on the end of the spark plug.

12 Follow the above procedure for the remaining spark plugs, replacing them one at a time to prevent mixing up the spark plug wires.

28 Spark plug wire, distributor cap and rotor check and replacement

Refer to illustrations 28.11a, 28.11b and 28.12

1 The spark plug wires should be checked whenever new spark plugs are installed.

2 Begin this procedure by making a visual check of the spark plug wires while the engine is running. In a darkened garage (make sure there is ventilation) start the engine and observe each plug wire. Be careful not to come into contact with any moving engine parts. If there is a break in the wire, you will see arcing or a small spark at the damaged area. If arcing is noticed, make a note to obtain new wires, then allow the engine to cool and check the distributor cap and rotor.

27.4a Spark plug manufacturers recommend using a wire type gauge when checking the gap – if the wire does not slide between the electrodes with a slight drag, adjustment is required

27.4b To change the gap, bend the *side* electrode only, as indicated by the arrows, and be very careful not to crack or chip the porcelain insulator surrounding the center electrode

27.6 When removing the spark plug wires, pull only on the boot and use a twisting/pulling motion – a special tool like this, available at auto parts stores, will ease the job

Chapter 1 Tune-up and routine maintenance

1 – 27

27.8 Use a socket and extension to unscrew the spark plug

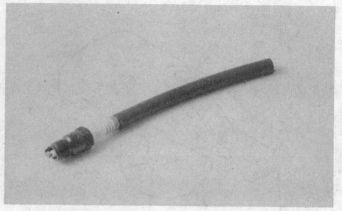

27.10 A length of 3/16-inch ID rubber hose will save time and prevent damaged threads when installing the spark plugs

28.11a Use a small screwdriver to pry off the distributor cap retaining clips

3   The spark plug wires should be inspected one at a time to prevent mixing up the order, which is essential for proper engine operation. Each original plug wire should be numbered to help identify its location. If the number is illegible, a piece of tape can be marked with the correct number and wrapped around the plug wire.
4   Disconnect the plug wire from the spark plug. A removal tool can be used for this purpose or you can grasp the rubber boot, twist the boot half a turn and pull the boot free. Do not pull on the wire itself (see illustration 27.6).
5   Check inside the boot for corrosion, which will look like a white crusty powder.
6   Push the wire and boot back onto the end of the spark plug. It should fit tightly onto the end of the plug. If it doesn't, remove the wire and use pliers to carefully crimp the metal connector inside the wire boot until the fit is snug.
7   Using a clean rag, wipe the entire length of the wire to remove built-up dirt and grease. Once the wire is clean, check for burns, cracks and other damage. Do not bend the wire sharply, because the conductor within the wire might break.
8   Disconnect the wire from the distributor. Again, pull only on the rubber boot. Check for corrosion and a tight fit. Press the wire back into the distributor.
9   Inspect the remaining spark plug wires, making sure that each one is securely fastened at the distributor and spark plug when the check is complete.
10   If new spark plug wires are required, purchase a set for your specific engine model. Pre-cut wire sets with the boots already installed are available. Remove and replace the wires one at a time to avoid mix-ups in the firing order.
11   Detach the distributor cap by prying off the two cap retaining clips (see illustration). Look inside it for cracks, carbon tracks and worn, burned or loose contacts (see illustration).
12   Pull the rotor off the distributor shaft and examine it for cracks and car-

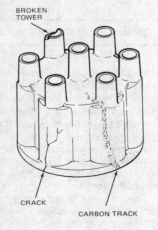
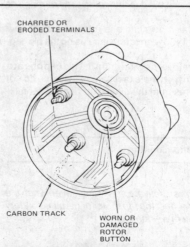

28.11b Shown here are some of the common defects to look for when inspecting the distributor cap (if in doubt about its condition, install a new one)

Chapter 1 Tune-up and routine maintenance

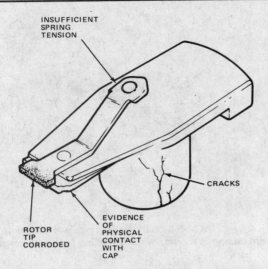

28.12 The ignition rotor should be checked for wear and corrosion as indicated here (if in doubt about its condition, buy a new one)

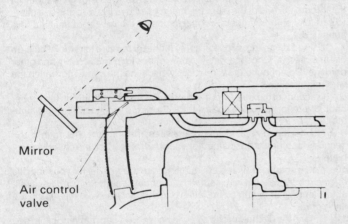

30.3 Use a mirror to observe movement of the air control valve

bon tracks **(see illustration)**. Replace the cap and rotor if any damage or defects are noted.

13 It is common practice to install a new cap and rotor whenever new spark plug wires are installed, but if you wish to continue using the old cap, clean the terminals first.

14 When installing a new cap, remove the wires from the old cap one at a time and attach them to the new cap in the exact same location – do not simultaneously remove all the wires from the old cap or firing order mix-ups may occur.

29 Carburetor choke check and cleaning

Refer to illustration 29.3

Warning: *The electric cooling fan can activate at any time, even when the ignition is in the Off position. Disconnect the fan motor or negative battery cable when working in the vicinity of the fan.*

Check

1 The choke operates only when the engine is cold, so this check should be performed before the engine has been started for the day.

29.3 The choke plate is located in the carburetor throat

2 Take off the top plate of the air cleaner assembly. It's held in place by a wing nut at the top and clips on the side. If any vacuum hoses must be disconnected, make sure you tag the hoses for reinstallation in their original positions. Place the top plate and wing nut aside, out of the way of moving engine components.

3 Look at the center of the air cleaner housing. You will notice a flat plate at the carburetor opening **(see illustration)**.

4 Press the accelerator pedal to the floor. The plate should close completely. Start the engine while you watch the plate at the carburetor. Don't position your face near the carburetor, as the engine could backfire, causing serious burns. When the engine starts, the choke plate should open slightly.

5 Allow the engine to continue running at an idle speed. As the engine warms up to operating temperature, the plate should slowly open, allowing more air to enter through the top of the carburetor.

6 After a few minutes, the choke plate should be fully open to the vertical position. Tap the accelerator to make sure the fast idle cam disengages.

7 You'll notice that the engine speed corresponds with the plate opening. With the plate fully closed, the engine should run at a fast idle speed. As the plate opens and the throttle is moved to disengage the fast idle cam, the engine speed will decrease.

Cleaning

8 With the engine off, use aerosol carburetor solvent (available at auto parts stores) to clean the contact surfaces of the choke plate shaft where it passes through the carburetor body. Also clean the fast idle cam and link mechanism.

9 Refer to Chapter 4 for specific information on adjusting and servicing the choke components.

30 Heated Air Intake (HAI) air cleaner check

Refer to illustration 30.3

Warning: *The electric cooling fan can activate at any time, even when the ignition is in the Off position. Disconnect the fan motor or negative battery cable when working in the vicinity of the fan.*

1 All engines are equipped with a Heated Air Intake (HAI) air cleaner which draws air to the carburetor from different locations, depending on engine temperature.

2 This is a visual check, requiring use of a small mirror.

3 When the engine is cold, locate the air control valve inside the air cleaner assembly. It's inside the long snorkel of the air cleaner housing **(see illustration)**.

4 There is a flexible air duct attached to the end of the snorkel, leading to an area behind the headlight. Disconnect it at the snorkel. This will enable you to look through the end of the snorkel and see the air control valve inside.

5 Start the engine and look through the snorkel at the valve, which should move up to block off the air cleaner snorkel. With the valve closed,

Common spark plug conditions

NORMAL
Symptoms: Brown to grayish-tan color and slight electrode wear. Correct heat range for engine and operating conditions.
Recommendation: When new spark plugs are installed, replace with plugs of the same heat range.

WORN
Symptoms: Rounded electrodes with a small amount of deposits on the firing end. Normal color. Causes hard starting in damp or cold weather and poor fuel economy.
Recommendation: Plugs have been left in the engine too long. Replace with new plugs of the same heat range. Follow the recommended maintenance schedule.

CARBON DEPOSITS
Symptoms: Dry sooty deposits indicate a rich mixture or weak ignition. Causes misfiring, hard starting and hesitation.
Recommendation: Make sure the plug has the correct heat range. Check for a clogged air filter or problem in the fuel system or engine management system. Also check for ignition system problems.

ASH DEPOSITS
Symptoms: Light brown deposits encrusted on the side or center electrodes or both. Derived from oil and/or fuel additives. Excessive amounts may mask the spark, causing misfiring and hesitation during acceleration.
Recommendation: If excessive deposits accumulate over a short time or low mileage, install new valve guide seals to prevent seepage of oil into the combustion chambers. Also try changing gasoline brands.

OIL DEPOSITS
Symptoms: Oily coating caused by poor oil control. Oil is leaking past worn valve guides or piston rings into the combustion chamber. Causes hard starting, misfiring and hesitation.
Recommendation: Correct the mechanical condition with necessary repairs and install new plugs.

GAP BRIDGING
Symptoms: Combustion deposits lodge between the electrodes. Heavy deposits accumulate and bridge the electrode gap. The plug ceases to fire, resulting in a dead cylinder.
Recommendation: Locate the faulty plug and remove the deposits from between the electrodes.

TOO HOT
Symptoms: Blistered, white insulator, eroded electrode and absence of deposits. Results in shortened plug life.
Recommendation: Check for the correct plug heat range, over-advanced ignition timing, lean fuel mixture, intake manifold vacuum leaks, sticking valves and insufficient engine cooling.

PREIGNITION
Symptoms: Melted electrodes. Insulators are white, but may be dirty due to misfiring or flying debris in the combustion chamber. Can lead to engine damage.
Recommendation: Check for the correct plug heat range, over-advanced ignition timing, lean fuel mixture, insufficient engine cooling and lack of lubrication.

HIGH SPEED GLAZING
Symptoms: Insulator has yellowish, glazed appearance. Indicates that combustion chamber temperatures have risen suddenly during hard acceleration. Normal deposits melt to form a conductive coating. Causes misfiring at high speeds.
Recommendation: Install new plugs. Consider using a colder plug if driving habits warrant.

DETONATION
Symptoms: Insulators may be cracked or chipped. Improper gap setting techniques can also result in a fractured insulator tip. Can lead to piston damage.
Recommendation: Make sure the fuel anti-knock values meet engine requirements. Use care when setting the gaps on new plugs. Avoid lugging the engine.

MECHANICAL DAMAGE
Symptoms: May be caused by a foreign object in the combustion chamber or the piston striking an incorrect reach (too long) plug. Causes a dead cylinder and could result in piston damage.
Recommendation: Repair the mechanical damage. Remove the foreign object from the engine and/or install the correct reach plug.

1-30   Chapter 1   Tune-up and routine maintenance

31.4 Before opening the drain valve located at the bottom of the radiator, push a short section of 3/8-inch diameter hose onto the fitting to prevent the coolant from splashing as it drains

air cannot enter through the end of the snorkel, but instead enters the air cleaner through the flexible duct attached to the exhaust manifold and the heat stove passage.

6   As the engine warms up to operating temperature, the valve should move down to allow air to be drawn through the snorkel end. Depending on outside temperature, this may take 10-to-15 minutes. To speed up this check you can reconnect the snorkel air duct, drive the vehicle, then check to see if the valve is completely open.

7   If the thermostatically controlled air cleaner isn't operating properly, see Chapter 6 for more information.

31 Cooling system servicing (draining, flushing and refilling)

Warning: *Antifreeze is a corrosive and poisonous solution, so be careful not to spill any of the coolant mixture on the vehicle's paint or your skin. If this happens, rinse immediately with plenty of clean water. Consult local authorities regarding proper disposal procedures for antifreeze before draining the cooling system. In many areas, reclamation centers have been established to collect used oil and coolant mixtures. The electric cooling fan can activate at any time, even when the ignition is in the Off position. Disconnect the fan motor or negative battery cable when working in the vicinity of the fan.*

1   Periodically, the cooling system should be drained, flushed and refilled to replenish the antifreeze mixture and prevent formation of rust and corrosion, which can impair the performance of the cooling system and cause engine damage. When the cooling system is serviced, all hoses and the radiator cap should be checked and replaced if necessary.

Draining
Refer to illustrations 31.4 and 31.5

2   Apply the parking brake and block the wheels. If the vehicle has just been driven, wait several hours to allow the engine to cool down before beginning this procedure.

3   Once the engine is completely cool, remove the radiator cap.

4   Remove the splash cover located beneath the radiator (some models). Then move a large container under the radiator drain to catch the coolant. Attach a 3/8-inch inner diameter hose to the drain fitting to direct the coolant into the container (some models are already equipped with a hose), then open the drain fitting (a pair of pliers may be required to turn it) **(see illustration)**.

5   After the coolant stops flowing out of the radiator, move the container under the engine block drain plug **(see illustration)**. Loosen the plug and allow the coolant in the block to drain.

6   While the coolant is draining, check the condition of the radiator hoses, heater hoses and clamps (refer to Section 14 if necessary).

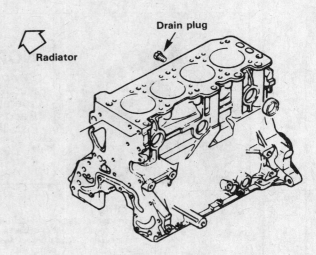

31.5 After the coolant stops flowing from the radiator, remove the engine block drain plug and allow the coolant to drain from the engine block

7   Replace any damaged clamps or hoses (refer to Chapter 3 for detailed replacement procedures).

Flushing
8   Once the system is completely drained, flush the radiator with fresh water from a garden hose until water runs clear at the drain. The flushing action of the water will remove sediments from the radiator but will not remove rust and scale from the engine and cooling tube surfaces.

9   These deposits can be removed by the chemical action of a cleaner. Follow the procedure outlined in the manufacturer's instructions. If the radiator is severely corroded, damaged or leaking, it should be removed (see Chapter 3) and taken to a radiator repair shop.

10   Remove the overflow hose from the coolant recovery reservoir. Drain the reservoir and flush it with clean water, then reconnect the hose.

Refilling
11   Close and tighten the radiator drain. Install and tighten the block drain plug.

12   Place the heater temperature control in the maximum heat position.

13   Slowly add new coolant (a 50/50 mixture of water and antifreeze) to the radiator until it's full. Add coolant to the reservoir up to the lower mark.

14   Leave the radiator cap off and run the engine in a well-ventilated area until the thermostat opens (coolant will begin flowing through the radiator and the upper radiator hose will become hot).

15   Turn the engine off and let it cool. Add more coolant mixture to bring the level back up to the lip on the radiator filler neck.

16   Squeeze the upper radiator hose to expel air, then add more coolant mixture if necessary. Replace the radiator cap.

17   Start the engine, allow it to reach normal operating temperature and check for leaks.

32 Rear wheel bearing check, repack and adjustment (drum brakes only)

Check
Refer to illustration 32.1

1   In most cases the rear wheel bearings will not need servicing until the brake shoes are changed. However, the bearings should be checked whenever the rear of the vehicle is raised for any reason. Several items,

Chapter 1 Tune-up and routine maintenance

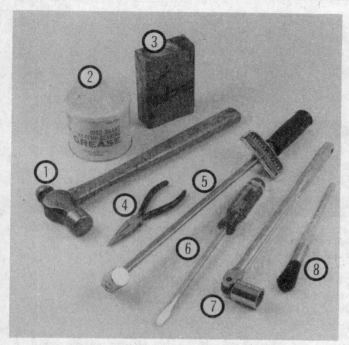

32.1 Tools and materials needed for front wheel bearing maintenance

1. **Hammer** – A common hammer will do just fine
2. **Grease** – High-temperature grease that is formulated specially for front wheel bearings should be used
3. **Wood block** – If you have a scrap piece of 2x4, it can be used to drive the new seal into the hub
4. **Needle-nose pliers** – Used to straighten and remove the cotter pin in the spindle
5. **Torque wrench** – This is very important in this procedure; if the bearing is too tight, the wheel won't turn freely – if it's too loose, the wheel will "wobble" on the spindle. Either way, it could mean extensive damage.
6. **Screwdriver** – Used to remove the seal from the hub (a long screwdriver would be preferred)
7. **Socket/breaker bar** – Needed to loosen the nut on the spindle if it's extremely tight
8. **Brush** – Together with some clean solvent, this will be used to remove old grease from the hub and spindle

32.6 Pry the dust cap out of the hub

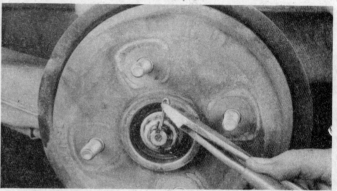

32.7a Wire cutters are useful for gripping the greasy cotter pin and pulling it out

32.7b Pull the nut lock off the spindle nut

32.8 Remove the washer with a small screwdriver (the spindle nut has been removed in this illustration)

including a torque wrench and special grease, are required for this procedure **(see illustration)**.
2 With the vehicle securely supported on jackstands, spin each wheel and check for noise, rolling resistance and freeplay.
3 Grasp the top of each tire with one hand and the bottom with the other. Move the wheel in-and-out on the spindle. If there's any noticeable movement, the bearings should be checked and then repacked with grease or replaced if necessary.

Repack

Refer to illustrations 32.6, 32.7a, 32.7b, 32.8, 32.9, 32.11 and 32.15

4 Remove the tire/wheel assembly.
5 If necessary, back off the parking brake adjuster (see Chapter 9).
6 Pry the dust cap out of the drum/hub assembly using a screwdriver or hammer and chisel **(see illustration)**.
7 Straighten the bent ends of the cotter pin, then pull the cotter pin out of the nut lock **(see illustration)**. Discard the cotter pin and use a new one during reassembly. Remove the nut lock **(see illustration)**. **Note:** *If no cotter pin is present, the vehicle is equipped with a self-locking nut.*
8 Remove the spindle nut and washer from the end of the spindle **(see illustration)**.

Chapter 1 Tune-up and routine maintenance

32.9 Remove the outer wheel bearing after the pulling the drum/hub out slightly to dislodge it

32.11 Pry the seal out of the hub with a screwdriver or hooked seal puller such as this one (available at auto parts stores)

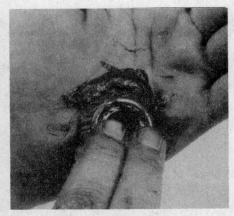

32.15 Work the grease into the bearing rollers from the back side of the bearing race

32.27 Tap the grease cap into place with a large punch and a hammer, working around the outer circumference

9 Pull the drum/hub assembly out slightly, then push it back into its original position. This should force the outer bearing off the spindle enough so it can be removed (see illustration).
10 Pull the drum/hub off the spindle.
11 Use a seal puller or screwdriver to pry the seal out of the rear of the drum/hub (see illustration). As this is done, note how the seal is installed.
12 Remove the inner wheel bearing from the drum/hub.
13 Use solvent to remove all traces of the old grease from the bearings, hub and spindle. A small brush may prove helpful; however make sure no bristles from the brush embed themselves inside the bearing rollers. Allow the parts to air dry.
14 Carefully inspect the bearings for cracks, heat discoloration, worn rollers, etc. Check the bearing races inside the hub for wear and damage. If the bearing races are defective, the hubs should be taken to a machine shop with the facilities to remove the old races and press new ones in. Note that the bearings and races come as matched sets and old bearings should never be installed on new races.
15 Use wheel bearing grease to pack the bearings. Work the grease completely into the bearings, forcing it between the rollers, cone and cage from the back side (see illustration).
16 Apply a thin coat of grease to the spindle at the outer bearing seat, inner bearing seat, shoulder and seal seat.
17 Put a small quantity of grease inboard of each bearing race inside the hub. Using your finger, form a dam at these points to provide extra grease availability and to keep thinned grease from flowing out of the bearing.
18 Place the grease-packed inner bearing into the rear of the drum/hub, and put a little more grease outboard of the bearing.
19 Place a new seal over the inner bearing and tap the seal evenly into place with a hammer and block of wood until it's flush with the hub.
20 Carefully place the drum/hub assembly onto the spindle and push the grease-packed outer bearing into position.

Adjustment – castellated nut
Refer to illustration 32.27
Note: *If equipped with a self-locking nut, install a new self-locking nut, torque to specs listed in this Chapter's specifications, then proceed to step 27.*
21 Install the washer and spindle nut. Tighten the nut only slightly (no more than 12 ft-lbs of torque).
22 Spin the drum/hub in a forward direction to seat the bearings and remove any grease or burrs which could cause excessive bearing play later.
23 Check to see that the tightness of the spindle nut is still approximately 12 ft-lbs.
24 Loosen the spindle nut until it's just loose, no more.
25 Tighten the nut until it's snug (4 to 7 ft-lbs). Install the nut lock, then install a new cotter pin through the hole in the spindle and the slots in the nut lock. If the slots don't line up, loosen the nut slightly until they do. The nut should not be loosened more than one-half flat to install the cotter pin.
26 Bend the ends of the cotter pin until they're flat against the nut. Cut off any extra length which could interfere with the dust cap.
27 Install the dust cap, tapping it into place with a hammer and a large punch (see illustration).
28 Install the tire/wheel assembly on the drum/hub and tighten the lug nuts.
29 Grasp the top and bottom of the tire and check the bearings in the manner described earlier in this Section.
30 Lower the vehicle.

33 Brake fluid replacement (Precis models only)

1 Because brake fluid absorbs moisture which could ultimately cause corrosion of the brake components, and air which could make the braking system less effective, the fluid should be replaced at the specified intervals. This job can be accomplished for a nominal fee by a properly equipped brake shop using a pressure bleeder. The task can also be done by the home mechanic with the help of an assistant. To bleed the air and old fluid and replace it with fresh fluid from sealed containers, refer to the brake bleeding procedure in Chapter 9.
2 If there is any possibility that incorrect fluid has been used in the system, drain all the fluid and flush the system with alcohol. Replace all piston seals and cups, as they will be affected and could possibly fail under pressure.

34 Automatic transaxle fluid and filter change

Refer to illustrations 34.7, 34.9 and 34.11
1 At the specified time intervals, the automatic transaxle fluid should be drained and replaced.

Chapter 1 Tune-up and routine maintenance

1 – 33

34.7 Use a box-end wrench to remove the automatic transaxle drain plug without rounding it off

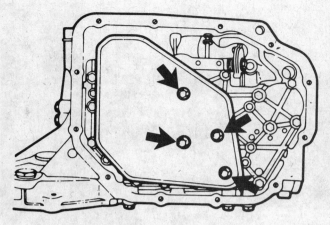

34.9 Remove the bolts (arrows) and detach the filter

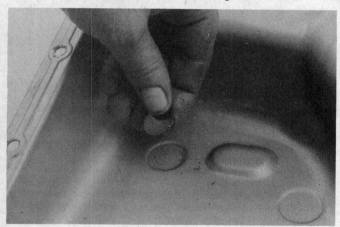

34.11 Be sure to reinstall any magnets in the fluid pan recesses

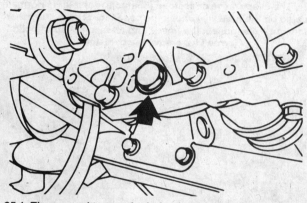

35.1 The manual transaxle drain plug (arrow) is located at the lower side of the case

2 Before beginning work, purchase the specified transmission fluid (see *Recommended lubricants and fluids* at the front of this Chapter).
3 Other tools necessary for this job include jackstands to support the vehicle in a raised position, a drain pan capable of holding at least eight pints, newspapers and clean rags.
4 The fluid should be drained immediately after the vehicle has been driven. Hot fluid is more effective than cold fluid at removing built-up sediment. **Warning:** *Fluid temperature can exceed 350-degrees F in a hot transaxle. Wear protective gloves.*
5 After the vehicle has been driven to warm up the fluid, raise it and place it on jackstands for access to the transaxle drain plug.
6 Move the necessary equipment under the vehicle, being careful not to touch any of the hot exhaust components.
7 Place the drain pan under the drain plug in the transaxle and remove the drain plug **(see illustration)**. Be sure the drain pan is in position, as fluid will come out with some force. Once the fluid is drained, reinstall the drain plug securely.
8 Remove the transaxle pan bolts, carefully pry the pan loose with a screwdriver and remove it.
9 Remove the filter retaining bolts, and detach the filter from the transaxle **(see illustration)**. Be careful when lowering the filter as it contains residual fluid.
10 Place the new filter in position and install the bolts. Tighten the bolts to the torque listed in this Chapter's Specifications.
11 Carefully clean the gasket surfaces of the fluid pan, removing all traces of old gasket material. Wash the pan in clean solvent and dry it with compressed air. If equipped, be sure to clean and reinstall any magnets **(see illustration)**.

12 Install a new gasket, place the fluid pan in position and install the bolts. Tighten the bolts to the torque listed in this Chapter's Specifications.
13 Lower the vehicle.
14 With the engine off, add new fluid to the transaxle through the dipstick tube (see *Recommended lubricants and fluids* for the recommended fluid type and capacity). Use a funnel to prevent spills. It is best to add a little fluid at a time, continually checking the level with the dipstick (Section 7). Allow the fluid time to drain into the pan.
15 Start the engine and shift the selector into all positions from Park through Low, then shift into Park and apply the parking brake.
16 With the engine idling, check the fluid level. Add fluid up to the HOT level on the dipstick.

35 Manual transaxle lubricant change

Refer to illustration 35.1

1 Remove the drain plug and drain the lubricant **(see illustration)**.
2 Reinstall the drain plug. Tighten it to the torque listed in this Chapter's Specifications.
3 Add new lubricant until it begins to run out of the filler hole (Section 25). See *Recommended lubricants and fluids* for the specified lubricant type.

36 Evaporative emissions control system check and canister replacement

Refer to illustration 36.2

1 The function of the evaporative emissions control system is to draw fuel vapors from the gas tank and fuel system, store them in a charcoal

36.2 The evaporative emissions (charcoal) canister is normally located on the firewall – check the hoses for cracks and other damage – disconnect the hoses and release the spring clip (arrow) to remove the canister

37.3 The timing marks are located on the drivebelt end of the engine (drivebelt is removed for clarity) – highlight the notch in the crankshaft pulley and the appropriate mark on the timing plate

canister and route them to the intake manifold during normal engine operation.

2 The most common symptom of a fault in the evaporative emissions system is a strong fuel odor in the engine compartment. If a fuel odor is detected, inspect the charcoal canister, located in the engine compartment **(see illustration)**. Check the canister and all hoses for damage and deterioration.

3 At the specified intervals, the charcoal canister must be replaced with a new one. Disconnect the hoses, release the spring clip and lift the canister from the engine compartment. Installation is the reverse of removal.

4 The evaporative emissions control system is explained in more detail in Chapter 6.

37 Ignition timing check and adjustment

Refer to illustrations 37.1, 37.3, 37.7 and 37.11

Note: *The ignition timing needs to be checked periodically on carbureted models only. On fuel-injected models, timing normally needs checking only if the distributor or crank angle sensor has been removed, the mounting bolt has loosened or the original factory setting has been tampered with.*

1 Some special tools are required for this procedure **(see illustration)**. The engine must be at normal operating temperature and the air conditioner must be Off. Make sure the idle speed is correct (Section 20).

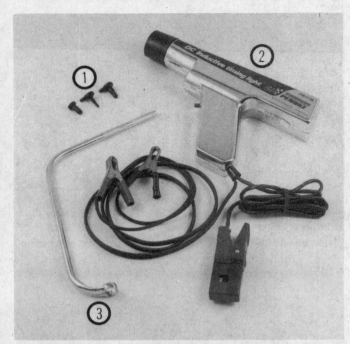

37.1 Tools needed to check and adjust the ignition timing

1 **Vacuum plugs** – *Vacuum hoses will, in most cases, have to be disconnected and plugged. Molded plugs in various shapes and sizes are available for this.*
2 **Inductive pick-up timing light** – *Flashes a bright concentrated beam of light when the number one spark plug fires. Connect the leads according to the instructions supplied with the light.*
3 **Distributor wrench** – *On some models, the hold-down bolt for the distributor is difficult to reach and turn with conventional wrenches or sockets. A special wrench like this must be used.*

2 Apply the parking brake and block the wheels to prevent movement of the vehicle. The transmission must be in Park (automatic) or Neutral (manual).

3 Locate the timing marks at the drivebelt end of the engine (they should be visible from above after the hood is opened) **(see illustration)**. The crankshaft pulley or vibration damper has a notch in it and a plate with raised numbers is attached to the timing cover. Clean the plate with solvent so the numbers are visible.

4 Use chalk or white paint to mark the notch in the pulley/vibration damper.

5 Highlight the point on the timing plate that corresponds to the ignition timing specification on the Vehicle Emission Control Information label.

6 Hook up the timing light by following the manufacturer's instructions (an inductive pick-up timing light is preferred). Generally, the power leads are attached to the battery terminals and the pick-up lead is attached to the number one spark plug wire. The number one spark plug is the one closest to the drivebelt end of the engine. **Caution:** *If an inductive pick-up timing light isn't available, don't puncture the spark plug wire to attach the timing light pick-up lead. Instead, use an adapter between the spark plug and plug wire. If the insulation on the plug wire is damaged, the secondary voltage will jump to ground at the damaged point and the engine will misfire.*

7 On fuel-injected models, locate the ignition timing adjustment connector and ground it, using a jumper wire **(see illustration)**. This will prevent the computer from attempting to advance the timing during adjustment.

8 Make sure the timing light wires are routed away from the drivebelts and fan, then start the engine.

9 Allow the idle speed to stabilize, then point the flashing timing light at the timing marks – be very careful of moving engine components!

Chapter 1 Tune-up and routine maintenance

1–35

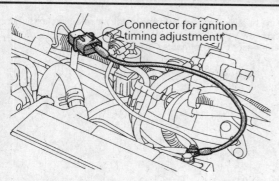

37.7 On fuel-injected models, there should be a connector, like the one shown here, in the engine compartment. Use a jumper wire to ground the connector terminal (Precis shown, others similar).

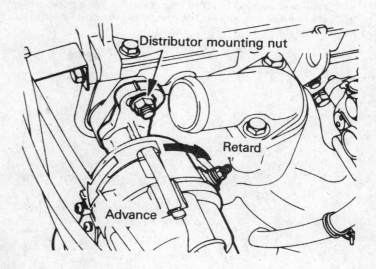

37.11 After loosening the mounting nut, rotate the distributor (or crank angle sensor) housing to adjust the ignition timing

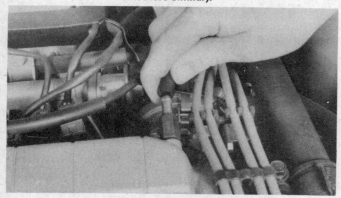

38.2a Pull the hose off the PCV valve . . .

10   The mark on the pulley/vibration damper will appear stationary. If it's aligned with the specified point on the timing plate, the ignition timing is correct.
11   If the marks aren't aligned, adjustment is required. Loosen the distributor or crank angle sensor mounting nut **(see illustration)** and turn the distributor or crank angle sensor very slowly until the marks are aligned.
12   Tighten the nut and recheck the timing.
13   Turn off the engine and remove the timing light (and adapter, if used).

38 Positive Crankcase Ventilation (PCV) valve check and replacement

Refer to illustrations 38.2a and 38.2b

1   The PCV valve is located in the valve cover.
2   Disconnect the hose, unscrew the PCV valve from the cover, then reconnect the hose **(see illustrations)**.
3   With the engine idling at normal operating temperature, place your finger over the valve opening. If there's no vacuum at the valve, check for a plugged hose or valve. Replace any plugged or deteriorated hoses.
4   Turn off the engine. Remove the PCV valve from the hose. Blow through the valve from the threaded end. If air will not pass through the valve in this direction, replace it with a new one.
5   When purchasing a replacement PCV valve, make sure it's for your particular vehicle and engine size. Compare the old valve with the new one to make sure they're the same.

39 Oxygen sensor replacement

Refer to illustration 39.1
Warning: *The electric cooling fan can activate at any time, even when the ignition is in the Off position. Disconnect the fan motor or negative battery cable when working in the vicinity of the fan.*

38.2b . . . the remove the valve with a wrench

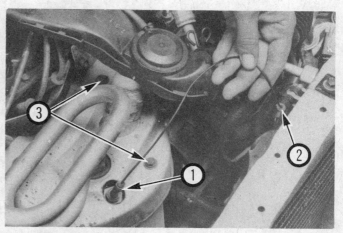

39.1 Prior to removing the oxygen sensor (1), unplug the wire at the connector (2), unscrew the bolts (3) and remove the exhaust manifold heat shield

1   Unplug the oxygen sensor wire at the connector located directly behind the top left corner of the radiator and remove the exhaust manifold heat shield bolts **(see illustration)**.

2 Lift the heat shield up for access and unscrew the oxygen sensor.
3 Screw the new oxygen sensor into the exhaust manifold. Tighten the sensor securely.
4 Place the heat shield in position and install the retaining bolts. Tighten the bolts securely.
5 Plug in the electrical connector.

Chapter 2 Part A Engine

Contents

Camshaft(s) – removal, inspection and installation 9	Oil pan – removal and installation 13
Crankshaft front oil seal – replacement 8	Oil pump – removal, inspection and installation 14
Cylinder compression check .... See Chapter 2B	Rear main oil seal – replacement 16
Cylinder head – removal and installation 12	Repair operations possible with the engine in the vehicle 2
Drivebelt check, adjustment and replacement .... See Chapter 1	Rocker arm assembly – removal, inspection and installation 4
Engine mounts – check and replacement 17	Spark plug replacement .... See Chapter 1
Engine oil and oil filter change .... See Chapter 1	Timing belt and sprockets – removal, inspection
Engine overhaul – general information .... See Chapter 2B	and installation 7
Engine – removal and installation .... See Chapter 2B	Top Dead Center (TDC) for number
Exhaust manifold – removal and installation 11	one piston – locating .... See Chapter 2B
Flywheel/driveplate – removal and installation 15	Valve cover – removal and installation 3
General information and engine identification 1	Valves – servicing .... See Chapter 2B
Hydraulic valve adjusters – removal, inspection	Valve springs, retainers and seals – replacement 5
and installation 6	Water pump – removal and installation .... See Chapter 3
Intake manifold – removal and installation 10	

Specifications

General

Firing order	1-3-4-2
Cylinder numbers (drivebelt end-to-transaxle end)	1-2-3-4

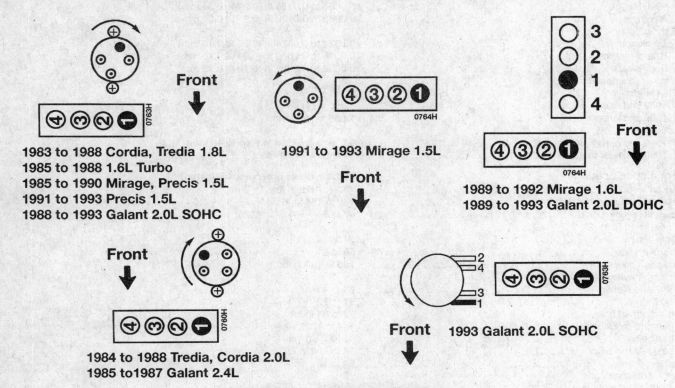

The blackened terminal shown on the distributor cap indicates the Number One spark plug wire position

Cylinder location and distributor rotation

1.5L engine

Camshaft
Camshaft endplay	0.002 to 0.008 in (0.05 to 0.20 mm)
Lobe height	
Standard	
Intake	
990 and later Precis and 1990 Mirage	1.5318 in (38.909 mm)
1991 and later Mirage	1.5256 in (38.75 mm)
All others	1.5 in (38.1 mm)
Exhaust	
1990 and later Precis and 1990 Mirage	1.5344 in (38.974 mm)
1991 and later Mirage	1.5394 in (39.10 mm)
All others	1.5039 in (38.2 mm)
Wear limit	0.002 in (0.05 mm)
Camshaft bearing oil clearance	
1990 Mirage	0.0015 to 0.0031 in (0.04 to 0.08 mm)
1991 and later Mirage	0.0024 to 0.0039 in (0.06 to 0.10 mm)
All others	0.002 to 0.0035 in (0.05 to 0.09 mm)

Cylinder head
Warpage limit	0.004 in (0.1 mm)

Timing belt
Timing belt deflection	9/32 to 11/32 in (7 to 9 mm)
Clearance between timing belt and seal line	Approx. 9/16 in (14 mm)

Oil pump
Clearances	
Outer gear-to-front case	0.0039 to 0.0079 in (0.10 to 0.20 mm)
Outer gear-to-crescent	
1990 Mirage	0.0087 to 0.0173 in (0.22 to 0.44 mm)
All others	0.0087 to 0.0134 in (0.22 to 0.34 mm)
Inner gear-to-crescent	
1990 Mirage	0.0083 to 0.0134 in (0.21 to 0.34 mm)
All others	0.0083 to 0.0126 in (0.21 to 0.32 mm)
Gear endplay	
Tip clearance	0.0024 to 0.0071 in (0.06 to 0.18 mm)
Side clearance	0.0016 to 0.0039 in (0.04 to 0.10 mm)
Body clearance	0.0039 to 0.0071 in (0.10 to 0.18 mm)
Pressure regulator spring	
Free height	
1989 and earlier models	1.850 in (47 mm)
1990 and later models	1.835 in (46.6 mm)
Load	
1989 and earlier models	9.5 lb @ 1.575 in (4.3 kg @ 40 mm)
1990 and later models	13.4 lb @ 1.579 in

Torque specifications
	Ft-lbs (unless otherwise indicated)
Valve cover bolts	13.2 to 16.8 in-lbs
Intake/exhaust manifold nuts/bolts	132 to 168 in-lbs
Camshaft sprocket bolt	47 to 54
Cylinder head bolts	
Cold engine	51 to 54
Warm engine	58 to 61
Crankshaft pulley bolts	108 to 132 in-lbs
Crankshaft pulley center bolt	
1987 and earlier	37 to 43
1988 on	51 to 72
Oil pump cover bolts	72 to 84 in-lbs
Oil pump-to-block bolts	104 to 132 in-lbs
Oil pan bolts	52 to 70 in-lbs
Oil seal retainer bolts	84 to 108 in-lbs
Flywheel/driveplate bolts	94 to 101
Rocker arm shaft bolts	
1991 and later Mirage	21 to 25
All others	15 to 19
Cylinder head rear cover bolts	70 to 86 in-lbs
Timing belt tensioner bolt	15 to 18

Chapter 2 Part A Engine

1.6L engine

Camshaft
Camshaft endplay
  SOHC .... 0.002 to 0.006 in (0.05 to 0.15 mm)
  DOHC .... 0.004 to 0.008 in (0.10 to 0.20 mm)
Lobe height (standard)
  SOHC
    Intake .... 1.432 in (36.7 mm)
    Exhaust .... 1.434 in (36.4 mm)
  DOHC
    Intake .... 1.3858 in (35.2 mm)
    Exhaust .... 1.3743 in (34.9 mm)
  Wear limit .... 0.020 in (0.5 mm)
Camshaft bearing oil clearance .... 0.002 to 0.0035 in (0.05 to 0.09 mm)

Cylinder head
Warpage limit .... 0.002 in (0.05 mm)

Timing belt
SOHC
  Timing belt deflection .... Not available
  Clearance between timing belt and seal line .... Approx. 15/64 in (6 mm)

Timing belt (continued)
DOHC
  Projection of tensioner rod (on the bench) .... 15/32 in (12.0 mm)
  Projection of tensioner rod (assembled) .... 5/32 to 3/16 in (3.8 to 4.5 mm)

Oil pump
Clearances
  SOHC
    Side clearance .... 0.0024 to 0.0047 in (0.06 to 0.12 mm)
    Tip clearance .... 0.0016 to 0.0047 in (0.04 to 0.12 mm)
    Body clearance .... 0.0039 to 0.0063 in (0.10 to 0.16 mm)
    Driveshaft-to-cover clearance .... 0.0008 to 0.0020 in (0.02 to 0.05 mm)
  DOHC
    Tip clearance
      Drive gear .... 0.0063 to 0.0083 in (0.16 to 0.21 mm)
      Driven gear .... 0.0051 to 0.0071 in (0.13 to 0.18 mm)
    Side clearance
      Drive gear .... 0.0031 to 0.0055 in (0.08 to 0.14 mm)
      Driven gear .... 0.0024 to 0.0047 in (0.06 to 0.12 mm)
Pressure regulator spring
  Free length .... 1.834 in (47 mm)
  Load .... 13.4 lb @ 1.579 in (6.1 kg @ 40 mm)

Torque specifications   Ft-lbs (unless otherwise indicated)
Valve cover bolts
  1990 and later Mirage .... 24 to 36 in-lbs
  All others .... 43 to 61 in-lbs
Intake/exhaust manifold nuts/bolts
  1990 and later Mirage (10 mm bolts) .... 22 to 30
  All others .... 11 to 14
Camshaft bearing cap (DOHC) .... 14 to 15
Camshaft sprocket bolt .... 58 to 72
Cylinder head bolts
  SOHC
    Cold engine .... 51 to 54
    Warm engine .... 58 to 61
  DOHC (cold engine) .... 65 to 72
Crankshaft pulley bolts
  1990 and later Mirage .... 168 to 264 in-lbs
  All others .... 108 to 132 in-lbs
Crankshaft pulley center bolt .... 80 to 94
Oil pump sprocket bolt/nut
  1990 and later Mirage (bolt) .... 36 to 43
  All others (nut) .... 25 to 28
Oil pump cover bolts
  1990 and later Mirage .... 132 to 156 in-lbs
  All others .... 72 to 84 in-lbs
Oil pan bolts .... 51 to 69 in-lbs
Oil pan nuts .... 43 to 60 in-lbs

Oil seal retainer bolts	84 to 108 in-lbs
Flywheel/driveplate bolts	94 to 101
Rocker arm shaft bolts	14 to 15
Cylinder head rear cover bolts	69 to 86 in-lbs
Timing belt tensioner bolt	15 to 18

1.8L turbo, 2.0L and 2.4L engines

Camshaft

Camshaft endplay	0.004 to 0.008 in (0.1 to 0.2 mm)
Lobe height (standard)	
1.8L Turbo and 2.0L Cordia/Tredia	1.6565 in (42.08 mm)
2.0L SOHC Galant	1.7529 in (44.525 mm)
2.4L	1.6693 in (42.4 mm)
2.0L DOHC	
Intake	1.3974 in (35.493 mm)
Exhaust	1.3858 in (35.200 mm)
Wear limit	0.020 in (0.5 mm)
Camshaft bearing oil clearance	0.002 to 0.0035 in (0.05 to 0.09 mm)

Cylinder head

Warpage limit	0.002 in (0.05 mm)

Timing belt

Timing belt deflection	Tension automatically adjusted
Clearance between timing belt and seal line	Approx. 9/16 in (14 mm)
Projection of tensioner rod (on the bench)	15/32 in (12 mm)
Projection of tensioner rod (assembled)	5/32 to 3/16 in (3.8 to 4.5 mm)

Oil pump

Clearances	
Tip clearance	
Drive gear	0.0063 to 0.0083 in (0.16 to 0.21 mm)
Driven gear	0.0051 to 0.0071 in (0.13 to 0.18 mm)
Side clearance	
Drive gear	0.0031 to 0.0055 in (0.08 to 0.14 mm)
Driven gear	0.0024 to 0.0047 in (0.06 to 0.12 mm)
Pressure regulator spring	
Free length	1.834 in (47 mm)
Load	13.4 lb @ 1.579 in (6.1 kg @ 40 mm)

Torque specifications

Ft-lbs (unless otherwise indicated)

Valve cover bolts	
1990 and later Galant DOHC 16-valve engine	24 to 36 in-lbs
1993 Galant SOHC 16-valve	26 to 30 in-lbs
Intake/exhaust manifold nuts/bolts	
1990 and later Galant DOHC 16-valve engine	
Exhaust	18 to 22
Intake (10 mm bolts)	22 to 30
Galant SOHC 16-valve engine	18 to 22
All others	11 to 14
Camshaft bearing cap bolts (DOHC 16-valve engine)	14 to 15
Camshaft sprocket bolt	58 to 72
Cylinder head bolts	
Cold engine	
1993 Galant DOHC 16-valve engine	
Step 1	58
Step 2	loosen all bolts
Step 3	14
Step 4	tighten all bolts 1/4-turn (90-degrees)
Step 5	tighten all bolts an additional 1/4-turn (90-degrees)
All others	65 to 72
Warm engine (all except 1993 Galant DOHC 16-valve engine)	73 to 80
Crankshaft pulley bolts	
1990 and later Galant DOHC 16-valve engine	14 to 22
All others	108 to 132 in-lbs
Crankshaft pulley center bolt	80 to 94
Oil pump sprocket nut	
1990 and later Galant	36 to 43
All others	25 to 28
Oil pump cover bolts	
1990 and later Galant	132 to 156 in-lbs
All others	72 to 84 in-lbs

Chapter 2 Part A Engine

Oil pan bolts	51 to 69 in-lbs
Oil pan nuts	43 to 60 in-lbs
Oil seal retainer bolts	84 to 108 in-lbs
Flywheel/driveplate bolts	94 to 101
Rocker arm shaft bolts	
1990 Galant SOHC 16-valve engine	14 to 20
All others	14 to 15
Silent Shaft sprocket nut	
1990 and later Galant (thick shim)	31 to 35
All others	22 to 28
Cylinder head rear cover bolts	69 to 86 in-lbs
Automatic tensioner bolt	14 to 20
Timing belt tensioner bolt	32 to 40

1   General information and engine identification

Refer to illustration 1.4

This Part of Chapter 2 is devoted to in-vehicle engine repair procedures. Information concerning engine removal and installation and engine block and cylinder head overhaul can be found in Part B of this Chapter.

The following repair procedures are based on the assumption that the engine is installed in the vehicle. If the engine has been removed from the vehicle and mounted on a stand, many of the steps outlined in this Part of Chapter 2 will not apply.

The Specifications included in this Part of Chapter 2 apply only to the procedures contained in this Part. Part B of Chapter 2 contains the Specifications necessary for cylinder head and engine block rebuilding.

There are basically two types of four-cylinder engines installed in the Mitsubishi models covered in this book: Single Overhead Camshaft (SOHC) and Double Overhead Camshaft (DOHC) engines. There are several different versions of the SOHC engine; each one has its own engine number – G64B, G62B, G63B, G15B, G32B, 4G63, G4DJ, 4G15 and G62B Turbo. There are two versions of the DOHC engine – 4G32 and 4G61. Refer to the Engine Identification Chart **(see illustration)** to determine the correct engine using the vehicle's year and model. If there are optional engines available for the same year, use the engine identification plate that is attached to the engine firewall to determine the correct engine type.

2   Repair operations possible with the engine in the vehicle

Many major repair operations can be accomplished without removing the engine from the vehicle.

Clean the engine compartment and the exterior of the engine with some type of degreaser before any work is done. It will make the job easier and help keep dirt out of the internal areas of the engine.

Depending on the components involved, it may be helpful to remove the hood to improve access to the engine as repairs are performed (refer to Chapter 11 if necessary). Cover the fenders to prevent damage to the paint. Special pads are available, but an old bedspread or blanket will also work.

If vacuum, exhaust, oil or coolant leaks develop, indicating a need for gasket or seal replacement, the repairs can generally be made with the engine in the vehicle. The intake and exhaust manifold gaskets, oil pan gasket, crankshaft oil seals and cylinder head gasket are all accessible with the engine in place.

Exterior engine components, such as the intake and exhaust manifolds, the oil pan (and the oil pump), the water pump, the starter motor, the alternator, the distributor and the fuel system components can be removed for repair with the engine in place.

Since the camshaft(s) and cylinder head can be removed without pulling the engine, valve component servicing can also be accomplished with the engine in the vehicle. Replacement of the timing belt(s) and sprockets is also possible with the engine in the vehicle.

Single Overhead Camshaft (SOHC) engines

Model	Engine no. (Displacement)	Valve type
Galant		
1985 to 1987	G64B (2.4L)	Adjustable w/ Jet valve
1989 and later	4G63 (2.0L)	Hydraulic w/o Jet valve
Precis		
1987 and 1988	G4DJ (1.5L)	Adjustable w/ Jet valve
1989 and later	G4DJ (1.5L)	Adjustable w/o Jet valve
Cordia/Tredia		
1983	G62B (1.8L)	Adjustable w/ Jet valve
1984 and 1985	G62B Turbo (1.8L)	Adjustable w/ Jet valve
1984 and 1985	G63B (2.0L)	Adjustable w/ Jet valve
1985 to 1988	G62B Turbo (1.8L)	Hydraulic w/ Jet valve
1985 to 1988	G63B (2.0L)	Hydraulic w/ Jet valve
Mirage		
1985 to 1987	G15B (1.5L)	Adjustable w/ Jet valve
1985 to 1987	G32B (1.6L)	Adjustable w/ Jet valve
1989 and later	4G15 (1.5L)	Adjustable w/o Jet valve

Double Overhead Camshaft (DOHC) engines

Model	Engine no. (Displacement)	Valve type
Galant		
1989 and later	4G32 (2.0L)	Hydraulic w/o Jet valve
Mirage		
1989 and later	4G61 (1.6L)	Hydraulic w/o Jet valve

1.4   Engine Identification Chart

In extreme cases caused by a lack of necessary equipment, repair or replacement of piston rings, pistons, connecting rods and rod bearings is possible with the engine in the vehicle. However, this practice is not recommended because of the cleaning and preparation work that must be done to the components involved.

3   Valve cover – removal and installation

Refer to illustrations 3.3, 3.5a, 3.5b and 3.5c

1   Disconnect the negative cable from the battery.
2   Remove the air cleaner assembly (see Chapter 4).
3   Detach the spark plug wires and cable brackets from the valve cover **(see illustration)**. **Note:** *On DOHC engines, remove the center cover to gain access to the spark plugs.*

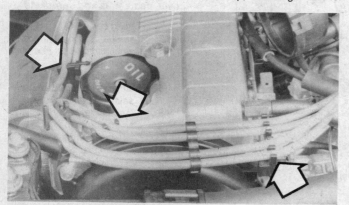

3.3   Carefully disconnect the spark plug wires from the clips

2A–6   Chapter 2 Part A   Engine

3.5a Remove the valve cover bolts (arrows) – 2.0L engine shown

4   Clearly label and then disconnect any emission hoses and cables which connect to or cross over the valve cover.
5   Remove the valve cover bolts **(see illustrations)** and lift the cover off. If the cover sticks to the cylinder head, tap on it with a soft-face hammer or place a block of wood against the cover and tap on the wood with a hammer.
6   Thoroughly clean the valve cover and remove all traces of old gasket material.
7   Install a new gasket on the cover, using RTV to hold it in place. Place the cover on the engine and install the cover bolts. **Note:** *Be sure to install a new semi-circular seal* **(see illustrations 3.5a and 3.5b)** *into the cylinder head. Apply a small amount of sealant to the bottom of the seal and, after it has been installed, to the top of the seal, at the seal-to-head joint.*
8   Tighten the bolts to the torque listed in this Chapter's Specifications. The remaining steps are the reverse of removal. When finished, run the engine and check for oil leaks.

---

4   Rocker arm assembly – removal, inspection and installation

Removal
Refer to illustration 4.3
1   Remove the valve cover (see Section 3).

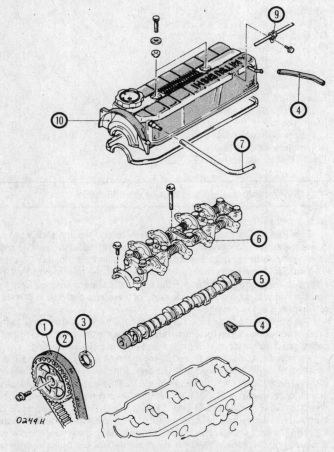

3.5b Exploded view of the SOHC valve cover, rocker arms and camshaft

1   Timing belt   6   Rocker arm assembly
2   Camshaft sprocket   7   PCV hose
3   Camshaft oil seal   8   Breather hose
4   Semi-circular seal   9   Cable clip
5   Camshaft   10 Valve cover

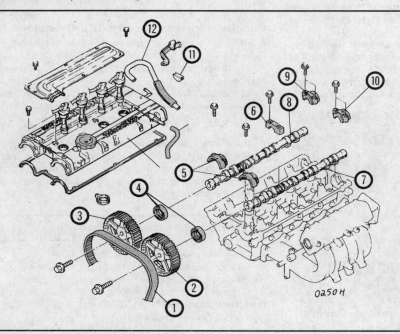

3.5c Exploded view of a DOHC valve cover, camshafts and sprockets

1   Timing belt
2   Camshaft sprocket (intake)
3   Camshaft sprocket (exhaust)
4   Camshaft seal
5   Camshaft bearing caps
6   Camshaft bearing cap
7   Camshaft (intake)
8   Camshaft (exhaust)
9   Camshaft bearing cap
10 Camshaft bearing cap
11 Semi-circular seal
12 Breather hose

Chapter 2 Part A Engine

4.3 Remove the rocker arm assembly bolts (arrows) – 2.0L Galant engine shown

2 Position the number one piston at Top Dead Center (see Chapter 2B).
3 Loosen the rocker arm shaft mounting bolts 1/4-turn at a time each until the spring pressure is relieved **(see illustration)**. Remove the bolts. If you're working on a DOHC engine, remove the camshafts following the procedure in Section 9.
4 Lift the rocker arms and shaft assembly, or the individual rocker arms, from the cylinder head. If you're working on an engine with hydraulic valve adjusters, be sure not to let the adjusters fall out of the rocker arms. Wrap pieces of tape around the ends of the rocker arms to keep the adjusters in place.

Inspection

Refer to illustrations 4.5 and 4.6

5 If you wish to disassemble and inspect the rocker arm assemblies (a good idea as long as you have them off), remove the retaining bolts and slip the rocker arms and springs off the shafts **(see illustration)**. Keep the parts in order so you can reassemble them in the same positions.
6 Thoroughly clean the parts and inspect them for wear and damage. On SOHC engines, check the rocker arm faces that contact the camshaft and the adjusting screw tips **(see illustration)**. On DOHC engines, visually check the roller for dents, damage or seizure. Replace any parts that are damaged or excessively worn. Also, make sure the oil holes in the shafts are not plugged.

Installation

Refer to illustration 4.7

7 On rocker arms with adjusting screws, loosen the locknuts and back off the adjusters **(see illustration)** until they only protrude 1 mm (0.040-inch).

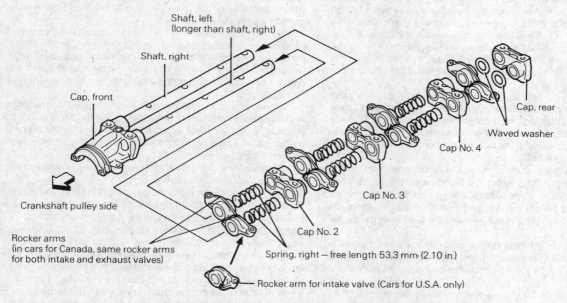

4.5 Exploded view of the rocker arms and shafts (typical)

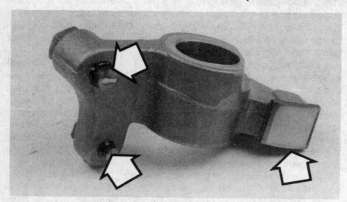

4.6 Check the contact faces and adjusting screw tips (arrows)

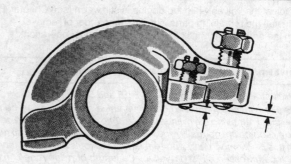

4.7 On adjustable-type rocker arms, back off the adjusters until they only protrude 1 mm (0.040 in)

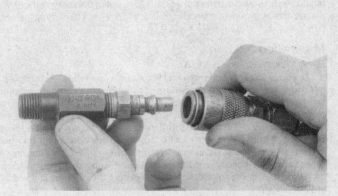

5.4 This is what the air hose adapter that threads into the spark plug hole looks like – they're commonly available from auto parts stores

5.5 If you use a lever-type tool such as this one, leave the intake shaft in place while you replace the exhaust seals and vice-versa

8 Lubricate all components with assembly lube or engine oil and reassemble the shafts (SOHC engines). When installing the rocker arms, shafts and springs, note the markings and the difference between the left and right side parts. On SOHC 8-valve engines, position the notches in the ends of the shafts up. On Galant SOHC 16-valve engines, position the shafts so the notches face out, so a line drawn between them would be parallel with the cylinder head gasket surface. Install the bolts through the shafts (16-valve) or bearing caps and shafts (8-valve) to hold the components in place until reassembly.

9 If removed, insert the hydraulic valve adjusters into the rocker arms and wrap tape around the ends of the rocker arms to prevent the adjusters from falling out. **Note:** *New adjusters are primed with diesel oil. To make sure the oil is not spilled from the adjuster while handling, avoid tilting the adjuster as much as possible.* If the oil is spilled from the adjuster, bleed the adjuster as described in Section 6.

10 Position the rocker arm assemblies on the cylinder head and install the mounting bolts finger tight. **Note:** *On DOHC engines, bearing cap numbers 2 through 5 are the same shape.* Check the markings on the caps to identify the correct journal number and intake/exhaust position. On models with hydraulic valve adjusters, remove the tape from the rocker arms.

11 Tighten the bolts in several stages until the specified torque is reached. **Note:** *On DOHC engines, tighten the bearing caps a little at a time, starting with the center journals and working toward the ends of the camshaft.*

12 On adjustable-type rocker arms, adjust the valve clearances (cold) as described in Chapter 1.

13 Temporarily install the valve cover and run the engine until it is fully warmed up.

14 Readjust the valves while the engine is still warm (see Chapter 1).

15 Reinstall the remaining parts in the reverse order of removal.

16 Run the engine and check for oil leaks and proper operation.

5 Valve springs, retainers and seals – replacement

Refer to illustrations 5.4, 5.5, 5.9, 5.10, 5.15 and 5.17

Note: *Broken valve springs and defective valve stem seals can be replaced without removing the cylinder heads. Two special tools and a compressed air source are normally required to perform this operation, so read through this Section carefully and rent or buy the tools before beginning the job. If compressed air isn't available, a length of nylon rope can be used to keep the valves from falling into the cylinder during this procedure.*

Note: *Several different engines are equipped with an auxiliary valve or jet valve. The jet valve is mounted directly beside each exhaust valve. Refer to Chapter 2B for the jet valve servicing procedure.*

1 Refer to Section 3 and remove the valve cover.

2 Remove the spark plug from the cylinder which has the defective component. If all of the valve stem seals are being replaced, all of the spark plugs should be removed.

3 Turn the crankshaft until the piston in the affected cylinder is at top dead center on the compression stroke (refer to Chapter 2B for instructions). If you're replacing all of the valve stem seals, begin with cylinder number one and work on the valves for one cylinder at a time. Move from cylinder-to-cylinder following the firing order sequence (see this Chapter's Specifications).

4 Thread an adapter into the spark plug hole **(see illustration)** and connect an air hose from a compressed air source to it. Most auto parts stores can supply the air hose adapter. **Note:** *Many cylinder compression gauges utilize a screw-in fitting that may work with your air hose quick-disconnect fitting.*

5 Remove the rocker arm shaft (see Section 4), unless you are using a lever-type tool **(see illustration)**.

6 Apply compressed air to the cylinder. **Warning:** *The piston may be forced down by compressed air, causing the crankshaft to turn suddenly. If the wrench used when positioning the number one piston at TDC is still attached to the bolt in the crankshaft nose, it could cause damage or injury when the crankshaft moves.*

7 The valves should be held in place by the air pressure. If the valve faces or seats are in poor condition, leaks may prevent air pressure from retaining the valves – refer to the alternative procedure below.

8 If you don't have access to compressed air, an alternative method can be used. Position the piston at a point a few degrees before TDC on the compression stroke, then feed a long piece of nylon rope through the spark plug hole until it fills the combustion chamber. Be sure to leave the end of the rope hanging out of the engine so it can be removed easily. Use a large ratchet and socket to rotate the crankshaft in the normal direction of rotation until slight resistance is felt.

9 Stuff shop rags into the cylinder head holes above and below the valves to prevent parts and tools from falling into the engine, then use a valve spring compressor to compress the spring. Remove the keepers with small needle-nose pliers or a magnet **(see illustration)**. **Note:** *A couple of different types of tools are available for compressing the valve springs with the head in place. One type grips the lower spring coils and presses on the retainer as the knob is turned, while the other type utilizes the rocker arm shaft for leverage. Both types work very well, although the lever type is usually less expensive.*

10 Remove the spring retainer and valve spring, then remove the guide seal **(see illustration)**. **Note:** *If air pressure fails to hold the valve in the closed position during this operation, the valve face and/or seat is probably damaged. If so, the cylinder head will have to be removed for additional repair operations.*

Chapter 2 Part A Engine

5.9 Use needle-nose pliers (shown) or a small magnet to remove the valve spring keepers – Be careful not to drop them down into the engine!

5.10 Remove the valve guide seal with a pair of pliers

5.15 Gently tap the seal into place with a hammer and deep socket

5.17 Apply a small dab of grease to each keeper before installation to hold it in place on the valve stem until the spring is released

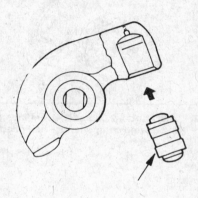

6.1 The hydraulic valve adjusters fit into the ends of the rocker arms

11 Wrap a rubber band or tape around the top of the valve stem so the valve won't fall into the combustion chamber, then release the air pressure. **Note:** *If a rope was used instead of air pressure, turn the crankshaft slightly in the direction opposite normal rotation.*
12 Inspect the valve stem for damage. Rotate the valve in the guide and check the end for eccentric movement, which would indicate that the valve is bent.
13 Move the valve up-and-down in the guide and make sure it doesn't bind. If the valve stem binds, either the valve is bent or the guide is damaged. In either case, the head will have to be removed for repair.
14 Reapply air pressure to the cylinder to retain the valve in the closed position, then remove the tape or rubber band from the valve stem. If a rope was used instead of air pressure, rotate the crankshaft in the normal direction of rotation until slight resistance is felt.
15 Lubricate the valve stem with engine oil and install a new guide seal **(see illustration)**.
16 Install the spring in position over the valve.
17 Install the valve spring retainer. Compress the valve spring and carefully position the keepers in the groove. Apply a small dab of grease to the inside of each keeper to hold it in place if necessary **(see illustration)**.
18 Remove the pressure from the spring tool and make sure the keepers are seated.
19 Disconnect the air hose and remove the adapter from the spark plug hole. If a rope was used in place of air pressure, pull it out of the cylinder.

20 Refer to Section 4 and install the rocker arms and shafts.
21 Install the spark plug(s) and connect the wire(s).
22 Refer to Section 3 and install the valve cover.
23 Start and run the engine, then check for oil leaks and unusual sounds coming from the valve cover area.

6 Hydraulic valve adjusters – removal, inspection and installation

Refer to illustration 6.1

Note: *Since the hydraulic valve adjusters are precision parts, do not allow any dirt or other contaminants to enter. Do not attempt to disassemble a hydraulic adjuster. Use only clean diesel oil to prime the adjusters.*

1 Remove the rocker arm assemblies and pull the adjusters from the rocker arms **(see illustration)**. Be sure to keep them in order so they can be returned to their original positions.
2 One at a time, submerge an adjuster in a container filled with diesel oil. Using a rigid wire (such as a straightened-out paper clip) inserted through the hole in the top of the adjuster, lightly hold down the steel ball and move the plunger up and down approximately four or five times. Remove the wire and push the plunger forcefully with your finger. If the plunger moves even a slight amount, the adjuster still has air inside. If the

2A – 10 Chapter 2 Part A Engine

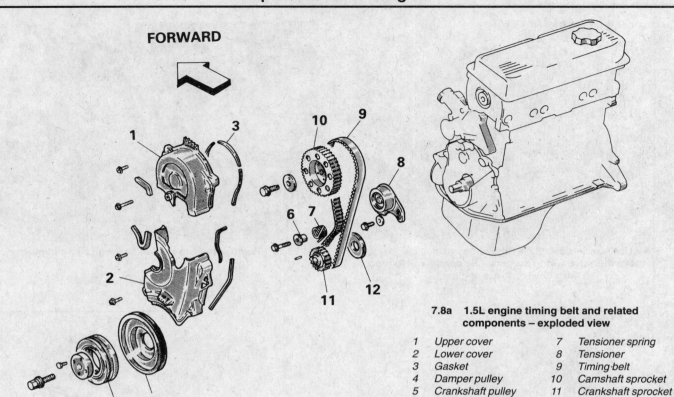

7.8a 1.5L engine timing belt and related components – exploded view

1 Upper cover
2 Lower cover
3 Gasket
4 Damper pulley
5 Crankshaft pulley
6 Tensioner spacer
7 Tensioner spring
8 Tensioner
9 Timing belt
10 Camshaft sprocket
11 Crankshaft sprocket
12 Belt guide

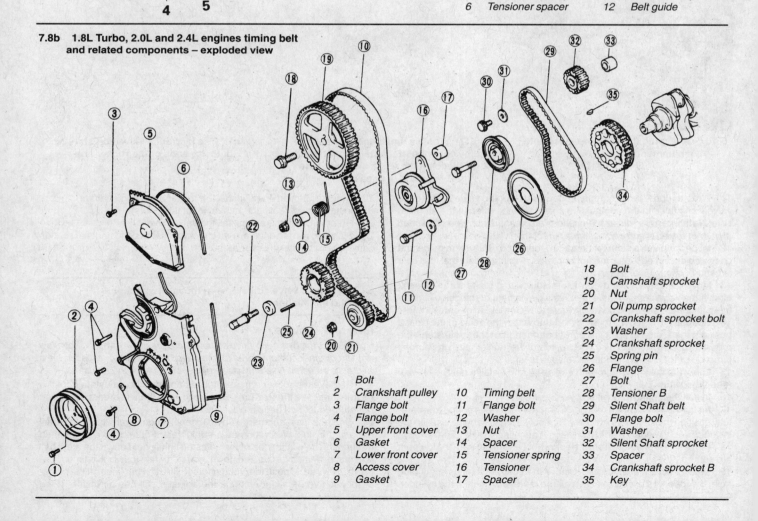

7.8b 1.8L Turbo, 2.0L and 2.4L engines timing belt and related components – exploded view

1 Bolt
2 Crankshaft pulley
3 Flange bolt
4 Flange bolt
5 Upper front cover
6 Gasket
7 Lower front cover
8 Access cover
9 Gasket
10 Timing belt
11 Flange bolt
12 Washer
13 Nut
14 Spacer
15 Tensioner spring
16 Tensioner
17 Spacer
18 Bolt
19 Camshaft sprocket
20 Nut
21 Oil pump sprocket
22 Crankshaft sprocket bolt
23 Washer
24 Crankshaft sprocket
25 Spring pin
26 Flange
27 Bolt
28 Tensioner B
29 Silent Shaft belt
30 Flange bolt
31 Washer
32 Silent Shaft sprocket
33 Spacer
34 Crankshaft sprocket B
35 Key

Chapter 2 Part A Engine

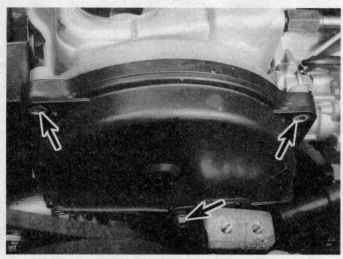

7.8c The upper timing belt cover is attached with three bolts (arrows) (2.0L engine shown)

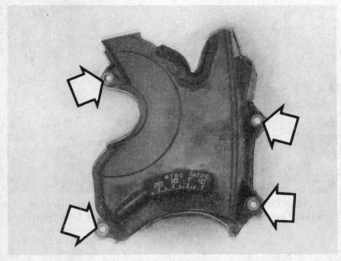

7.8d Arrows point to the locations of the lower timing belt cover bolts (cover removed for clarity) (1.5L engine shown)

7.8e Arrows point to the locations of the lower timing belt cover bolts (cover removed for clarity) (2.0L engine shown)

7.9 If you plan to reuse the belt, paint an arrow on it to indicate direction of rotation (clockwise)

plunger still moves, repeat the previous steps and try to bleed the air out again. If the adjuster does not respond, replace it with a new one. **Caution:** *Once the hydraulic adjuster has been properly bled, be sure to hold it in an upright position and don't allow the diesel oil to spill out.*

3 Assemble the adjusters into the rocker arms. Install the valve cover and related components.

4 Occasionally, the adjuster might not get completely bled and could be noisy when the engine is first started. If this happens, the air can be bled by slowly raising the speed of the engine from idle to 3,000 rpm for one minute. If the adjuster(s) do not become silent, replace the defective ones.

7 Timing belt and sprockets – removal, inspection and installation

Caution: *Do not try to turn the crankshaft with the camshaft sprocket bolt and do not rotate the crankshaft counterclockwise. Also, don't turn the crankshaft after the timing belt has been removed.*

1 Position the number one piston at Top Dead Center (see Chapter 2B).
2 Disconnect the negative cable from the battery.
3 Remove the air cleaner assembly and associated hoses (see Chapter 4).
4 Set the parking brake and block the rear wheels. Raise the front of the vehicle and support it securely on jackstands.
5 Remove the left engine mount (see Section 17). **Note:** *Make sure the engine is supported with a piece of wood and a floor jack placed under the oil pan. The wood will prevent the floor jack from denting or damaging the oil pan.*

SOHC engines

Removal

Refer to illustrations 7.8a, 7.8b, 7.8c, 7.8d, 7.8e, 7.9, 7.10a, 7.10b, 7.10c, 7.11, 7.12, 7.14, 7.15a, 7.15b and 7.16

6 Loosen the water pump pulley bolts and remove the drivebelts (see Chapter 1).
7 Unbolt and remove the water pump pulley.
8 Remove the bolts that secure the timing belt upper cover **(see illustrations)** and lift the cover off. Remove the bolts that secure the lower cover **(see illustrations)** and lift it off.
9 If you plan to reuse the timing belt, paint an arrow on it **(see illustration)** to indicate the direction of rotation (clockwise).

7.10a Loosen the belt tensioner bolt (arrow) and pry the tensioner towards the water pump (belt removed for clarity) (1.5L engine shown)

7.10b Some engines use an Allen-head bolt (arrow) to retain the tensioner (2.0L engine shown)

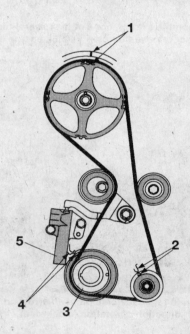

7.10c Locations of the timing marks on SOHC 16-valve engines. Notice how the auto tensioner is attached to the engine with two bolts; its plunger presses against the tensioner arm, pressing the tensioner pulley against the belt

1. Camshaft sprocket timing marks
2. Oil pump sprocket timing marks
3. Crankshaft sprocket
4. Crankshaft sprocket timing marks
5. Automatic tensioner

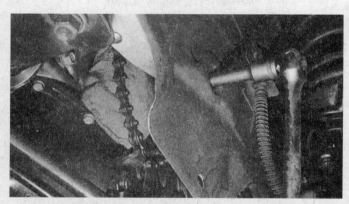

7.11 Wrap a protective cloth around the pulley and attach a chain wrench – loosen the crankshaft pulley center bolt with a socket, extension and breaker bar (1.5L engine shown)

7.12 Remove the flywheel/driveplate cover for access and wedge a flat-blade screwdriver in the ring gear teeth at the right corner of the engine block (axle removed for clarity)

10 On 8-valve engines, loosen the adjusting bolt and move the timing belt tensioner towards the water pump as far as possible **(see illustrations)**. Temporarily secure the tensioner by tightening the bolt. On 16-valve engines, unbolt and remove the tensioner, then loosen the tensioner arm attaching bolt and allow the tensioner arm to pivot down, relieving tension on the belt **(see illustration)**. See Steps 48 and 49 under DOHC engines to inspect the automatic tensioner used on 16-valve engines.

11 Remove the splash pan from beneath the drivebelt end of the engine, then remove the large center bolt from the crankshaft pulley. **Note:** *Some models secure the crankshaft pulley with four bolts. It's very tight, so, to break it loose, wrap a rag around the pulley and attach a chain wrench. Slip a socket onto the bolt and insert an extension through the hole in the inner fender. Turn the extension with a breaker bar* **(see illustration)**.

12 If you are unable to loosen the bolt due to the chain wrench slipping, you can prevent the crankshaft from turning by having an assistant wedge a flat-blade screwdriver in the flywheel/driveplate ring gear teeth **(see illustration)**. To do this, you must first remove the flywheel/driveplate cover (described in the transaxle removal procedures in Chapter 7).

13 Make sure the tensioner bolts are loose and slip the timing belt off the sprockets and set it aside.

Chapter 2 Part A Engine

2A – 13

7.14 Slip a large screwdriver through the sprocket to prevent the camshaft from turning – be sure to pad the gasket surface (arrow) to prevent damage to the head (1.5L engine shown)

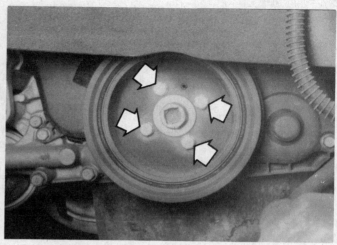

7.15a Remove the crankshaft pulley bolts (arrows) (2.0L engine shown)

7.15b When removing the belt guide, note how it's installed – the chamfered side (side with the tooth marks) faces out (1.5L engine shown)

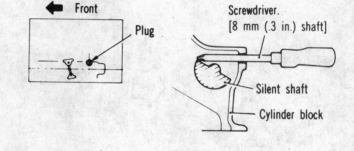

7.16 A screwdriver is used to keep the Silent Shaft from turning as the oil pump is loosened

7.19 Turn the tensioner pulley by hand to detect roughness and excess play (1.5L engine shown)

14   If you intend to remove the camshaft, unscrew the camshaft sprocket bolt and slide the sprocket off – a large screwdriver inserted through a hole in the sprocket will keep it from turning while you remove the bolt **(see illustration)**. Also, remove the oil pump sprocket at this time and inspect the oil pump seal for leaks or apparent damage.

15   If you intend to replace the crankshaft front oil seal, unscrew the four bolts for the crankshaft pulley **(see illustration)** and slide it off; then slide off the crankshaft sprocket and the belt guide flange **(see illustration)** located behind the crankshaft sprocket. When removing the flange, note the way it's installed (the chamfered side faces out).

16   On 1.8L turbo, 2.0L and 2.4L engines, remove the plug at the left side of the cylinder block and insert a screwdriver or a long punch to hold the Silent Shaft stationary **(see illustration)**. The screwdriver or punch should have an approximate diameter of 5/16-inch and a length of 2 3/8-inches.

17   Loosen the bolt that retains the Silent Shaft sprocket.

18   Loosen the Silent Shaft tensioner **(see illustration 7.8c)** and remove the Silent Shaft belt (timing belt B). **Caution:** *Do not attempt to loosen the sprocket bolt by holding the sprocket stationary with pliers.*

Inspection

Refer to illustrations 7.19 and 7.20

19   Rotate the tensioner pulley by hand and move it side-to-side to detect roughness and excess play **(see illustration)**. Visually inspect the sprockets for any signs of damage and wear. Replace parts as necessary.

Chapter 2 Part A Engine

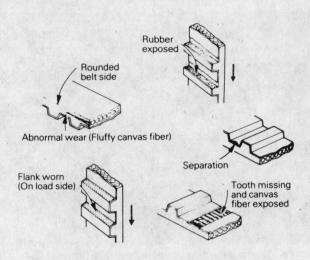

7.20 Carefully inspect the timing belt for the conditions shown here

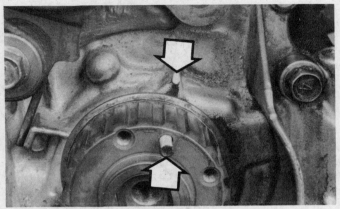

7.22a Line up the dowel on the crankshaft sprocket with the stationary mark on the oil pump (arrows) (1.5L engine shown)

7.22b Align the triangular mark on the camshaft sprocket with the stationary mark on the cylinder head (arrows) – (1.5L engine shown)

20 Inspect the timing belt (see illustration) for cracks, separation, wear, missing teeth and oil contamination. Replace the belt if it's in questionable condition.

Installation

Refer to illustrations 7.22a, 7.22b, 7.22c, 7.22d, 7.22e, 7.23, 7.24, 7.25a, 7.25b, 7.26, 7.29, 7.30a, 7.30b and 7.32

21 Reinstall the timing belt sprockets, if they were removed. Note that the camshaft sprocket is indexed by a dowel (1.5L engine only) or by punch marks. Slip the belt guide flange onto the crankshaft before installing the lower sprocket – the chamfered side of the flange faces out. The crankshaft sprocket has two different size flats which match those on the crankshaft (see illustrations 7.8b, 7.8c and 7.8d).

22 Align the timing marks located on the camshaft, crankshaft and oil pump sprockets with the marks on the cylinder head and the front case (see the accompanying illustrations and illustration 7.10c). Note: *Different engines have different types of alignment marks. Some alignment marks are notches or raised arrows and some are punch indentations. Identify the exact type and location of the alignment marks on each sprocket before continuing the procedure. When aligning the oil pump sprocket marks, it is critical that the Silent Shaft (if equipped) has the weighted portion at the bottom of the shaft (it is possible to align the marks with the Silent Shaft weight at the top; if you do this accidentally, severe engine vibration will result). Before installing the timing belt, slightly rock the oil pump sprocket by hand and watch carefully that the sprocket has the tendency to remain stationary (return to approximately the marks-aligned position) when the sprocket is rotated. This means the sprocket is*

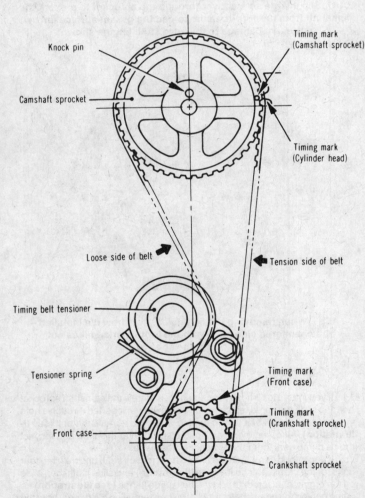

7.22c Camshaft and crankshaft sprocket alignment marks on the 1.5L engine

CORRECTLY timed. If the sprocket has the tendency to rotate clockwise when spun lightly, the shaft is INCORRECTLY timed. If there is any doubt about whether or not the silent shaft is in the correct position, insert a screwdriver through the hole in the side of the cylinder block (see illustration 7.16). Make sure the screwdriver extends approximately two inches into the hole and also make sure the sprocket cannot be rotated with the screwdriver in place; now you can be sure the timing is correct.

Chapter 2 Part A Engine

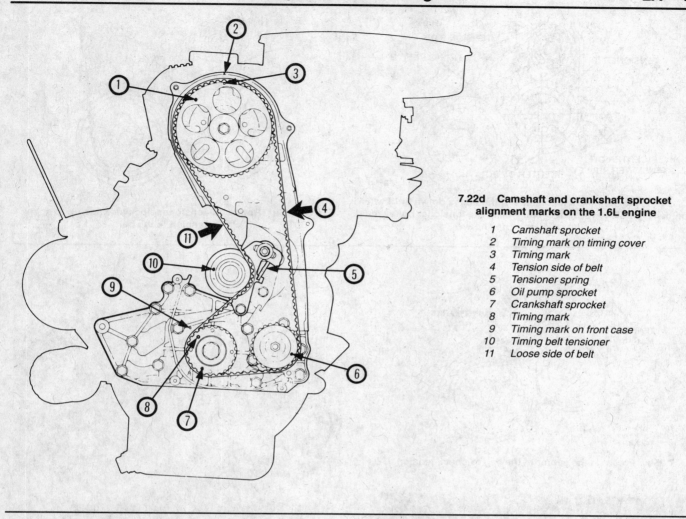

7.22d Camshaft and crankshaft sprocket alignment marks on the 1.6L engine

1. Camshaft sprocket
2. Timing mark on timing cover
3. Timing mark
4. Tension side of belt
5. Tensioner spring
6. Oil pump sprocket
7. Crankshaft sprocket
8. Timing mark
9. Timing mark on front case
10. Timing belt tensioner
11. Loose side of belt

23 On 1.8L Turbo, 2.0L and 2.4L engines, install timing belt B. Align the timing marks on each sprocket with the timing marks on the front case **(see illustration)**.

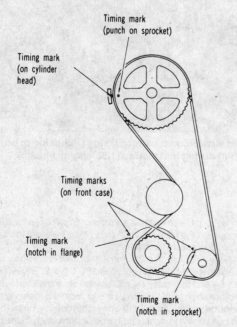

7.22e Camshaft and crankshaft sprocket alignment marks on the 1.8L Turbo, 2.0L and 2.4L engines

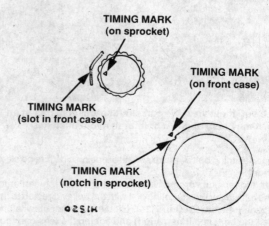

7.23 Align the marks as shown before installing timing belt B – 1.8L Turbo, 2.0L and 2.4L engines

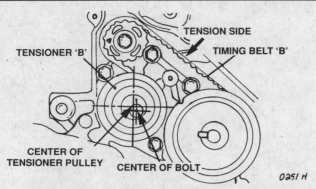

7.24 Make sure the tension side of the belt is drawn tight when initially installing the timing belt — also, the tensioner pulley must be positioned as shown

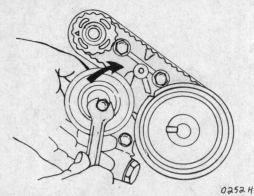

7.25a Rotate the tensioner clockwise and up to apply tension to the belt

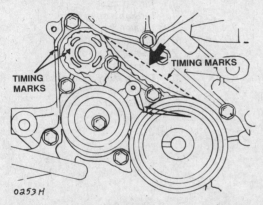

7.25b Press on the center of the belt to check tension

7.26 Maintain tension on the firewall side (right side of photo) of the belt during installation (1.5L engine shown)

7.29 Since you must view the camshaft sprocket marks at an angle, it may help to use a pointer to avoid mistakes

24 After installing timing belt B, make sure the tension side has no slack **(see illustration)**.

25 Make sure the tension sprocket for timing belt B has the center located just to the left side of the mounting bolt with the pulley directed to the front of the engine **(see illustration 7.24)**. Lift the tensioner up with one finger to tighten the belt **(see illustration)** and tighten the tensioner bolt and the Silent Shaft bolt to the torque listed in this Chapter's Specifications. **Note:** *Use the index finger and press firmly on the timing belt* **(see illustration)**. The belt deflection should be 13/64 to 9/32-inch (5.0 to 7.0 mm).

26 On all engines, slip the timing belt onto the crankshaft sprocket. While maintaining tension on the rear (firewall) side of the belt, slip the belt onto the camshaft sprocket **(see illustration)**.

27 On 8-valve engines, release the tensioner hold-down bolt to apply spring tension against the belt. Retighten the bolt. On 16-valve engines, follow Steps 53, 54 and 55 under *DOHC engines* to install the tensioner.

28 Install the crankshaft pulley, taking care to align the locating pin with the small hole in the pulley. Install the crankshaft pulley bolts (both the pulley securing bolts and the center bolt) and tighten them to the torques listed in this Chapter's Specifications. When tightening the bolts, hold the crankshaft in place using one of the methods discussed in Steps 11 and 12.

Chapter 2 Part A Engine

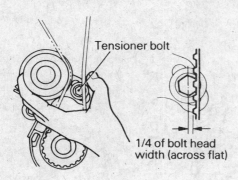

7.30a The belt is tensioned properly if it can be pressed in to 1/4 of the adjustment bolt head width (1.5L engine only)

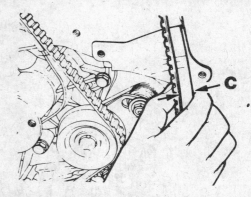

7.30b Squeeze the timing belt and case and observe the distance between the belt and Line C to measure the belt deflection

29 Using the bolt in the center of the crankshaft pulley, turn the crankshaft clockwise through two complete revolutions (720-degrees). Recheck the alignment of the timing marks (see illustration). If the marks do not align properly, loosen the tensioner, slip the belt off the camshaft sprocket, align the marks, reinstall the belt, and check the alignment again.

30 On 1.5L engines, check for proper timing belt tension by pushing on the belt with thumb pressure. It should deflect to 1/4 of adjuster bolt head width (see illustration). If it is too tight or too loose, loosen the tensioner bolts and adjust the tensioner. On 1.8L Turbo, 2.0L 8-valve, 2.4L and 1.6L engines, squeeze the timing belt and the cover (see illustration), and measure the distance between the timing belt and Line C. The distance should be approximately 9/16-inch on the 1.8L Turbo, 2.0L 8-valve and 2.4L engines. On 1.6L engines, the distance should be 15/64-inch. On 16-valve engines, follow Step 60 under DOHC engines to properly adjust the timing belt tension.

31 Tighten the tensioner bolts to the torque listed in this Chapter's Specifications, starting with the adjustment bolt; then tighten the bolt which goes through the tension spring.

32 Reinstall the remaining parts in the reverse order of removal. Note that the timing belt cover bolts come in different lengths (see illustration).

33 Start the engine, set the ignition timing (see Chapter 1) and road test the vehicle.

DOHC engines

Note: *The Mitsubishi DOHC engines, 4G32 and 4G61, are similar except that the engine installed in the Mirage (4G61 – 1989 and 1990) does not have a Silent Shaft.*

Removal

Refer to illustrations 7.41 and 7.45

34 Follow steps 1 through 5 of this Section. Be sure to support the engine with a block of wood placed between the floor jack and the engine oil pan.

35 Remove the clamp for the power steering pressure hose (if equipped) and the clamp for the air conditioning hose, if equipped.

36 Remove the drivebelts (see Chapter 1) and the tensioner pulley bracket.

37 Remove the water pump pulley (see Chapter 3).

38 Remove the crankshaft pulley (see Chapter 2B). Remove the bolts from the crankshaft pulley. They might be very tight, so, to break them loose, wrap a rag around the pulley and attach a chain wrench. Slip a socket onto the bolt(s) and install an extension through the hole in the inner fender. Turn the extension with a breaker bar (see illustration 7.11). If you are unable to loosen the bolt due to the chain wrench slipping, you can prevent the crankshaft from turning by having an assistant wedge a flat-blade screwdriver in the flywheel/driveplate ring gear teeth (see illustration 7.12). To do this, you must first remove the flywheel/driveplate cover (described in the transaxle removal procedures in Chapter 7).

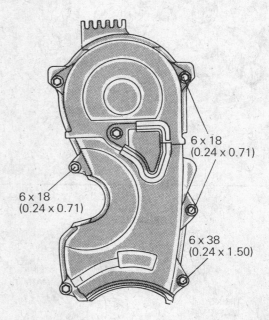

7.32 Be sure to return the timing belt cover bolts to their proper locations (1.5L engine shown)

39 Remove the retaining bolts from the upper and lower timing belt covers and remove the covers.

40 Remove the valve cover (see Section 3).

41 Remove the automatic tensioner (see illustration).

42 Make a mark on the timing belt in the direction of rotation so it may be reinstalled in the same direction in the event the timing belt is reused (see illustration 7.9). Remove the timing belt, the tensioner arm and the idle pulley (see illustration 7.41). Note: *Be sure that the timing marks are correctly aligned before removing the timing belt* (see illustration 7.51).

43 Remove the camshaft sprockets. Using an adjustable wrench or an open end wrench, hold the camshaft at the hexagon (located between the number 2 and 3 journals) and remove the camshaft bolt(s).

44 Remove the oil pump sprocket. Remove the plug on the side of the engine block and insert a Phillips screwdriver (see illustration 7.16) to prevent the shaft from turning. Remove the oil pump sprocket nut with a socket and a breaker bar.

2A–18   Chapter 2 Part A   Engine

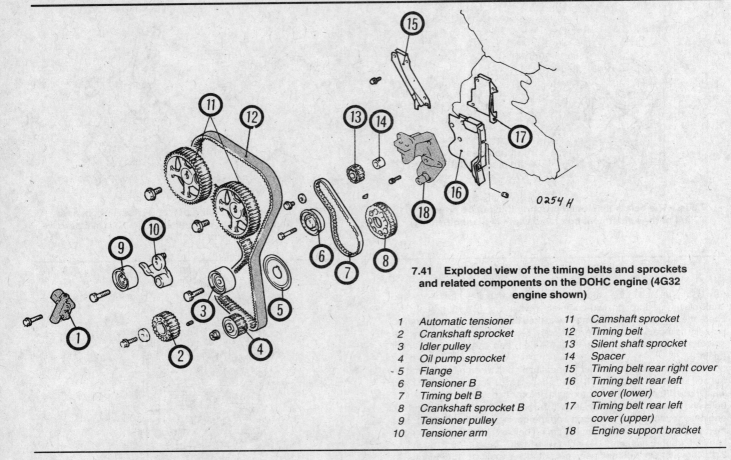

7.41 Exploded view of the timing belts and sprockets and related components on the DOHC engine (4G32 engine shown)

1. Automatic tensioner
2. Crankshaft sprocket
3. Idler pulley
4. Oil pump sprocket
5. Flange
6. Tensioner B
7. Timing belt B
8. Crankshaft sprocket B
9. Tensioner pulley
10. Tensioner arm
11. Camshaft sprocket
12. Timing belt
13. Silent shaft sprocket
14. Spacer
15. Timing belt rear right cover
16. Timing belt rear left cover (lower)
17. Timing belt rear left cover (upper)
18. Engine support bracket

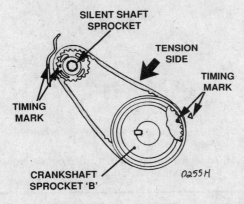

7.45 Timing marks on the DOHC timing belt B

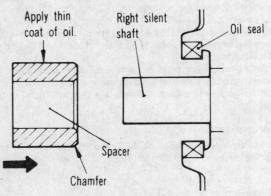

7.50 The chamfered edge of the spacer must face in

45   On the 4G32 DOHC engine, remove timing belt B. Be sure the alignment marks are positioned correctly before removing the belt **(see illustration)**. **Note:** *The timing marks for timing belt B are located in different places on the engine and sprockets for the DOHC and the SOHC engines, but the procedure is the same.*

Inspection

46   Rotate the tensioner pulley by hand and move it side-to-side to detect roughness and excess play. Visually inspect the sprockets for any signs of damage and wear. Replace parts as necessary. Also, replace the pulley if there is a lubricant leak.

47   Inspect the timing belt **(see illustration 7.20)** for cracks, separation, wear, missing teeth and oil contamination. Replace the belt(s) if they are in questionable condition.

48   Check the automatic tensioner for leaks or any obvious damage to the body. Also, check the rod end for wear or damage. Measure the rod protrusion for the correct length – it should extend 15/32-inch (12 mm) beyond the body of the tensioner.

49   Place the tensioner in a vise that is equipped with soft jaws (or put a shop rag over the jaws to prevent damage to the tensioner). If the rod is easily retracted, replace it with a new unit, The tensioner should have a fair amount of strength or resistance. **Caution:** *Be sure the tensioner is in a level position when it is in the vise. Also, place a washer over the plug on the bottom of the tensioner to prevent the vise from contacting the plug.*

Installation

Refer to illustrations 7.50, 7.51 and 7.60

50   Reinstall the timing belt sprockets, if they were removed. Tighten the bolts to the values listed in this Chapter's Specifications. **Note:** *Be sure to properly install the spacer on the crankshaft sprocket. The chamfered edge must face toward the oil seal* **(see illustration)**.

Chapter 2 Part A Engine

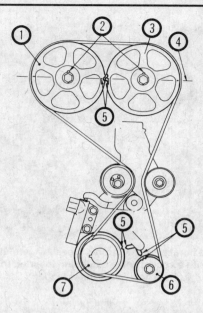

7.51 Identify each timing mark or notch and make sure the sprockets are positioned correctly

1. Exhaust camshaft sprocket
2. Dowel pin holes (must be in the 12 o'clock position)
3. Intake camshaft sprocket
4. Machined surface of cylinder head
5. Timing marks
6. Oil pump sprocket
7. Crankshaft sprocket

51 Align the timing marks located on the camshaft, crankshaft and oil pump sprockets **(see illustration)**. **Note:** *The engines are equipped with different types of alignment marks. Some alignment marks are notches and some are punch indentations or raised arrows. Identify the exact type and location of the alignment marks on each sprocket before continuing the procedure.* When aligning the oil pump sprocket marks, it is critical that the Silent Shaft has the weighted portion at the top of the shaft. Before installing the timing belt, slightly turn the oil pump sprocket by hand and watch carefully that the sprocket has the tendency to rotate clockwise (the direction of the engine) as if the shaft weight falls to one side. The timing mark should move when the sprocket is INCORRECTLY timed. When the sprocket is CORRECTLY timed, the shaft will have the tendency to remain stationary when it is turned slightly, as the weight is at the bottom of the shaft.
52 On 4G32 DOHC engines, install timing belt B. Follow steps 23 through 25 of this Section. **Note:** *The timing marks on DOHC engines equipped with timing belt B are located in different places but the procedure is the same* **(see illustration 7.45)**. *Also, when adjusting the belt tension on timing belt B, adjust it to 13/64 to 9/32-inch (5.0 to 7.0 mm).*
53 Prepare the automatic tensioner for installation. First, if the rod is in its fully extended position, reset the tensioner. Clamp the tensioner in a soft-jawed vise and slowly force the rod into the tensioner. Make sure the plug at the bottom of the tensioner is surrounded with a thick washer to prevent the jaws from contacting it. Push the rod in with the vise until the set hole in the rod is aligned with the hole in the tensioner body. Insert a small piece of wire (such as a straightened-out paper clip) through the holes to hold the tensioner in the retracted position.
54 Install the automatic tensioner onto the engine, leaving the wire attached to it.
55 Install the tensioner pulley onto the tensioner arm. Position the two small holes in the tensioner pulley hub just to the left of the center bolt. Tighten the center bolt finger tight. Don't remove the wire from the tensioner yet.
56 Turn the two camshaft sprockets until the dowel pins are located at the top **(see illustration 7.51)**. Then, align the timing marks facing each other with the upper surface of the cylinder head. **Note:** *When the exhaust*

7.60 Use a special tool (Mitsubishi part no. MD998752) to apply the correct amount of torque to the tensioner

camshaft is released, it will tend to rotate one tooth in the counterclockwise direction. This shift must be taken into account when installing the timing belt onto the sprockets. Also, the camshaft sprockets are identical and are provided with two timing marks. When the sprocket is mounted on the exhaust camshaft, use the timing mark on the right with the dowel pin hole on the top. On the intake camshaft sprocket, use the one on the left with the dowel pin hole on the top.
57 Align the crankshaft sprocket timing mark and the oil pump sprocket timing mark with their pointers **(see illustration 7.51)**.
58 Remove the plug on the side of the block and insert a Phillips screwdriver or a long punch through the hole **(see illustration 7.16)**. If the tool CAN be inserted into the hole as deep as 2 1/2-inches or more, the timing marks are aligned correctly. If the tool CANNOT be inserted more than 1-inch, the oil pump sprocket must be reset. When it's positioned properly, reinstall the screwdriver and keep it there until the timing belt is installed.
59 Install the timing belt in the following sequence:
  a) Install the timing belt around the tensioner pulley and crankshaft sprocket and hold the timing belt to the tensioner pulley with your left hand.
  b) Pulling the belt with your right hand, install it around the oil pump sprocket.
  c) Install the belt around the idler pulley.
  d) Install the belt around the intake camshaft sprocket.
  e) Turn the exhaust camshaft sprocket one tooth to align it's timing mark with the upper surface of the cylinder head **(see illustration 7.51)**. Pull the belt with both hands and install it around the exhaust camshaft.
  f) Gently raise the tensioner pulley so the belt does not sag, then temporarily tighten the center bolt.
60 Adjust the timing belt tension in the following sequence:
  a) Turn the crankshaft 1/4 turn counterclockwise, then clockwise to move the number 1 cylinder to TDC.
  b) Loosen the tensioner center bolt and attach Mitsubishi special tool no. MD998752 (or equivalent) to a torque wrench **(see illustration)**. **Note:** *The torque wrench must be capable of measuring small increments between 0 and 30 in-lbs.* Apply between 23 and 25 in-lbs. to the tensioner.
  c) While holding tension on the timing belt tensioner, tighten the center bolt to the torque listed in this Chapter's Specifications.
  d) Screw Mitsubishi special tool no. MD998738 (or equivalent) into the engine left support bracket until its end makes contact with the tensioner arm. Continue to screw the tool into the pulley and when tension is relieved from the automatic tensioner, remove the wire or clip that was inserted into the tensioner. An alternate method would be to pry the tensioner pulley towards the front of the vehicle (don't pry against the belt) until the automatic tensioner is compressed, then remove the wire.
  e) Remove the special tool.
  f) Rotate the crankshaft two complete turns and wait 15 minutes. Measure how far the tensioner plunger protrudes from the tensioner body (the distance between the tensioner arm and the automatic tensioner body). It should be between 5/32 and 3/16-inch (3.8 to 4.5 mm).

Chapter 2 Part A Engine

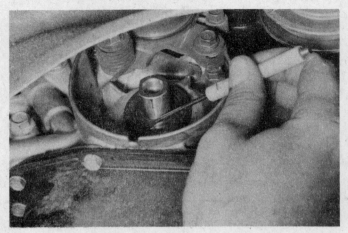

8.2 Working from below the left inner fender, carefully pry the seal out with a small screwdriver

8.4 Install the new crankshaft front oil seal by carefully tapping it into place with a socket and a extension inserted through the access hole in the inner fender

61 If the tensioner protrusion is not as specified, repeat the belt adjustment procedure.
62 Install the timing covers.
63 The remaining steps are the reverse of removal.

8 Crankshaft front oil seal – replacement

Refer to illustrations 8.2 and 8.4

1 Remove the timing belt and the crankshaft sprocket (see Section 7).
2 Wrap the tip of a small screwdriver with tape. Working from below the left inner fender, use the screwdriver to pry the seal out of its bore **(see illustration)**. Take care to prevent damaging the crankshaft and the seal bore.
3 Thoroughly clean and inspect the seal bore and sealing surface on the crankshaft. Minor imperfections can be removed with emery cloth. If there is a groove worn in the crankshaft sealing surface (from contact with the seal), installing a new seal will probably not stop the leak. Such wear normally indicates the internal engine components are also worn. Consider overhauling the engine.
4 Lubricate the new seal with engine oil and drive the seal into place with a hammer and socket **(see illustration)**.
5 Reinstall the timing belt and related components as described in Section 7.
6 Run the engine, checking for oil leaks.

9 Camshaft(s) – removal, inspection and installation

Removal

Refer to illustrations 9.8, 9.10, 9.11, 9.12, 9.13 and 9.14

1 Position the number one piston at Top Dead Center (see Chapter 2B).
2 Disconnect the negative cable from the battery.
3 Remove the valve cover (see Section 3).
4 Remove the distributor (SOHC engine) or crank angle sensor (DOHC engine) (see Chapter 5).

SOHC engines

5 Drain the cooling system (see Chapter 1) and disconnect the upper radiator hose from the thermostat housing.
6 Remove the fuel pump (carbureted models only) (see Chapter 4).
7 Remove the timing belt and camshaft sprocket(s) (see Section 7).
8 Wrap some tape around the end of a small screwdriver and use it to pry out the camshaft oil seal **(see illustration)**. Be careful not to damage the seal bore or the sealing surface on the camshaft.
9 Detach the rocker arm assembly (see Section 4).
10 On the 1.5L engine, remove the cylinder head end cover and the camshaft thrust case bolts **(see illustration)**.

9.8 Carefully pry out the seal without scratching the metal surfaces (1.5L engine shown)

9.10 On the 1.5L engine, remove the lower bolts (A) first and set the bracket aside, then the upper bolts (B) and finally the thrust case bolt (C)

Chapter 2 Part A Engine

2A – 21

9.11 On the 1.5L engine, grip the camshaft between the third and fourth exhaust lobes and remove the thrust case-to-camshaft bolt

9.12 On the 1.5L engine, pull the thrust case out of the cylinder head – if it's stuck, slide the camshaft back and forth slightly to push it out

9.13 Guide the camshaft out of the cylinder head – be careful not to damage the bearings in the head (1.5L engine shown)

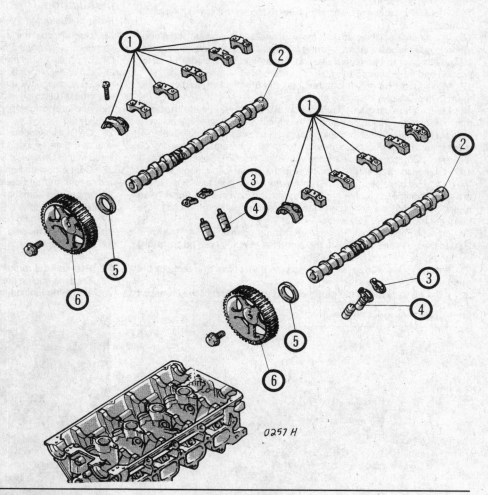

9.14 Exploded view of the camshafts and related components on the DOHC engine

1. Bearing caps
2. Camshaft
3. Rocker arms
4. Lifters
5. Camshaft oil seal
6. Camshaft sprocket

11 Hold the camshaft to keep it from turning and remove the camshaft thrust case bolt **(see illustration)**.
12 Pull the camshaft thrust case out of the cylinder head **(see illustration)**.
13 To gain clearance, unbolt the windshield washer reservoir and set it aside. Slip the camshaft out of the cylinder head toward the transaxle end of the engine **(see illustration)**.

DOHC engines

14 Remove the camshaft bearing caps, loosening the bolts a little at a time to prevent distorting the camshaft(s). Store them in order so they can be returned to their original locations. Carefully lift the camshaft(s) out of the cylinder head **(see illustration)**.

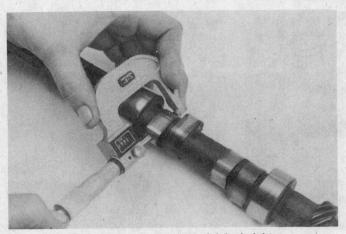

9.17 Measure the camshaft lobe heights

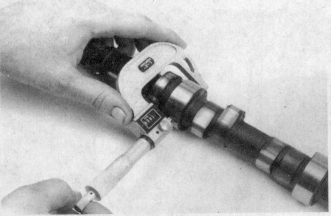

9.18 Measure the camshaft journal diameters

Inspection

Refer to illustrations 9.17, 9.18 and 9.20

15 Thoroughly clean the camshaft(s), the rear cover plate and the gasket surfaces.
16 Visually inspect the camshaft for wear and/or damage to the distributor drive gear, lobe surfaces, bearing journals and seal contact surfaces. Visually inspect the camshaft bearing surfaces in the cylinder head for scoring and other damage.
17 Measure the camshaft lobe heights **(see illustration)** and compare them to this Chapter's Specifications.
18 Measure the camshaft bearing journal diameters **(see illustration)**, then measure the inside diameter of the camshaft bearing surfaces in the cylinder head, using a telescoping gauge. Subtract the journal measurement from the bearing measurement to obtain the camshaft bearing oil clearance. Compare this clearance with this Chapter's Specifications.
19 Replace the camshaft if it fails any of the above inspections. **Note:** *If the distributor drive gear is faulty, replace the driven gear also.* If the lobes are worn, replace the rocker arms along with the camshaft. Cylinder head replacement may be necessary if the camshaft bearing surfaces in the head are damaged or excessively worn.
20 On 1.5L engines, install the camshaft thrust case on the camshaft and tighten the bolt securely. Check the camshaft endplay **(see illustration)** and compare it to this Chapter's Specifications.
21 If the endplay is excessive, replace the thrust case and recheck the endplay.
22 If the endplay is still too great, check the outer end of the camshaft transaxle-end journal for wear. If it is badly worn, replace the camshaft.

Installation

Refer to illustration 9.27

23 Liberally coat the journals and thrust portions of the camshaft with assembly lube or engine oil.
24 Carefully install the camshaft(s) in the cylinder head.
25 On SOHC models, insert the camshaft thrust case with the threaded hole up and align the threaded hole with the bolt hole in the cylinder head. Lock the thrust case in position with the bolt. Install the rear cover with a new gasket and tighten the bolts to the torque listed in this Chapter's Specifications.
26 On DOHC models, install the camshaft bearing caps. Tighten them a little at a time, working from the center journals on out, until the torque listed in this Chapter's Specifications is reached.
27 Coat a new camshaft oil seal with engine oil and press it into place with a hammer and deep socket **(see illustration)**.
28 Install the camshaft sprocket(s) and tighten the bolts to the torque listed in this Chapter's Specifications.
29 Install the timing belt (see Section 7).
30 On SOHC models, install the rocker arm assembly (see Section 4).
31 Temporarily set the valve clearances prior to engine start-up (see Chapter 1).
32 Reinstall the remaining parts in the reverse order of removal.
33 Start the engine and allow it to warm up (176 to 203-degrees F) while you adjust the ignition timing (see Chapter 1).

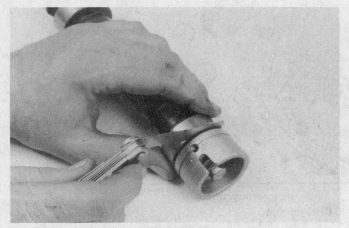

9.20 Measure camshaft endplay with a feeler gauge (1.5L engine shown)

9.27 Tap the seal gently into place with a hammer and a large socket (1.5L engine shown)

Chapter 2 Part A Engine

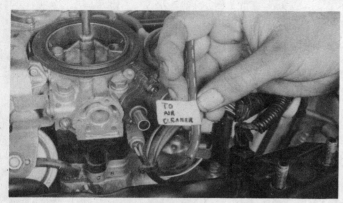

10.4a Label the wires and hoses with masking tape and a marking pen

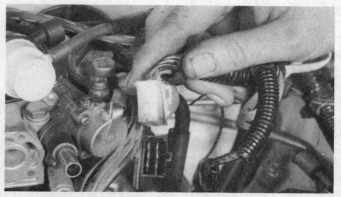

10.4b Unplug the solenoid control valve connector (1.5L engine shown)

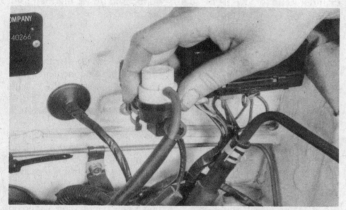

10.4c Unsnap the emission control valves from the brackets (1.5L engine shown)

10.12 Remove the bolts (arrows) that hold down the fuel rail

34 Readjust the valve clearances as described in Chapter 1.
35 Reinstall the valve cover and run the engine while checking for oil leaks.

10 Intake manifold – removal and installation

1 Relieve the fuel system pressure (see Chapter 4), then disconnect the negative cable from the battery.
2 Drain the cooling system (see Chapter 1).

Removal
Carbureted engines
Refer to illustrations 10.4a, 10.4b and 10.4c
3 Remove the air cleaner (see Chapter 4).
4 Clearly label **(see illustration)** and disconnect all hoses, wires, brackets and emission lines which run to the carburetor and intake manifold. Several components may be slipped out of brackets and laid over the carburetor **(see illustrations)**.
5 Disconnect the fuel lines from the carburetor and cap the fittings to prevent leakage (see Chapter 4).
6 Disconnect the throttle cable from the carburetor (see Chapter 4).
7 Detach the cable which runs from the carburetor to the transmission (automatic transmission models only) and the cruise control cable, on vehicles so equipped.
8 Unbolt the intake manifold and remove it from the engine. If it sticks, tap the manifold with a soft-face hammer. **Caution:** *Do not pry between gasket sealing surfaces or tap on the carburetor.*

Fuel-injected engines
Refer to illustrations 10.12, 10.13, 10.15, 10.16a and 10.16b
9 Remove the air intake hose (see Chapter 4).

10.13 Remove the bolts that retain the plenum to the intake manifold (arrows)

10 Clearly label and disconnect all hoses, wires, brackets and emission lines which run to the fuel injection system and intake manifold.
11 Remove the distributor (see Chapter 5) and the accelerator cable(s) (see Chapter 4).
12 Remove the bolts that hold the fuel rail to the intake manifold **(see illustration)**, then remove the fuel rail (see Chapter 4).
13 On SOHC engines, remove the bolts that retain the air intake plenum to the intake manifold **(see illustration)**. Remove the plenum.
14 Raise the vehicle and support it securely on jackstands.

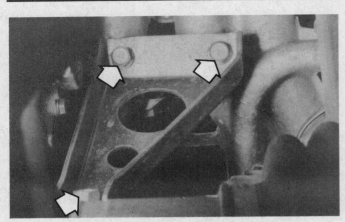

10.15 Remove the bolts (arrows) that retain the intake manifold brace

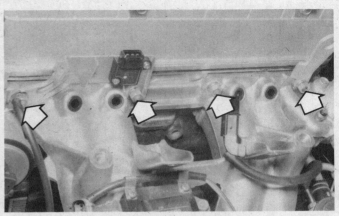

10.16a Remove the bolts that hold the intake manifold to the cylinder head (arrows) – there are two located under the manifold, not visible in this photo

15 Remove the bolts that retain the lower intake manifold bracket to the engine **(see illustration)**.
16 Unbolt the intake manifold and remove it from the engine **(see illustration)**. If it sticks, tap the manifold with a soft-face hammer or carefully pry it from the head **(see illustration)**. **Caution:** *Do not pry between gasket sealing surfaces or tap on the fuel injectors.*

Installation
Refer to illustration 10.17

17 Thoroughly clean the manifold and cylinder head mating surfaces, removing all traces of gasket material **(see illustration)**.
18 Install the manifold, using a new gasket. Tighten the nuts in several stages, working from the center out, until the torque listed in this Chapter's Specifications is reached.
19 Reinstall the remaining parts in the reverse order of removal.
20 Add coolant, run the engine and check for leaks and proper operation.

11 Exhaust manifold – removal and installation

Refer to illustrations 11.3, 11.6, 11.7, 11.8 and 11.12
Warning: *Allow the engine to cool completely before beginning this procedure.*

Removal
1 Disconnect the negative cable from the battery. On Galant SOHC 16-valve and DOHC non-turbo models, remove the air conditioning condenser fan motor; on Galant DOHC turbo models, remove the radiator (see Chapter 3) and turbo air outlet.
2 Set the parking brake and block the rear wheels. Raise the vehicle and support it securely on jackstands.
3 Working from under the vehicle, remove the nuts that secure the exhaust system to the bottom of the exhaust manifold **(see illustration)**. Apply penetrating oil to the threads to make removal easier.
4 On carbureted engines, remove the air cleaner assembly (see Chapter 4).
5 Unplug the oxygen sensor wire (see chapter 6).
6 If equipped with an AIR injection tube, apply penetrating oil and unscrew the flare nuts on the AIR injection tube **(see illustration)**. Remove the tube from the manifold.
7 Remove the three bolts that secure the heat shield **(see illustration)** to the exhaust manifold. Lift the heat shield off.
8 Apply penetrating oil to the threads and remove the exhaust manifold mounting nuts **(see illustration)**, brackets and emission components.
9 Slip the manifold off the studs and remove it from the engine compartment.

Installation
10 Clean and inspect all threaded fasteners and repair as necessary.

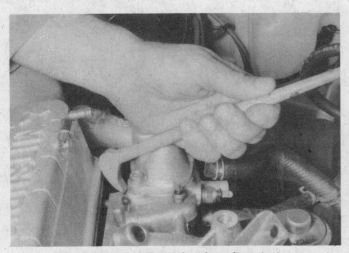

10.16b Make sure the prybar doesn't contact any gasket surfaces

10.17 Carefully scrape all traces of gasket material off the intake manifold without damaging the surface

Chapter 2 Part A Engine

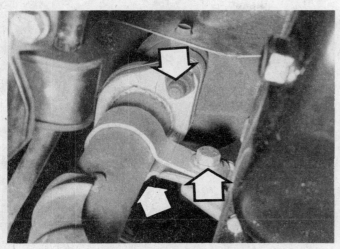

11.3 Working under the vehicle, apply penetrating oil to the threads and remove the nuts (arrows) – note that one is hidden from view

11.6 Apply penetrating oil and unscrew the flare nuts (arrows) (1.5L engine shown)

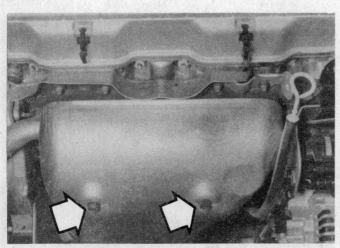

11.7 Remove the bolts from the heat shield (arrows) (2.0L engine shown)

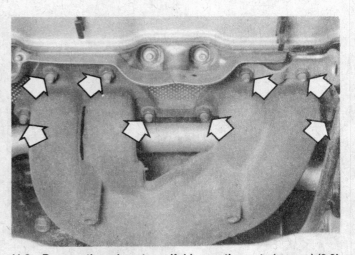

11.8 Remove the exhaust manifold mounting nuts (arrows) (2.0L engine shown)

11 Remove any traces of gasket material from the mating surfaces and inspect them for wear and cracks.
12 Place a new gasket over the studs (see illustration), install the manifold and tighten the nuts in several stages, working from the center out, to the torque listed in this Chapter's Specifications.
13 Reinstall the remaining parts in the reverse order of removal.
14 Run the engine and check for exhaust leaks.

12 Cylinder head – removal and installation

Caution: *Allow the engine to cool completely before following this procedure.*

Removal
Refer to illustrations 12.10, 12.11 and 12.12

1 Position the number one piston at Top Dead Center (see Chapter 2B).
2 Disconnect the negative cable from the battery.
3 Drain the cooling system and remove the spark plugs (see Chapter 1).
4 Remove the intake manifold (see Section 10).
5 Remove the exhaust manifold (see Section 11).

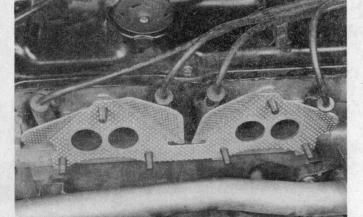

11.12 The exhaust gasket is installed with the shields facing out (1.5L engine shown)

2A–26 Chapter 2 Part A Engine

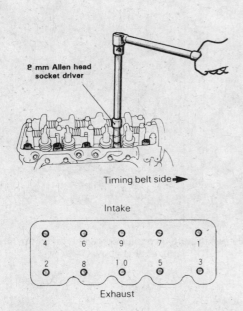

12.10 Cylinder head bolt loosening sequence – turn this illustration upside-down if the camshaft sprocket is on the opposite end of the cylinder head

12.11 Pry on an area of the cylinder head that won't be damaged when pressure is applied

12.12 Use a gasket scraper to remove the cylinder head gasket

12.13a Install the new head gasket dry (no sealer) as shown – it can only go on one way due to the dowels (arrows) (1.5L engine shown)

6 Remove the distributor (see Chapter 5), including the cap and wires.
7 On carbureted engines, remove the fuel pump (see Chapter 4).
8 Remove the timing belt (see Section 7). **Note:** *On some models it is possible to remove the cylinder head without removing the timing belt. Hang the sprocket with a piece of wire attached to the hood of the vehicle – make sure the belt does not lose tension after it has been disconnected from the vehicle.*
9 Remove the valve cover (see Section 3).
10 Using an 8 mm Allen head socket, loosen the cylinder head bolts, 1/4-turn at a time, in the sequence shown **(see illustration)** until they can be removed by hand. **Note:** *DOHC engines require a special tool (Mitsubishi part no. MD998051-01) to remove the cylinder head bolts.*
11 Carefully lift the cylinder head straight up and place the head on wood blocks to prevent damage to the sealing surfaces. If the head sticks to the engine block, dislodge it by placing a block of wood against the head casting and tapping the wood with a hammer or by prying the head with a pry bar placed carefully on a casting protrusion **(see illustration)**. Cylinder head disassembly and inspection procedures are covered in Chapter 2, Part B. It's a good idea to have the head checked for warpage, even if you're just replacing the gasket.
12 Remove all traces of old gasket material from the block and head **(see illustration)**. Do not allow anything to fall into the engine. Clean and inspect all threaded fasteners and be sure the threaded holes in the block are clean and dry.

Installation

Refer to illustrations 12.13a, 12.13b and 12.14

Note: *On 1993 Galant SOHC and DOHC models, check the cylinder head bolt shank length before reassembly. If it exceeds 3.91 inches (99.4 mm), replace the bolt.*

13 Place a new gasket **(see illustration)** and the cylinder head in position. **Note:** *Position the identification mark facing up and on the timing belt end of the engine block* **(see illustration)**.
14 The cylinder head bolts should be tightened in several stages following the proper sequence **(see illustration)** to the torque listed in this Chapter's Specifications.
15 Reinstall the timing belt (see Section 7).

Chapter 2 Part A Engine

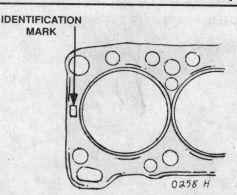

12.13b The identification mark should face up and be positioned next to the camshaft sprocket

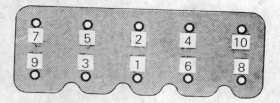

12.14 Cylinder head bolt tightening sequence – turn this illustration upside-down if the camshaft sprocket is on the opposite end of the cylinder head

13.5a Arrows point to oil pan bolts visible from below – others (not visible here) are located around the perimeter

13.5b Tap the oil pan to break the seal

16 Reinstall the remaining parts in the reverse order of removal.
17 Be sure to refill the cooling system and check all fluid levels. Rotate the crankshaft clockwise slowly by hand through two complete revolutions. Recheck the camshaft timing marks (see Section 7).
18 Start the engine and set the ignition timing (see Chapter 1). Run the engine until normal operating temperature is reached. Check for leaks and proper operation. Shut off the engine. Remove the valve cover and retorque the cylinder head bolts, unless the gasket manufacturer states otherwise. Recheck the valve adjustment if the engine is not equipped with hydraulic valve adjusters.

6 Thoroughly clean the oil pan and sealing surfaces. Remove all traces of old gasket material. Check the oil pan sealing surface for distortion. Straighten or replace as necessary.

Installation
7 If the oil pan was sealed with RTV only (no gasket), apply a 4 mm bead of RTV sealer to the oil pan flange (see illustration). If you are using a gasket, apply a thin coat of RTV sealer to the oil pan flange and affix the gasket.
8 Place the oil pan into position and install the bolts finger tight. Working

13 Oil pan – removal and installation

Refer to illustrations 13.5a, 13.5b and 13.7

Removal
1 Disconnect the negative cable from the battery.
2 Raise the vehicle and support it securely on jackstands.
3 Remove the splash pan under the drivebelt end of the engine, then remove the dipstick and drain the engine oil (see Chapter 1).
4 Unbolt the front exhaust pipe from the exhaust manifold (see Section 11).
5 Remove the bolts (see illustration) and lower the oil pan from the vehicle. If the pan is stuck, tap it with a soft-face hammer (see illustration) or place a block of wood against the pan and tap the wood with a hammer.

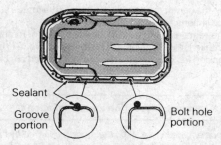

13.7 If the oil pan was sealed with RTV only, apply RTV sealant in the groove and to the inside of the bolt holes

14.3 The oil pickup tube is attached to the bottom of the oil pump by two bolts (arrows) (1.5L engine shown)

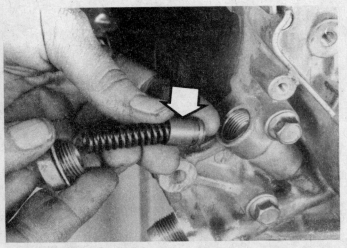

14.4 Remove the pressure regulator plug, spring and plunger – if the plunger (arrow) sticks in the bore, remove it with a magnet

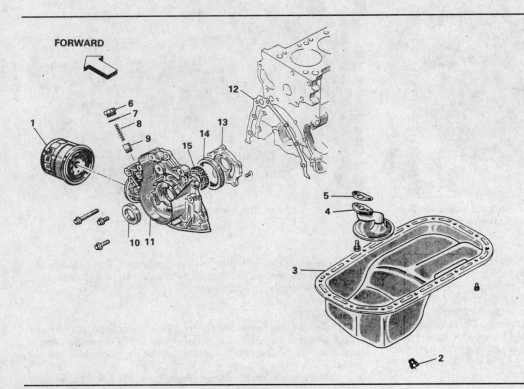

14.6a Exploded view of the oil pump and related components – 1.5L engine

1. Oil filter
2. Drain plug
3. Oil pan
4. Oil pickup tube and screen
5. Gasket
6. Plug
7. Gasket
8. Pressure regulator spring
9. Pressure regulator plunger
10. Crankshaft front oil seal
11. Front case
12. Front case gasket
13. Oil pump cover
14. Outer gear
15. Inner gear

side-to-side from the center out, tighten the bolts to the torque listed in this Chapter's Specifications.

9 Reinstall the remaining parts in the reverse order of removal.

10 Refill the crankcase with the proper quantity and grade of oil and run the engine, checking for leaks. Road test the vehicle and check for leaks again.

14 Oil pump – removal, inspection and installation

Note: *The various engines installed in the models covered by this manual incorporate three different types of oil pumps. The 1.5L engine utilizes a gear assembly that is rotated by the crankshaft. The oil pumps on all other engines are turned by the timing belt.*

Removal

Refer to illustrations 14.3, 14.4, 14.6a, 14.6b, 14.6c and 14.7

1 Remove the timing belt and crankshaft sprocket (see Section 7).
2 Remove the oil pan (see Section 13).
3 Unbolt the oil pickup tube (screen) from the bottom of the pump housing **(see illustration)**.
4 Remove the pressure regulator plug, spring and plunger **(see illustration)**.
5 Remove the timing belt tensioner (see Section 7).
6 Remove the oil pump-to-block bolts noting the different lengths and locations of the bolts. Carefully separate the front case from the engine **(see illustrations)**.

Chapter 2 Part A Engine

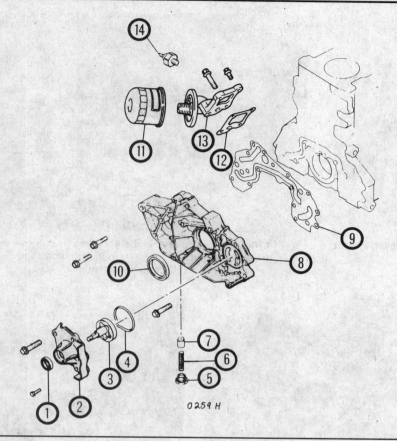

14.6b Exploded view of the oil pump and related components – 1.6L engines

1. Oil seal
2. Oil pump cover
3. Oil pump rotor assembly
4. Gasket
5. Plug
6. Relief spring
7. Relief plunger
8. Front case
9. Front case gasket
10. Crankshaft oil seal
11. Oil filter
12. Gasket
13. Oil filter adapter
14. Oil pressure switch

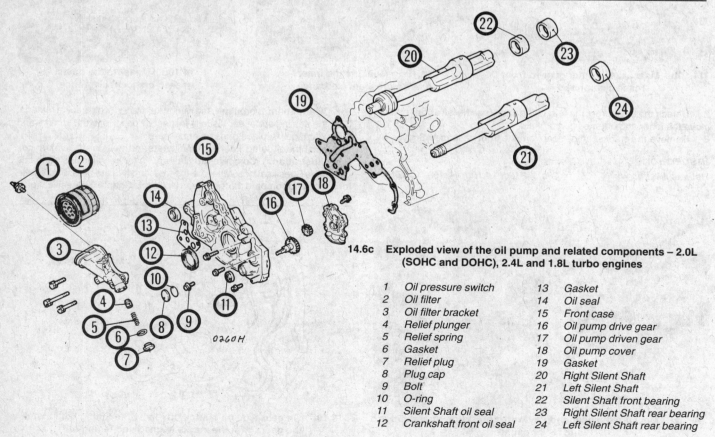

14.6c Exploded view of the oil pump and related components – 2.0L (SOHC and DOHC), 2.4L and 1.8L turbo engines

1. Oil pressure switch
2. Oil filter
3. Oil filter bracket
4. Relief plunger
5. Relief spring
6. Gasket
7. Relief plug
8. Plug cap
9. Bolt
10. O-ring
11. Silent Shaft oil seal
12. Crankshaft front oil seal
13. Gasket
14. Oil seal
15. Front case
16. Oil pump drive gear
17. Oil pump driven gear
18. Oil pump cover
19. Gasket
20. Right Silent Shaft
21. Left Silent Shaft
22. Silent Shaft front bearing
23. Right Silent Shaft rear bearing
24. Left Silent Shaft rear bearing

Chapter 2 Part A Engine

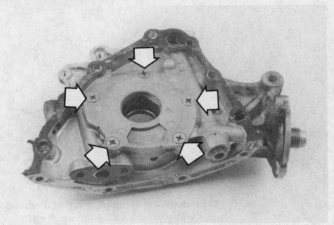

14.7 On the 1.5L oil pump, remove the screws (arrows) and lift off the oil pump cover

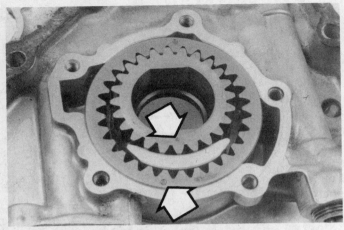

14.10a Install the inner and outer gears as shown – the matchmarks (arrows) must be in the positions shown (1.5L engine shown)

14.10b Measure the outer gear-to-front case clearance

14.10c Measure the inner gear-to-crescent clearance

14.10d Measure the outer gear-to-crescent clearance

7   Detach the cover from the rear of the case **(see illustration)** and remove the inner and outer oil pump gears.
8   Remove the crankshaft oil seal from the front case (see Section 8).

Inspection

Refer to illustrations 14.10a, 14.10b, 14.10c, 14.10d, 14.10e, 14.10f and 14.10g

9   Clean all parts thoroughly and remove all traces of old gasket material from the sealing surfaces. Visually inspect all parts for wear, cracks and other damage. Replace parts as necessary.
10 Install the oil pump outer and inner gears and measure the clearances **(see illustrations)**. Compare the clearances to the values listed in this Chapter's Specifications. Measure the free length of the pressure regulator spring and compare the measurement to this Chapter's Specifications.

14.10e Measure the gear endplay with a precision straightedge and feeler gauge

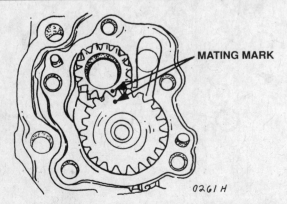

14.10f On gear type oil pumps driven by the timing belt, install the gears with the marks aligned and facing out

Chapter 2 Part A Engine

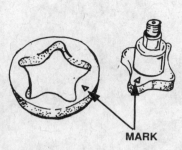

14.10g On rotor type oil pumps, install the rotor into the rotor housing with the marks facing up

14.13 When installing the pump on the engine, note that the crankshaft has machined flats which the oil pump inner gear must fit over (where applicable)

Replace parts as necessary. Pack the pump cavity with petroleum jelly and install the cover. Tighten the bolts to the torque listed in this Chapter's Specifications. **Note:** *Be sure to install the gears or rotor(s) with the mating marks in the correct place* **(see illustration).**

Installation

Refer to illustration 14.13

11 Install a new crankshaft front oil seal (see Section 8).
12 Install the pressure regulator valve components and tighten the plug securely.
13 Using a new gasket, position the pump on the engine **(see illustration).** Install the bolts in their proper locations, according to length. Tighten the bolts to the torque listed in this Chapter's Specifications.
14 Reinstall the remaining parts in the reverse order of removal.
15 Add oil, start the engine and check for oil pressure and leaks.

15 Flywheel/driveplate – removal and installation

Refer to illustrations 15.3a and 15.3b

Removal

1 Raise the vehicle and support it securely on jackstands, then refer to Chapter 7 and remove the transaxle. If it's leaking, now would be a very good time to replace the front seal (the torque converter must be removed to gain access to this seal).
2 Remove the pressure plate and clutch disc (manual transaxle equipped vehicles) (see Chapter 8). Now is a good time to check/replace the clutch components.
3 The bolt holes are either staggered or symmetrical on the flywheel/driveplate bolt pattern. To ensure correct alignment during reinstallation, mark the position of the flywheel/driveplate to the crankshaft before removal **(see illustrations).**
4 Remove the bolts that secure the flywheel/driveplate to the crankshaft. If the crankshaft turns, wedge a screwdriver in the ring gear teeth to jam the flywheel.
5 Remove the flywheel/driveplate from the crankshaft. Since the flywheel is fairly heavy, be sure to support it while removing the last bolt.
6 Clean the flywheel to remove grease and oil. Inspect the surface for cracks, rivet grooves, burned areas and score marks. Light scoring can be removed with emery cloth. Check for cracked and broken ring gear teeth. Lay the flywheel on a flat surface and use a straightedge to check for warpage. **Note:** *Ring gears are available separately and can be installed by an automotive machine shop.*
7 Clean and inspect the mating surfaces of the flywheel/driveplate and the crankshaft. If the crankshaft rear seal is leaking, replace it before reinstalling the flywheel/driveplate (see Section 16).

15.3a The flywheel/driveplate bolt holes are staggered so they line up only one way on the 1.5L engine

15.3b On engines with a symmetrical bolt pattern, be sure to mark the flywheel/driveplate to the crankshaft

2A–32 Chapter 2 Part A Engine

16.2a The quick (but not recommended) way to replace the rear main oil seal on the 1.5L engine is to simply pry the old one out – other models may require removal of the rear main seal retainer

16.2b Lubricate the crankshaft journal and the lip of the new seal with engine oil and tap the new seal into place – the seal lip is stiff and can be easily damaged during installation if you're not careful

16.5 After removing the retainer from the engine, support it on wood blocks and drive out the old seal with a punch and hammer

16.6 Drive the new seal into the retainer with a block of wood or a section of pipe, if you have one large enough – make sure you don't cock the seal in the retainer bore

Installation

8   Position the flywheel/driveplate against the crankshaft. If the bolt holes aren't staggered, align the previously applied matchmarks. Before installing the bolts, apply thread locking compound to the threads.
9   Wedge a screwdriver in the ring gear teeth to keep the flywheel/driveplate from turning as you tighten the bolts to the torque listed in this Chapter's Specifications.
10 The remainder of installation is the reverse of the removal procedure.

16 Rear main oil seal – replacement

Refer to illustrations 16.2a, 16.2b, 6.5 and 16.6

1   The transaxle must be removed from the vehicle for this procedure (see Chapter 7).
2   The seal can be replaced without removing the oil pan or seal retainer. However, this method is not recommended because the lip of the seal is quite stiff and it's possible to cock the seal in the retainer bore or damage it during installation. If you want to take the chance, pry out the old seal **(see illustration)**. Apply a film of clean oil to the crankshaft seal journal and the lip of the new seal and carefully tap the new seal into place **(see illustration)**. The lip is stiff so carefully work it onto the seal journal of the crankshaft with a smooth object like the end of an extension as you tap the seal into place. Don't rush it or you may damage the seal.
3   The following method is recommended but requires removal of the oil pan (see Section 13) and the seal retainer.
4   After the oil pan has been removed, remove the bolts, detach the seal retainer and peel off all the old gasket material.
5   Position the seal and retainer assembly on a couple of wood blocks on a workbench and drive the old seal out from the back side with a punch and hammer **(see illustration)**.
6   Drive the new seal into the retainer with a block of wood **(see illustration)** or a section of pipe slightly smaller in diameter than the outside diameter of the seal.
7   Lubricate the crankshaft seal journal and the lip of the new seal with clean engine oil. Position a new gasket on the engine block.
8   Slowly and carefully push the seal onto the crankshaft. The seal lip is stiff, so work it onto the crankshaft with a smooth object such as the end of an extension as you push the retainer against the block.
9   Install and tighten the retainer bolts to the torque listed in this Chapter's Specifications. The bottom sealing flange of the retainer must not ex-

Chapter 2 Part A Engine

2A – 33

17.8a On 2.0L engines, the left mount is located adjacent to the upper timing belt cover – it is attached to the engine with two nuts and two bolts (arrows)

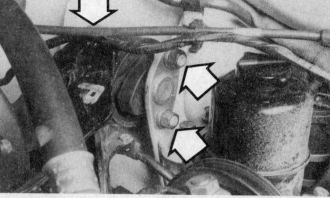

17.8b On 1.5L engines, the left engine mount is attached to the bracket with a through-bolt that is secured by two retaining nuts on this side and one hidden by the mount (arrows)

tend below the bottom sealing flange (oil pan rail) of the block.
10 The remaining steps are the reverse of removal.
11 Run the engine and check for oil leaks.

17 Engine mounts – check and replacement

1 Engine mounts seldom require attention, but broken or deteriorated mounts should be replaced immediately or the added strain placed on the driveline components may cause damage or wear.

Check

2 During the check, the engine must be raised slightly to remove the weight from the mounts.
3 Raise the vehicle and support it securely on jackstands, then position a jack under the engine oil pan. Place a large block of wood between the jack head and the oil pan, then carefully raise the engine just enough to take the weight off the mounts. **Warning:** *DO NOT place any part of your body under the engine when it's supported only by a jack!*
4 Check the mounts to see if the rubber is cracked, hardened or separated from the metal backing. Sometimes the rubber will split right down the center. Check that the rod on the front mount is not bent.
5 Check for relative movement between the mount plates and the engine or frame (use a large screwdriver or pry bar to attempt to move the mounts). If movement is noted, lower the engine and tighten the mount fasteners.
6 Rubber preservative may be applied to the mounts to slow deterioration.

Replacement

Refer to illustrations 17.8a, 17.8b, 17.8c, 17.8d and 17.8e

7 Disconnect the negative battery cable from the battery, then raise the vehicle and support it securely on jackstands (if not already done).
8 Remove the fasteners and detach the mount from the frame and engine **(see illustrations)**. Do not disconnect more than one mount at a time, except during engine removal.
9 The mounts are available as inserts, which are pressed into place. Obtain new inserts and take them to an automotive machine shop or dealer service department to be pressed into the existing bracket.
10 Installation is the reverse of removal. Use thread locking compound on the mount bolts and be sure to tighten them securely.

17.8c The transaxle mount on some models is secured to the transaxle by a through-bolt and nut (arrows) – it is secured to the body by three bolts accessible from inside the right fender

17.8d The rear mount is located above the steering rack – remove the mounting bolts and through-bolt (arrows)

17.8e The front mount assembly is held in place by two through bolts – one at the bottom and one at the top (the bottom one is shown here)

Notes:

Chapter 2 Part B
General engine overhaul procedures

Contents

Compression check	3
Crankshaft – inspection	19
Crankshaft – installation and main bearing oil clearance check	23
Crankshaft – removal	14
Cylinder head – cleaning and inspection	10
Cylinder head – disassembly	9
Cylinder head – reassembly	12
Cylinder honing	17
Engine block – cleaning	15
Engine block – inspection	16
Engine overhaul – disassembly sequence	8
Engine overhaul – general information	2
Engine overhaul – reassembly sequence	21
Engine rebuilding alternatives	7
Engine – removal and installation	6
Engine removal – methods and precautions	5
General information	1
Initial start-up and break-in after overhaul	26
Main and connecting rod bearings – inspection	20
Pistons/connecting rods – inspection	18
Pistons/connecting rods – installation and rod bearing oil clearance check	25
Pistons/connecting rods – removal	13
Piston rings – installation	22
Rear main oil seal installation	24
Top Dead Center (TDC) for number one piston – locating	4
Valves – servicing	11

Specifications

General
Cylinder compression pressure (at 250 rpm)	192 psi
Oil pressure (engine warm)	11 psi minimum at idle

1.5L engine

Cylinder head warpage
Standard	0.002 in (0.05 mm)
Service limit	
1990 and later Mirage	0.008 in (0.2 mm)
All others	0.004 in (0.1 mm)

Valves and related components
Valve margin width	
Intake	
Standard	0.039 in (1.0 mm)
Service limit	0.020 in (0.5 mm)
Exhaust	
Standard	0.059 in (1.5 mm)
Service limit	0.039 in (1.0 mm)
Valve stem diameter (intake and exhaust)	0.26 in (6.6 mm)
Valve stem-to-guide clearance	
Intake (1989 and later Mirage) only	
Standard	0.0008 to 0.0028 in (0.02 to 0.05 mm)
Service limit	0.004 in (0.10 mm)
Intake (all others)	
Standard	0.0012 to 0.0024 in (0.03 to 0.06 mm)
Service limit	0.004 in (0.1 mm)
Exhaust (all)	
Standard	0.0020 to 0.0035 in (0.05 to 0.09 mm)
Service limit	0.006 in (0.15 mm)
Valve spring	
Out-of-square limit	3 degrees
Pressure	
1991 and later Mirage	
Intake	51 lbs at 1.075 inch
Exhaust	64 lbs at 1.075 inch
All others (intake and exhaust)	53 lbs at 1.075 inch
Installed height	1.417 in (36 mm)

Valves and related components (continued)

Free length
  1991 and later Mirage
    Intake .... 1.815 in (46.1 mm)
    Exhaust .... 1.843 in (46.8 mm)
  All others (intake and exhaust) .... 1.756 in (44.6 mm)

Jet valves
Stem diameter .... 0.1693 in (4.30 mm)
Face/seat angle 45-degrees
Spring
  Free length .... 1.165 in (29.60 mm)
  Load .... 5.5 lbs at 0.846 in (3.5 kg at 21.5 mm)

Crankshaft and connecting rods
Connecting rod journal
  Diameter .... 1.6535 in (42 mm)
  Out-of-round/taper limits .... 0.0004 in (0.01 mm)
  Bearing oil clearance .... 0.0006 to 0.0017 in (0.014 to 0.044 mm)
Connecting rod endplay (side clearance) .... 0.004 to 0.010 in (0.10 to 0.25 mm)
Main bearing journal
  Diameter .... 1.8898 in (48 mm)
  Out-of-round/taper limits .... 0.0004 in (0.01 mm)
  Runout limit .... 0.0012 in (0.03 mm)
  Bearing oil clearance .... 0.0008 to 0.0028 in (0.02 to 0.07 mm)
Crankshaft endplay .... 0.002 to 0.007 in (0.05 to 0.18 mm)

Cylinder bore
Diameter (nominal) .... 2.972 in (75.5 mm)
Out-of-round/taper limit .... 0.0008 in (0.02 mm)

Pistons and rings
Piston diameter (nominal)* .... 2.972 in (75.5 mm)
Piston-to-bore clearance .... 0.0008 to 0.0016 in (0.02 to 0.04 mm)
Piston ring end gap
  Compression rings
    Standard .... 0.008 to 0.014 in (0.20 to 0.35 mm)
    Service limit .... 0.031 in (0.8 mm)
  Oil ring
    Standard .... 0.008 to 0.028 in (0.20 to 0.70 mm)
    Service limit .... 0.039 in (1.0 mm)
Piston ring side clearance
  No. 1 (top) compression ring
    Standard .... 0.0012 to 0.0028 in (0.03 to 0.07 mm)
    Service limit .... 0.006 in (0.15 mm)
  No. 2 compression ring
    Standard .... 0.0008 to 0.0024 in (0.02 to 0.06 mm)
    Service limit .... 0.005 in (0.12 mm)

Measured 5/64-inch up from bottom of skirt

Torque specifications*
Ft-lbs
Main bearing cap bolts .... 36 to 39
Connecting rod cap nuts
  1990 and later Mirage
    Step 1 15
    Step 2 .... Rotate an additional 1/4-turn (90-degrees)
  All others .... 23 to 25
Jet valves .... 13 to 16

Note: Refer to Part A for additional torque specifications.

1.6L engine

Cylinder head warpage
Standard .... 0.002 in (0.05 mm)
Service limit
  1990 and later Mirage .... 0.008 in (0.2 mm)
  All others .... 0.004 in (0.1 mm)

Valves and related components
Valve margin width
  Intake (DOHC)
    Standard .... 0.039 in (1.0 mm)
    Service limit .... 0.028 in (0.7 mm)

Chapter 2 Part B General engine overhaul procedures

Intake (SOHC)
  Standard .... 0.059 in (1.5 mm)
  Service limit .... 0.039 in (1.0 mm)
Exhaust (all)
  Standard .... 0.059 in (1.5 mm)
  Service limit .... 0.039 in (1.0 mm)
Valve stem diameter
  SOHC (intake and exhaust) .... 0.32 in (8.0 mm)
  DOHC
    Intake .... 0.260 in (6.6 mm)
    Exhaust .... 0.257 in (6.5 mm)
Valve stem-to-guide clearance
  Intake (SOHC)
    Standard .... 0.0012 to 0.0024 in (0.03 to 0.06 mm)
    Service limit .... 0.004 in (0.10 mm)
  Intake (DOHC)
    Standard .... 0.0008 to 0.0019 in (0.02 to 0.05 mm)
    Service limit .... 0.004 in (0.10 mm)
  Exhaust (all)
    Standard .... 0.0020 to 0.0035 in (0.05 to 0.09 mm)
    Service limit .... 0.006 in (0.15 mm)
Valve spring
  Out-of-square limit .... 4 degrees
  Pressure .... 62 lbs at 1.47 in (24 kg at 27.3 mm)
  Installed height
    SOHC .... 1.469 in (37.3 mm)
    DOHC .... Not available
  Free length
    SOHC .... 1.756 in (44.6 mm)
    DOHC .... 1.902 in (48.3 mm)

Jet valves (SOHC only)
Stem diameter .... 0.1693 in (4.30 mm)
Face/seat angle 45-degrees
Spring
  Free length .... 1.165 in (29.60 mm)
  Load .... 7.7 lbs at 0.846 in (34.3 N at 21.5 mm)

Crankshaft and connecting rods
Connecting rod journal
  Diameter .... 1.77 in (45 mm)
  Out-of-round/taper limit .... 0.0006 in (0.015 mm)
  Bearing oil clearance .... 0.0008 to 0.0019 in (0.02 to 0.05 mm)
Connecting rod endplay (side clearance) ... 0.004 to 0.010 in (0.10 to 0.25 mm)
Main bearing journal
  Diameter .... 2.24 in (57.0 mm)
  Out-of-round/taper limits .... 0.0006 in (0.015 mm)
  Runout limit .... 0.0002 in (0.005 mm)
  Bearing oil clearance .... 0.0008 to 0.0020 in (0.02 to 0.05 mm)
Crankshaft endplay .... 0.002 to 0.007 in (0.05 to 0.18 mm)

Cylinder bore
Diameter (nominal)
  SOHC .... 3.028 in (76.9 mm)
  DOHC .... 3.240 in (82.3 mm)
Out-of-round/taper limits
  SOHC .... 0.0008 in (0.02 mm)
  DOHC .... 0.0004 in (0.01 mm)

Pistons and rings
Piston diameter (nominal)*
  SOHC .... 3.028 in (76.9 mm)
  DOHC .... 3.239 in (82.3 mm)
Piston-to-bore clearance
  SOHC
    1985 and 1986 .... 0.0008 to 0.0016 in (0.02 to 0.04 mm)
    1987 and 1988 .... 0.0012 to 0.0020 in (0.03 to 0.05 mm)
  DOHC
    1989 .... 0.0012 to 0.0020 in (0.03 to 0.05 mm)
    1990 and later .... 0.0008 to 0.0016 in (0.02 to 0.04 mm)

2B-4 Chapter 2 Part B General engine overhaul procedures

Pistons and rings (continued)
Piston ring end gap
  Compression rings (SOHC)
    Number 1 (top)
      Standard .... 0.010 to 0.016 in (0.25 to 0.40 mm)
      Service limit .... 0.031 in (0.08 mm)
    Number 2
      Standard .... 0.008 to 0.014 in (0.20 to 0.35 mm)
      Service limit .... 0.031 in (0.08 mm)
  Oil ring (SOHC)
    Standard .... 0.008 to 0.028 in (0.20 to 0.70 mm)
    Service limit .... 0.039 in (0.10 mm)
  Compression rings (DOHC)
    Number 1 (top)
      Standard .... 0.010 to 0.016 in (0.25 to 0.40 mm)
      Service limit .... 0.031 in (0.80 mm)
    Number 2
      Standard .... 0.014 to 0.020 in (0.35 to 0.50 mm)
      Service limit .... 0.031 in (0.08 mm)
  Oil ring (DOHC)
    Standard .... 0.008 to 0.028 in (0.20 to 0.70 mm)
    Service limit .... 0.039 in (0.10 mm)
Piston ring side clearance
  SOHC
    No.1 (top) compression ring
      Standard .... 0.0012 to 0.0028 in (0.03 to 0.07 mm)
      Service limit .... 0.004 in (0.10 mm)
    No. 2 compression ring
      Standard .... 0.0008 to 0.0024 in (0.02 to 0.06 mm)
      Service limit .... 0.004 in (0.10 mm)
  DOHC (both compression rings)
    Standard .... 0.0012 to 0.0028 in (0.03 to 0.07 mm)
    Service limit .... 0.004 in (0.10 mm)
Measured 5/64-inch up from bottom of skirt

Torque specifications*   Ft-lbs
SOHC
  Main bearing cap bolts .... 36 to 39
  Connecting rod cap nuts .... 23 to 25
  Relief valve plug .... 27 to 29
  Jet valves .... 13 to 16
DOHC
  Main bearing cap bolts .... 47 to 51
  Connecting rod cap nuts .... 36 to 38
  Oil relief valve plug .... 29 to 36

Note: Refer to Part A for additional torque specifications.

1.8L, 2.0L and 2.4L engines

Cylinder head warpage
Standard .... 0.002 in (0.05 mm)
Service limit .... 0.008 in (0.2 mm)

Valves and related components
Valve margin width
  SOHC
    Intake
      Standard .... 0.047 in (1.2 mm)
      Service limit .... 0.028 in (0.7 mm)
    Exhaust
      Standard .... 0.079 in (2.0 mm)
      Service limit .... 0.059 in (1.5 mm)
  DOHC
    Intake
      Standard .... 0.040 in (1.0 mm)
      Service limit .... 0.028 in (0.7 mm)
    Exhaust
      Standard .... 0.059 in (1.5 mm)
      Service limit .... 0.040 in (1.0 mm)

Chapter 2 Part B General engine overhaul procedures

Valve stem diameter
  SOHC (intake and exhaust) .... 0.3100 in (8.00 mm)
  DOHC
    Intake .... 0.2590 in (6.58 mm)
    Exhaust .... 0.2575 in (6.55 mm)
Valve stem-to-guide clearance
  SOHC
    Intake
      Standard .... 0.0012 to 0.0024 in (0.03 to 0.06 mm)
      Service limit .... 0.004 in (0.10 mm)
    Exhaust
      Standard .... 0.0020 to 0.0035 in (0.05 to 0.09 mm)
      Service limit .... 0.006 in (0.15 mm)
  DOHC
    Intake
      Standard .... 0.0008 to 0.0019 in (0.02 to 0.05 mm)
      Service limit .... 0.004 in (0.10 mm)
    Exhaust
      Standard .... 0.002 to 0.0033 in (0.05 to 0.09 mm)
      Service limit .... 0.006 in (0.15 mm)
Valve spring
  Out-of-square limit .... 3 degrees
  Pressure
    1990 and later Galant DOHC .... 66 lbs (300 N)
    All others .... 72 lbs (322 N)
  Installed height
    2.4L, 1.8L and 2.0L (except Galant) .... 1.591 in (40.4 mm)
    2.0L Galant (SOHC and DOHC) .... Not available
  Free length
    SOHC Cordia/Tredia
      1.8L (1983 through 1986) .... 1.870 in (47.5 mm)
      1.8L (1987 and 1988) .... 1.960 in (49.8 mm)
      2.0L (1984 and 1985) .... 1.870 in (47.5 mm)
      2.0L (1986 through 1988) .... 1.960 in (49.8 mm)
    SOHC Galant
      2.4L (1985 through 1987) .... 1.960 in (49.8 mm)
      2.0L (1989 on) .... 1.960 in (49.8 mm)
    DOHC Galant
      2.0L (before March 1988) .... 1.803 in (45.8 mm)
      2.0L (after April 1988) .... 1.902 in (48.3 mm)

Jet valves
Stem diameter .... 0.1693 in (4.30 mm)
Face/seat angle 45-degrees
Spring
  Free length .... 1.165 in (29.60 mm)
  Load .... 7.7 lbs at 0.846 in (34.3 N at 21.5 mm)

Crankshaft and connecting rods
Connecting rod journal
  Diameter .... 1.7720 in (45.0 mm)
  Out-of-round/taper limits .... 0.0004 in (0.01 mm)
  Bearing oil clearance .... 0.0008 to 0.0020 in (0.02 to 0.05 mm)
Connecting rod endplay (side clearance) .... 0.004 to 0.010 in (0.10 to 0.25 mm)
Main bearing journal
  Diameter .... 2.2440 in (57.0 mm)
  Out-of-round limits .... 0.0006 in (0.016 mm)
  Taper limit .... 0.0002 in (0.005 mm)
  Bearing oil clearance .... 0.0008 to 0.0020 in (0.02 to 0.05 mm)
Crankshaft endplay .... 0.002 to 0.007 in (0.05 to 0.18 mm)

Cylinder bore
Diameter (nominal)
  Cordia\Tredia
    1983 through 1985 1.8L .... 3.346 in (85.0 mm)
    1986 through 1988 1.8L .... 3.173 in (80.6 mm)
    1984 through 1988 2.0L .... 3.346 in (85.0 mm)
  Galant
    2.4L .... 3.406 in (86.5 mm)
    2.0L (SOHC and DOHC) .... 3.347 in (85.1 mm)
Out-of-round/taper limits .... 0.0008 in (0.02 mm)

Pistons and rings

Piston diameter (nominal)*
- 2.4L .... 3.406 in (86.5 mm)
- 2.0L and 1.8L non-turbo .... 3.346 in (85.0 mm)
- 1.8L turbo .... 3.173 in (80.6 mm)

Piston-to-bore clearance
- 2.4L .... 0.0008 to 0.0016 in (0.02 to 0.04 mm)
- 2.0L and 1.8L non-turbo .... 0.0004 to 0.0012 in (0.01 to 0.03 mm)
- 1.8L turbo .... 0.0012 to 0.0020 in (0.03 to 0.05 mm)

Piston ring end gap
- 1.8L non-turbo and 2.0L SOHC (except 2.0L Galant)
  - Number 1 (top) compression ring
    - Standard .... 0.010 to 0.018 in (0.25 to 0.45 mm)
    - Service limit .... Not available
  - Number 2 compression ring
    - Standard .... 0.008 to 0.016 in (0.20 to 0.40 mm)
    - Service limit .... Not available
  - Oil ring
    - Standard .... 0.008 to 0.028 in (0.20 to 0.70 mm)
    - Service limit .... Not available
- 1.8L turbo SOHC
  - Number 1 compression ring
    - Standard .... 0.012 to 0.018 in (0.30 to 0.45 mm)
    - Service limit .... Not available
  - Number 2 compression ring
    - Standard .... 0.008 to 0.014 in (0.20 to 0.35 mm)
    - Service limit .... Not available
  - Oil ring
    - Standard .... 0.008 to 0.028 in (0.20 to 0.70 mm)
    - Service limit .... Not available
- 2.4L and 2.0L DOHC
  - Number 1 compression ring
    - Standard .... 0.010 to 0.016 in (0.25 to 0.40 mm)
    - Service limit .... 0.032 in (0.80 mm)
  - Number 2 compression ring (2.4L only)
    - Standard .... 0.008 to 0.016 in (0.20 to 0.40 mm)
    - Service limit .... 0.032 in (0.80 mm)
  - Number 2 compression ring (2.0L only)
    - Standard .... 0.014 to 0.020 in (0.35 to 0.50 mm)
    - Service limit .... 0.032 in (0.80 mm)
  - Oil ring
    - Standard .... 0.008 to 0.028 in (0.20 to 0.70 mm)
    - Service limit .... 0.039 in (0.10 mm)
- 2.0L SOHC Galant
  - Number 1 compression ring
    - Standard .... 0.010 to 0.016 in (0.25 to 0.40 mm)
    - Service limit .... 0.031 in (0.80 mm)
  - Number 2 compression ring
    - Standard .... 0.008 to 0.014 in (0.20 to 0.35 mm)
    - Service limit .... 0.031 in (0.80 mm)
  - Oil ring
    - Standard .... 0.008 to 0.028 in (0.20 to 0.70 mm)
    - Service limit .... 0.039 in (0.10 mm)

Piston ring side clearance
- 2.4L (1985 through 1987)
  - No. 1 (top) compression ring
    - Standard .... 0.0020 to 0.0040 in (0.05 to 0.09 mm)
    - Service limit .... Not available
  - No. 2 compression ring
    - Standard .... 0.0010 to 0.0020 in (0.02 to 0.06 mm)
    - Service limit .... Not available
- 2.4L (1988 only) and 2.0L SOHC Galant
  - No. 1 (top) compression ring
    - Standard .... 0.0012 to 0.0028 in (0.03 to 0.07 mm)
    - Service limit .... 0.004 in (0.10 mm)
  - No. 2 compression ring
    - Standard .... 0.0008 to 0.0024 in (0.02 to 0.06 mm)
    - Service limit .... 0.004 in (0.10 mm)

Chapter 2 Part B General engine overhaul procedures

SOHC 2.0L and 1.8L (except 2.0L Galant)
  No. 1 (top) compression ring
    Standard .... 0.0018 to 0.0033 in (0.045 to 0.085 mm)
    Service limit .... Not available
  No. 2 compression ring
    Standard .... 0.0008 to 0.0024 in (0.02 to 0.06 mm)
    Service limit .... Not available
2.0L DOHC engine
  Both compression rings
    Standard .... 0.0012 to 0.0028 in (0.03 to 0.07 mm)
    Service limit .... 0.004 in (0.10 mm)
* Measured 5/64-inch up from bottom of skirt

Torque specifications*   **Ft-lbs**
Main bearing cap bolts
  SOHC .... 36 to 39
  DOHC .... 47 to 51
Connecting rod cap nuts
  SOHC .... 23 to 25
  DOHC .... 36 to 38
Jet valves .... 13 to 16
Oil relief valve plug .... 29 to 36

* *Note: Refer to Part A for additional torque specifications.*

2.4a The oil pressure sending unit (arrow) is located below the exhaust manifold between the engine mount bracket and the oil pump (1.5L engine)

2.4b An oil pressure gauge can be installed in the hole where the oil pressure sending unit is normally located

1 General information

Included in this portion of Chapter 2 are the general overhaul procedures for the cylinder head and internal engine components.

The information ranges from advice concerning preparation for an overhaul and the purchase of replacement parts to detailed, step-by-step procedures covering removal and installation of internal engine components and the inspection of parts.

The following Sections have been written based on the assumption that the engine has been removed from the vehicle. For information concerning in-vehicle engine repair, as well as removal and installation of the external components necessary for the overhaul, see Part A of this Chapter and Section 8 of this Part. For information on determining models and engine numbers, refer to the Engine Identification Chart in Chapter 2A, Section 1.

The Specifications included in this Part are only those necessary for the inspection and overhaul procedures which follow. Refer to Part A for additional Specifications.

2 Engine overhaul – general information

Refer to illustrations 2.4a and 2.4b

It's not always easy to determine when, or if, an engine should be completely overhauled, as a number of factors must be considered.

High mileage is not necessarily an indication that an overhaul is needed, while low mileage doesn't preclude the need for an overhaul. Frequency of servicing is probably the most important consideration. An engine that's had regular and frequent oil and filter changes, as well as other required maintenance, will most likely give many thousands of miles of reliable service. Conversely, a neglected engine may require an overhaul very early in its life.

Excessive oil consumption is an indication that piston rings, valve seals and/or valve guides are in need of attention. Make sure that oil leaks aren't responsible before deciding that the rings and/or guides are bad. Perform a compression check to determine the extent of the work required (see Section 3).

Check the oil pressure with a gauge installed in place of the oil pressure sending unit **(see illustrations)** and compare it to the Specifications in this Chapter. If it's extremely low, the bearings and/or oil pump are probably worn out. For additional diagrams to help locate the oil pressure sender, refer to illustrations 13.5a, 13.5b and 13.5c in Chapter 2A.

Loss of power, rough running, knocking or metallic engine noises, excessive valve train noise and high fuel consumption rates may also point to the need for an overhaul, especially if they're all present at the same time. If a complete tune-up doesn't remedy the situation, major mechanical work is the only solution.

An engine overhaul involves restoring the internal parts to the specifications of a new engine. During an overhaul, the piston rings are replaced and the cylinder walls are reconditioned (rebored and/or honed). If a re-

3.6 A compression gauge with a threaded fitting for the spark plug hole is preferred over the type that requires hand pressure to maintain the seal – be sure to open the throttle and choke valves as far as possible during the compression check!

4.4 The crankshaft may be turned by placing a socket on the crankshaft pulley center bolt and slipping an extension through the left inner fender panel (wheel and tire removed for clarity)

4.6a Align the notch in the pulley (arrow) with the T on the timing plate (drivebelts and water pump pulley removed for clarity) (1.5L shown)

4.6b Some engines have the timing marks set off to the side of the timing belt cover (lower timing belt cover removed to show detail)

bore is done by an automotive machine shop, new oversize pistons will also be installed. The main bearings, connecting rod bearings and camshaft bearings are generally replaced with new ones and, if necessary, the crankshaft may be reground to restore the journals. Generally, the valves are serviced as well, since they're usually in less-than-perfect condition at this point. While the engine is being overhauled, other components, such as the distributor, starter and alternator, can be rebuilt as well. The end result should be a like new engine that will give many trouble free miles. **Note:** *Critical cooling system components such as the hoses, drivebelts, thermostat and water pump MUST be replaced with new parts when an engine is overhauled. The radiator should be checked carefully to ensure that it isn't clogged or leaking (see Chapter 3). Also, we don't recommend overhauling the oil pump – always install a new one when an engine is rebuilt.*

Before beginning the engine overhaul, read through the entire procedure to familiarize yourself with the scope and requirements of the job. Overhauling an engine isn't difficult, but it is time consuming. Plan on the vehicle being tied up for a minimum of two weeks, especially if parts must be taken to an automotive machine shop for repair or reconditioning. Check on availability of parts and make sure that any necessary special tools and equipment are obtained in advance. Most work can be done with typical hand tools, although a number of precision measuring tools are required for inspecting parts to determine if they must be replaced. Often an automotive machine shop will handle the inspection of parts and offer advice concerning reconditioning and replacement. **Note:** *Always wait until the engine has been completely disassembled and all components, especially the engine block, have been inspected before deciding what service*

and repair operations must be performed by an automotive machine shop. Since the block's condition will be the major factor to consider when determining whether to overhaul the original engine or buy a rebuilt one, never purchase parts or have machine work done on other components until the block has been thoroughly inspected. As a general rule, time is the primary cost of an overhaul, so it doesn't pay to install worn or substandard parts.

As a final note, to ensure maximum life and minimum trouble from a rebuilt engine, everything must be assembled with care in a spotlessly clean environment.

3 Compression check

Refer to illustration 3.6

1 A compression check will tell you what mechanical condition the upper end (pistons, rings, valves, head gaskets) of your engine is in. Specifically, it can tell you if the compression is down due to leakage caused by worn piston rings, defective valves and seats or a blown head gasket. **Note:** *The engine must be at normal operating temperature and the battery must be fully charged for this check. Also, the choke valve must be all the way open to get an accurate compression reading (if the engine's warm, the choke should be open).*

2 Begin by cleaning the area around the spark plugs before you remove them (compressed air should be used, if available, otherwise a small brush or even a bicycle tire pump will work). The idea is to prevent dirt from getting into the cylinders as the compression check is being done.

3 Remove all of the spark plugs from the engine (Chapter 1).

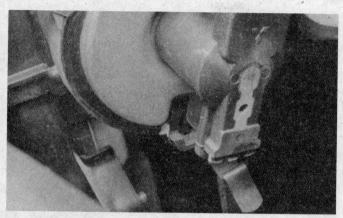

4.7 When the number one piston is at Top Dead Center (TDC) on the compression stroke, the distributor rotor should point straight down (Notice there is a paint mark on the cylinder head and distributor that is used to designate the position of the number one piston for purposes of correct assembly)

4 Block the throttle wide open.
5 Detach the coil wire from the center of the distributor cap and ground it on the engine block. Use a jumper wire with alligator clips on each end to ensure a good ground.
6 Install the compression gauge in the number one spark plug hole **(see illustration)**.
7 Crank the engine over at least seven compression strokes and watch the gauge. The compression should build up quickly in a healthy engine. Low compression on the first stroke, followed by gradually increasing pressure on successive strokes, indicates worn piston rings. A low compression reading on the first stroke, which doesn't build up during successive strokes, indicates leaking valves or a blown head gasket (a cracked head could also be the cause). Deposits on the undersides of the valve heads can also cause low compression. Record the highest gauge reading obtained.
8 Repeat the procedure for the remaining cylinders and compare the results to the Specifications in this Chapter.
9 Add some engine oil (about three squirts from a plunger-type oil can) to each cylinder, through the spark plug hole, and repeat the test.
10 If the compression increases after the oil is added, the piston rings are definitely worn. If the compression doesn't increase significantly, the leakage is occurring at the valves or head gasket. Leakage past the valves may be caused by burned valve seats and/or faces or warped, cracked or bent valves.
11 If two adjacent cylinders have equally low compression, there's a strong possibility that the head gasket between them is blown. The appearance of coolant in the combustion chambers or the crankcase would verify this condition.
12 If one cylinder is 20 percent lower than the others, and the engine has a slightly rough idle, a worn exhaust lobe on the camshaft could be the cause.
13 If the compression is unusually high, the combustion chambers are probably coated with carbon deposits. If that's the case, the cylinder head should be removed and decarbonized.
14 If compression is way down or varies greatly between cylinders, it would be a good idea to have a leak-down test performed by an automotive repair shop. This test will pinpoint exactly where the leakage is occurring and how severe it is.

4 Top Dead Center (TDC) for number one piston – locating

Refer to illustrations 4.4, 4.6a, 4.6b and 4.7
Note: *The following procedure is based on the assumption that the spark plug wires and distributor are correctly installed. If you are trying to locate TDC to install the distributor correctly, piston position must be determined by feeling for compression at the number one spark plug hole, then aligning the ignition timing marks as described in Step 6.*

1 Top Dead Center (TDC) is the highest point in the cylinder that each piston reaches as it travels up-and-down when the crankshaft turns. Each piston reaches TDC on the compression stroke and again on the exhaust stroke, but TDC generally refers to piston position on the compression stroke.
2 Positioning the piston(s) at TDC is an essential part of many procedures such as rocker arm removal, camshaft and timing belt/sprocket removal and distributor removal.
3 Before beginning this procedure, be sure to place the transmission in Neutral and apply the parking brake or block the rear wheels. Also, disable the ignition system by detaching the coil wire from the center terminal of the distributor cap and grounding it on the block with a jumper wire (non-DIS models). Remove the spark plugs (see Chapter 1).
4 In order to bring any piston to TDC, the crankshaft must be turned using one of the methods outlined below. When looking at the drivebelt end of the engine, normal crankshaft rotation is clockwise.
  a) The preferred method is to turn the crankshaft with a socket and ratchet attached to the bolt threaded into the drivebelt end of the crankshaft **(see illustration)**.
  b) A remote starter switch, which may save some time, can also be used. Follow the instructions included with the switch. Once the piston is close to TDC, use a socket and ratchet as described in the previous paragraph.
  c) If an assistant is available to turn the ignition switch to the Start position in short bursts, you can get the piston close to TDC without a remote starter switch. Make sure your assistant is out of the vehicle, away from the ignition switch, then use a socket and ratchet as described in Paragraph a) to complete the procedure.
5 On non-DIS models, detach the cap from the distributor and set it aside (see Chapter 1 if necessary). On DIS models, remove the valve cover (see Chapter 2A).
6 Turn the crankshaft (see Paragraph 3 above) until the notch in the crankshaft pulley is aligned with the T on the timing plate (located at the front of the engine) **(see illustrations)**.
7 On non-DIS models, look at the distributor rotor – it should be pointing straight down **(see illustration)**. On DIS models, check the rocker arms for the number one cylinder – they should be loose, not applying any pressure to the valves.
8 If the rotor is 180-degrees off, or the rocker arms are not loose, the number one piston is at TDC on the exhaust stroke. Go to Step 9.
9 To get the piston to TDC on the compression stroke, turn the crankshaft one complete turn (360-degrees) clockwise. The rotor should now be pointing straight down or the rocker arms should be loose. When the rotor is pointing at the number one spark plug wire terminal in the distributor cap or the rocker arms are loose and the ignition timing marks are aligned, the number one piston is at TDC on the compression stroke.
10 After the number one piston has been positioned at TDC on the compression stroke, TDC for any of the remaining pistons can be located by turning the crankshaft and following the firing order. On non-DIS models, with the distributor cap installed, use a felt-tip pen or chalk to make a mark on the distributor body directly beneath each of the terminals on the distributor cap. Then number the marks to correspond with the cylinder numbers. As you turn the crankshaft, the rotor will also turn. When it's pointing directly at one of the marks on the distributor, the piston for that particular cylinder is at TDC on the compression stroke.

5 Engine removal – methods and precautions

If you've decided that an engine must be removed for overhaul or major repair work, several preliminary steps should be taken.
Locating a suitable place to work is extremely important. Adequate work space, along with storage space for the vehicle, will be needed. If a shop or garage isn't available, at the very least a flat, level, clean work surface made of concrete or asphalt is required.
Cleaning the engine compartment and engine before beginning the removal procedure will help keep tools clean and organized.

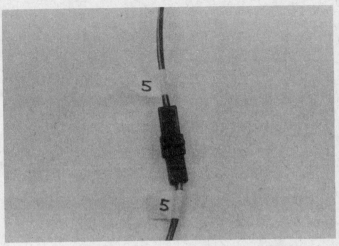

6.7 Label each wire before unplugging the connector

6.20 Attach a chain to the hoist brackets on the engine with large bolts and washers, then connect the hoist hook to the chain

6.23 On some models, it may be necessary to remove the master cylinder and unbolt the transaxle mount bracket (arrows) to remove the engine/transaxle assembly (1.5L shown)

6.24 Remove the long through-bolt from the transaxle mount

An engine hoist or A-frame will also be necessary. Make sure the equipment is rated in excess of the combined weight of the engine and accessories. Safety is of primary importance, considering the potential hazards involved in lifting the engine out of the vehicle.

If the engine is being removed by a novice, a helper should be available. Advice and aid from someone more experienced would also be helpful. There are many instances when one person cannot simultaneously perform all of the operations required when lifting the engine out of the vehicle.

Plan the operation ahead of time. Arrange for or obtain all of the tools and equipment you'll need prior to beginning the job. Some of the equipment necessary to perform engine removal and installation safely and with relative ease are (in addition to an engine hoist) a heavy duty floor jack, complete sets of wrenches and sockets as described in the front of this manual, wooden blocks and plenty of rags and cleaning solvent for mopping up spilled oil, coolant and gasoline. If the hoist must be rented, make sure that you arrange for it in advance and perform all of the operations possible without it beforehand. This will save you money and time.

Plan for the vehicle to be out of use for quite a while. A machine shop will be required to perform some of the work which the do-it-yourselfer can't accomplish without special equipment. These shops often have a busy schedule, so it would be a good idea to consult them before removing the engine in order to accurately estimate the amount of time required to rebuild or repair components that may need work.

Always be extremely careful when removing and installing the engine.

Serious injury can result from careless actions. Plan ahead, take your time and a job of this nature, although major, can be accomplished successfully.

6 Engine – removal and installation

Refer to illustrations 6.7, 6.20, 6.23, 6.24 and 6.26
Note: *Read through the entire Section before beginning this procedure. The engine and transaxle are removed as a unit and then separated outside the vehicle.*

Removal

1 If the vehicle is equipped with air conditioning, have the system discharged by a dealer service department or a service station.
2 Place protective covers on the front fenders.
3 Remove the hood (see Chapter 11).
4 Disconnect and remove the battery (see Chapter 5).
5 Remove the air cleaner assembly (see Chapter 4). On vehicles equipped with a turbocharger, remove the turbocharger intake hose (see Chapter 4).
6 Drain and remove the radiator (see Chapter 3). **Note:** *On vehicles equipped with an automatic transaxle, remove the cooler hoses from the radiator.*

Chapter 2 Part B General engine overhaul procedures

6.26 Slowly lift the engine up while pushing the transaxle down

Chapter 2, Part A) and the engine rear plate and mount the engine on a stand.

Installation

31 Check the engine/transaxle mounts. If they're worn or damaged, replace them.
32 On manual transaxle equipped vehicles, inspect the clutch components (see Chapter 8) and apply a very small amount of high temperature grease to the transaxle input shaft splines.
33 On automatic transaxle equipped vehicles, inspect the converter seal and bushing.
34 Carefully rejoin the transaxle and engine following the procedure outlined in Chapter 7. **Caution:** *Do not use the bolts to force the engine and transaxle into alignment. It may crack or damage major components.*
35 Install the transaxle-to-engine bolts and tighten them securely.
36 Attach the hoist to the engine and carefully lower the engine/transaxle assembly into the vehicle.
37 Install the mount bolts and tighten them securely.
38 Reinstall the remaining components and fasteners in the reverse order of removal.
39 Add coolant, oil, power steering and transmission fluid/lubricant as needed (see Chapter 1).
40 Run the engine and check for proper operation and leaks. Shut off the engine and recheck the fluid levels.

7 **Engine rebuilding alternatives**

The do-it-yourselfer is faced with a number of options when performing an engine overhaul. The decision to replace the engine block, piston/connecting rod assemblies and crankshaft depends on a number of factors, with the number one consideration being the condition of the block. Other considerations are cost, access to machine shop facilities, parts availability, time required to complete the project and the extent of prior mechanical experience on the part of the do-it-yourselfer.

Some of the rebuilding alternatives include:

Individual parts – If the inspection procedures reveal that the engine block and most engine components are in reusable condition, purchasing individual parts may be the most economical alternative. The block, crankshaft and piston/connecting rod assemblies should all be inspected carefully. Even if the block shows little wear, the cylinder bores should be surface honed.

Short block – A short block consists of an engine block with a crankshaft and piston/connecting rod assemblies already installed. All new bearings are incorporated and all clearances will be correct. The existing camshaft, valve train components, cylinder head(s) and external parts can be bolted to the short block with little or no machine shop work necessary.

Long block – A long block consists of a short block plus an oil pump, oil pan, cylinder head(s), rocker arm cover(s), camshaft and valve train components, timing sprockets and chain or gears and timing cover. All components are installed with new bearings, seals and gaskets incorporated throughout. The installation of manifolds and external parts is all that's necessary.

Give careful thought to which alternative is best for you and discuss the situation with local automotive machine shops, auto parts dealers and experienced rebuilders before ordering or purchasing replacement parts.

7 Carefully label, then disconnect all vacuum lines, coolant and emissions hoses and wire harness connectors. Masking tape and felt-tip pens work well for marking items **(see illustration)**. If necessary, take instant photos or sketch the locations to ensure correct reinstallation.
8 Disconnect the fuel lines from the carburetor or fuel injection system (see Chapter 4) and cap them to prevent leakage.
9 Detach the throttle cable (see Chapter 4).
10 Raise the vehicle and support it securely on jackstands.
11 On manual transaxle-equipped models, refer to Chapters 7 and 8 and detach the clutch cable and shift cables or shift control rod and extension rod. **Note:** *On models with a hydraulic release system, disconnect the slave cylinder without disconnecting the hydraulic line (see Chapter 8).*
12 On automatic transaxle-equipped vehicles, detach the shift control cable from the transaxle, then remove the bellhousing cover and the driveplate-to-converter bolts (see Chapter 7).
13 Detach the speedometer cable from the transaxle.
14 If equipped, remove the air conditioning compressor (see Chapter 3).
15 Remove the power steering pump and reservoir (if equipped) from the brackets without disconnecting the hoses and set them aside.
16 Remove the splash shield (if not already done) located at the drivebelt end of the engine.
17 Drain the engine oil and transaxle fluid and remove the oil filter (see Chapter 1).
18 Disconnect the exhaust pipe from the exhaust manifold (see Chapter 2, Part A).
19 Remove the driveaxles from the transaxle (see Chapter 8). Stuff clean rags into the openings to prevent the entry of foreign material.
20 Attach a short chain to the engine brackets and hook up the hoist **(see illustration)**. Take up the slack until there is tension on the chain.
21 Support the transaxle with a floor jack. Place a block of wood on the jack pad to protect the transaxle. Warning: Do not place any part of your body under the engine/transaxle when it's supported only by a hoist or other lifting device.
22 Remove the plugs from the right front inner fender.
23 Check for clearance and, if necessary, remove the brake master cylinder (see Chapter 9) and transaxle mount bracket **(see illustration)**.
24 Remove the mount through-bolts **(see illustration)**.
25 Confirm that all of the cables, hoses, wires and other items are disconnected from the engine.
26 Carefully push the transaxle down while lifting the engine up to clear obstructions **(see illustration)**.
27 Lift the engine and transaxle high enough to clear the front of the vehicle and slowly move the hoist away.
28 Lower the hoist and set the transaxle on blocks – leave the hoist hooked up.
29 Remove the bolts and separate the engine from the transaxle. Refer to Chapter 7 if necessary.
30 Remove the clutch components and flywheel (or driveplate) (see

8 **Engine overhaul – disassembly sequence**

Refer to illustrations 8.3a, 8.3b, 8.3c, 8.3d, 8.3e, 8.3f and 8.5

1 It's much easier to disassemble and work on the engine if it's mounted on a portable engine stand. A stand can often be rented quite cheaply from an equipment rental yard. Before the engine is mounted on a stand, the flywheel/driveplate should be removed from the engine.
2 If a stand isn't available, it's possible to disassemble the engine with it blocked up on the floor. Be extra careful not to tip or drop the engine when working without a stand.

8.3a 1.5L engine – drivebelt/timing belt end

8.3b 1.5L engine – intake manifold side

Chapter 2 Part B General engine overhaul procedures

8.3c 1.5L engine – exhaust manifold side

8.3d 2.0L SOHC engine – drivebelt/timing belt end

8.3e 2.0L SOHC engine – intake manifold side

3 If you're going to obtain a rebuilt engine, all external components **(see illustrations)** must come off first, to be transferred to the replacement engine, just as they will if you're doing a complete engine overhaul yourself. These include:

    Alternator and brackets
    Emissions control components
    Distributor, spark plug wires and spark plugs
    Thermostat and housing cover
    Water pump
    Carburetor or fuel injection components
    Intake/exhaust manifolds
    Oil filter

    Engine mounts
    Clutch and flywheel/driveplate
    Engine rear plate

Note: *When removing the external components from the engine, pay close attention to details that may be helpful or important during installation. Note the installed position of gaskets, seals, spacers, pins, brackets, washers, bolts and other small items.*

4 If you're obtaining a short block, which consists of the engine block, crankshaft, pistons and connecting rods all assembled, then the cylinder head, oil pan and oil pump will have to be removed as well. See Engine rebuilding alternatives for additional information regarding the different possibilities to be considered.

8.3f 2.0L SOHC engine – exhaust manifold side

5   If you're planning a complete overhaul, the engine must be disassembled and the internal components (**see illustration**) removed in the following order:

Valve cover
Intake and exhaust manifolds
Rocker arms and shafts
Timing belt cover
Timing belt and sprockets
Camshaft
Cylinder head
Oil pan
Oil pump
Rear main oil seal housing
Piston/connecting rod assemblies
Crankshaft and main bearings

6   Before beginning the disassembly and overhaul procedures, make sure the following items are available. Also, refer to Engine overhaul – reassembly sequence for a list of tools and materials needed for engine reassembly.

Common hand tools
Small cardboard boxes or plastic bags for storing parts
Gasket scraper
Ridge reamer
Micrometers
Telescoping gauges
Dial indicator set
Valve spring compressor
Cylinder surfacing hone
Piston ring groove cleaning tool
Electric drill motor
Tap and die set
Wire brushes
Oil gallery brushes
Cleaning solvent

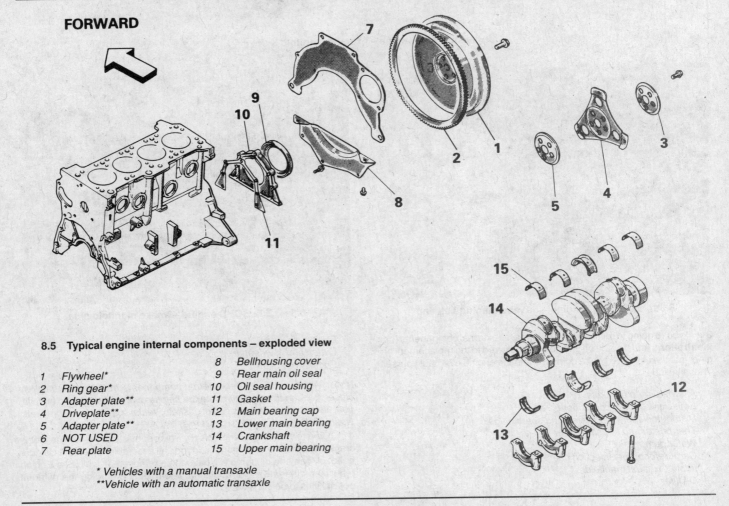

8.5 Typical engine internal components – exploded view

1	Flywheel*	8	Bellhousing cover
2	Ring gear*	9	Rear main oil seal
3	Adapter plate**	10	Oil seal housing
4	Driveplate**	11	Gasket
5	Adapter plate**	12	Main bearing cap
6	NOT USED	13	Lower main bearing
7	Rear plate	14	Crankshaft
		15	Upper main bearing

* Vehicles with a manual transaxle
** Vehicle with an automatic transaxle

Chapter 2 Part B General engine overhaul procedures

9.2 A small plastic bag, with an appropriate label, can be used to store the valve train components so they can be kept together and reinstalled in the original location

9.3 Use a valve spring compressor to compress the spring, then remove the keepers from the valve stem

9.4 If the valve won't pull through the guide, deburr the edge of the stem end and the area around the top of the keeper groove with a file

9.6 Unscrew the jet valves with Mitsubishi special tool no. MD998310 or a deep socket

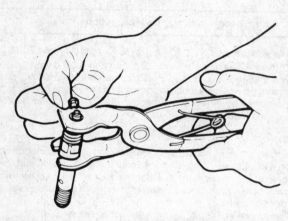

9.7 The jet valve spring should be compressed with special tool no. MD998309

9 Cylinder head – disassembly

Refer to illustrations 9.2, 9.3, 9.4, 9.6 and 9.7

Note: *New and rebuilt cylinder heads are commonly available for most engines at dealerships and auto parts stores. Due to the fact that some specialized tools are necessary for the disassembly and inspection procedures, and replacement parts may not be readily available, it may be more practical and economical for the home mechanic to purchase a replacement head rather than taking the time to disassemble, inspect and recondition the original.*

1 Cylinder head disassembly involves removal of the intake and exhaust valves and related components. If they're still in place, remove the rocker arm shafts. Label the parts or store them separately so they can be reinstalled in their original locations.

2 Before the valves are removed, arrange to label and store them, along with their related components, so they can be kept separate and reinstalled in the same valve guides they are removed from **(see illustration)**.

3 Compress the springs on the first valve with a spring compressor and remove the keepers **(see illustration)**. Carefully release the valve spring compressor and remove the retainer, the spring and the spring seat (if used).

4 Pull the valve out of the head, then remove the oil seal from the guide. If the valve binds in the guide (won't pull through), push it back into the head and deburr the area around the keeper groove with a fine file or whetstone **(see illustration)**.

5 Repeat the procedure for the remaining valves. Remember to keep all the parts for each valve together so they can be reinstalled in the same locations.

6 If equipped, remove the jet valves **(see illustration)**. **Caution:** *When the jet valve socket is used, make certain that the wrench is not tilted with respect to the center of the jet valve. If the tool is tilted, the valve stem might be bent by the force exerted on the valve spring retainer, resulting in defective jet valve operation or a damaged tool.*

7 When disassembling the jet valve, compress the spring with the Mitsubishi special tool (MD998309) **(see illustration)**, remove the valve spring retainer lock, the retainer and the spring. **Caution:** *Do not mix up the jet valve parts after disassembly or gas leakage and malfunctioning may result.*

8 Pull off the valve stem seals with pliers and discard them.

9 Once the valves and related components have been removed and stored in an organized manner, the head should be thoroughly cleaned and inspected. If a complete engine overhaul is being done, finish the engine disassembly procedures before beginning the cylinder head cleaning and inspection process.

2B–16 Chapter 2 Part B General engine overhaul procedures

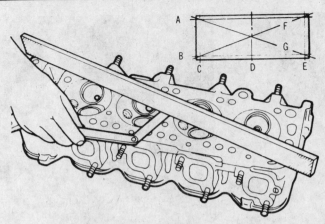

10.12 Check the cylinder head gasket surface for warpage by trying to slip a feeler gauge under the straightedge (see this Chapter's Specifications for the maximum warpage allowed and use a feeler gauge of that thickness)

10.14 A dial indicator can be used to determine the valve stem-to-guide clearance (move the valve stem as indicated by the arrows)

10 Cylinder head – cleaning and inspection

Refer to illustrations 10.12, 10.14, 10.15, 10.16, 10.17 and 10.18

1   Thorough cleaning of the cylinder head and related valve train components, followed by a detailed inspection, will enable you to decide how much valve service work must be done during the engine overhaul. **Note:** *If the engine was severely overheated, the cylinder head is probably warped (see Step 12).*

Cleaning

2   Scrape all traces of old gasket material and sealing compound off the head gasket, intake manifold and exhaust manifold sealing surfaces. Be very careful not to gouge the cylinder head. Special gasket removal solvents that soften gaskets and make removal much easier are available at auto parts stores.
3   Remove all built-up scale from the coolant passages.
4   Run a stiff wire brush through the various holes to remove deposits that may have formed in them.
5   Run an appropriate size tap into each of the threaded holes to remove corrosion and thread sealant that may be present. If compressed air is available, use it to clear the holes of debris produced by this operation. Warning: Wear eye protection when using compressed air!
6   Clean the valve adjuster threads in each rocker arm with a wire brush.
7   Clean the cylinder head with solvent and dry it thoroughly. Compressed air will speed the drying process and ensure that all holes and recessed areas are clean. **Note:** *Decarbonizing chemicals are available and may prove very useful when cleaning cylinder heads and valve train components. They are very caustic and should be used with caution. Be sure to follow the instructions on the container.*
8   Clean the rocker arms, springs, wave washers and shafts with solvent and dry them thoroughly (don't mix them up during the cleaning process). Compressed air will speed the drying process and can be used to clean out the oil passages.
9   Clean all the valve springs, spring seats, keepers and retainers with solvent and dry them thoroughly. Do the components from one valve at a time to avoid mixing up the parts.
10   Scrape off any heavy deposits that may have formed on the valves, then use a motorized wire brush to remove deposits from the valve heads and stems. Warning: Wear eye protection! Again, make sure the valves don't get mixed up.

Inspection

Note: *Be sure to perform all of the following inspection procedures before concluding that machine shop work is required. Make a list of the items that need attention.*

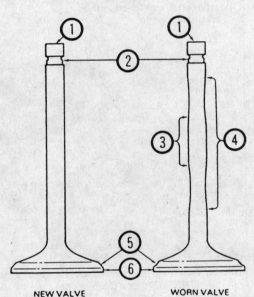

10.15 Check for valve wear at the points shown here

1	Valve tip	4	Stem (most worn area)
2	Keeper groove	5	Valve face
3	Stem (least worn area)	6	Margin

Cylinder head

11   Inspect the head very carefully for cracks, evidence of coolant leakage and other damage. If cracks are found, check with an automotive machine shop concerning repair. If repair isn't possible, a new cylinder head should be obtained.
12   Using a straightedge and feeler gauge, check the head gasket mating surface for warpage **(see illustration)**. If the warpage exceeds the limit listed in this Chapter's Specifications, it can be resurfaced at an automotive machine shop.
13   Examine the valve seats in each of the combustion chambers. If they're pitted, cracked or burned, the head will require valve service that's beyond the scope of the home mechanic.
14   Check the valve stem-to-guide clearance by measuring the lateral movement of the valve stem with a dial indicator attached securely to the head **(see illustration)**. The valve must be in the guide and approximately 1/16-inch off the seat. The total valve stem movement indicated by the gauge needle must be divided by two to obtain the actual clearance. After

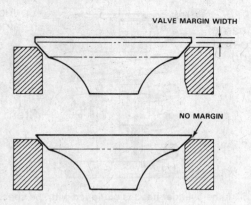

10.16 The margin width on each valve must be as specified (if no margin exists, the valve cannot be reused)

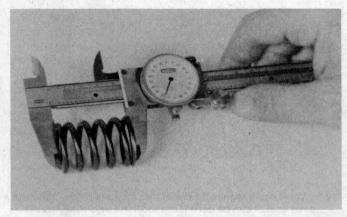

10.17 Measure the free length of each valve spring with a dial or vernier caliper

this is done, if there's still some doubt regarding the condition of the valve guides they should be checked by an automotive machine shop (the cost should be minimal).

Valves

15 Carefully inspect each valve face for uneven wear, deformation, cracks, pits and burned areas. Check the valve stem for scuffing and galling and the neck for cracks. Rotate the valve and check **(see illustration)** for any obvious indication that it's bent. Look for pits and excessive wear on the end of the stem. The presence of any of these conditions indicates the need for valve service by an automotive machine shop.

16 Measure the margin width on each valve **(see illustration)**. Any valve with a margin narrower than listed in this Chapter's Specifications will have to be replaced with a new one.

Valve components

17 Check each valve spring for wear (on the ends) and pits. Measure the free length and compare it to the Specifications **(see illustration)** listed in this Chapter. Any springs that are shorter than specified have sagged and should not be reused. The tension of all springs should be checked with a special fixture before deciding that they're suitable for use in a rebuilt engine (take the springs to an automotive machine shop for this check).

18 Stand each spring on a flat surface and check it for squareness **(see illustration)**. If any of the springs are distorted or sagged, replace all of them with new parts.

19 Check the spring retainers and keepers for obvious wear and cracks. Any questionable parts should be replaced with new ones, as extensive damage will occur if they fail during engine operation.

Rocker arm components

20 Refer to Chapter 2, Part A, for the rocker arm and shaft inspection procedures.

Jet valves

21 Make sure each jet valve slides smoothly in the jet valve body with no play. Do not interchange parts between jet valves. If any parts are worn or damaged, replace the entire jet valve assembly.

22 Check the valve head and valve seat for damage and evidence of seizure. Check the spring for distortion and cracks.

23 Any damaged or excessively worn parts must be replaced with new ones.

24 If the inspection process indicates that the valve components are in generally poor condition and worn beyond the limits specified, which is usually the case in an engine that's being overhauled, reassemble the valves in the cylinder head and refer to Section 10 for valve servicing recommendations.

11 Valves – servicing

1 Because of the complex nature of the job and the special tools and equipment needed, servicing of the valves, the valve seats and the valve

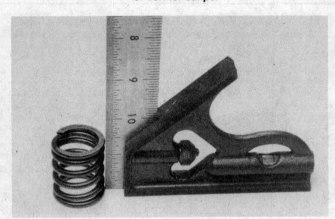

10.18 Check each valve spring for squareness

guides, commonly known as a valve job, should be done by a professional.

2 The home mechanic can remove and disassemble the head, do the initial cleaning and inspection, then reassemble and deliver it to a dealer service department or an automotive machine shop for the actual service work. Doing the inspection will enable you to see what condition the head and valvetrain components are in and will ensure that you know what work and new parts are required when dealing with an automotive machine shop.

3 The dealer service department, or automotive machine shop, will remove the valves and springs, recondition or replace the valves and valve seats, recondition the valve guides, check and replace the valve springs, spring retainers and keepers (as necessary), replace the valve seals with new ones, reassemble the valve components and make sure the installed spring height is correct. The cylinder head gasket surface will also be resurfaced if it's warped.

4 After the valve job has been performed by a professional, the head will be in like new condition. When the head is returned, be sure to clean it again before installation on the engine to remove any metal particles and abrasive grit that may still be present from the valve service or head resurfacing operations. Use compressed air, if available, to blow out all the oil holes and passages.

12 Cylinder head – reassembly

Refer to illustrations 12.3, 12.5, 12.6, 12.8, 12.9 and 12.12

1 Regardless of whether or not the head was sent to an automotive repair shop for valve servicing, make sure it's clean before beginning reassembly.

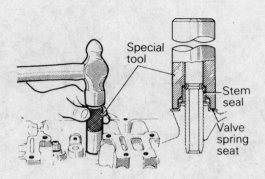

12.3 Valve seals require a special tool for installation (although a deep socket can be used if the tool isn't available) – don't hammer on the seals once they're seated!

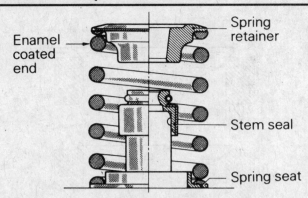

12.5 If the cylinder head is equipped with the special enamel-coated valve springs, install the valve springs with the enamel-coated ends away from the cylinder head, as shown here

12.6 Apply a small dab of grease to each keeper as shown here before installation – it will hold them in place on the valve stem as the spring is released

12.8 Be sure to check the valve spring installed height (the distance from the top of the seat/shims to the top of the spring)

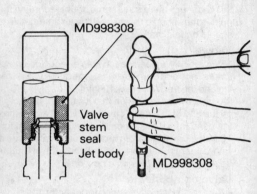

12.9 A special tool is available for jet valve stem seal replacement

2 If the head was sent out for valve servicing, the valves and related components will already be in place. Begin the reassembly procedure with Step 8.

3 Install new seals on each of the intake valve guides. Using a hammer and a deep socket or seal installation tool, gently tap each seal into place until it's completely seated on the guide (**see illustration**). Don't twist or cock the seals during installation or they won't seal properly on the valve stems.

4 Beginning at one end of the head, lubricate and install the first valve. Apply moly-base grease or clean engine oil to the valve stem.

5 Drop the spring seat over the valve guide and set the valve spring and retainer in place (**see illustration**).

6 Compress the spring with a valve spring compressor and carefully install the keepers in the upper groove, then slowly release the compressor and make sure the keepers seat properly. Apply a small dab of grease to each keeper to hold it in place if necessary (**see illustration**).

7 Repeat the procedure for the remaining valves. Be sure to return the components to their original locations – don't mix them up!

8 Check the installed valve spring height with a ruler graduated in 1/32-inch increments or a dial caliper. If the head was sent out for service work, the installed height should be correct (but don't automatically assume that it is). The measurement is taken from the top of each spring seat to the bottom of the retainer (**see illustration**). If the height is greater than listed in this Chapter's Specifications, shims can be added under the springs to correct it. **Caution:** *Don't, under any circumstances, shim the springs to the point where the installed height is less than specified.*

9 Using special tool number MD998308, drive the jet valve stem seal into place on the valve body (**see illustration**). Do not reuse the old seals and don't try to install the seals with any other type of tool.

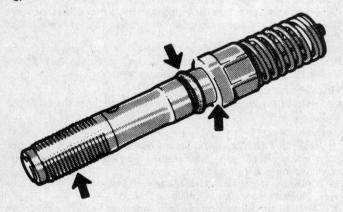

12.12 Be sure to use new O-rings when reinstalling the jet valves and apply engine oil to the O-rings, threads and seat area (arrows)

10 Apply engine oil to the jet valve stem when installing it in the valve body. Take care not to damage the valve stem seal lip. Make sure the jet valve stem slides smoothly in the body.

11 Compress the spring with special tool number MD998309 and install it together with the valve spring retainer. Install the retainer lock. Be careful not to damage the valve stem seal with the bottom of the retainer.

Chapter 2 Part B General engine overhaul procedures

13.1 A ridge reamer is required to remove the ridge from the top of each cylinder – do this before removing the pistons!

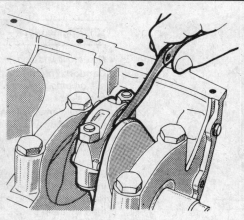

13.3 Check the connecting rod endplay with a feeler gauge as shown

13.6 To prevent damage to the crankshaft journals and cylinder walls, slip sections of hose over the rod bolts before removing the pistons

12 Install a new O-ring in the jet valve body groove and apply a thin coat of engine oil to it **(see illustration)**. Apply oil to the threads and seat as well.
13 Reinstall the jet valves and tighten them to the specified torque.
14 Apply moly-base grease to the rocker arm faces, the camshaft and the rocker shafts, then install the camshaft, rocker arms and shafts (refer to Part A).

13 Pistons/connecting rods – removal

Refer to illustrations 13.1, 13.3 and 13.6
Note: *Prior to removing the piston/connecting rod assemblies, remove the cylinder head and the oil pan by referring to the appropriate Sections in Chapter 2, Part A.*

1 Use your fingernail to feel if a ridge has formed at the upper limit of ring travel (about 1/4-inch down from the top of each cylinder). If carbon deposits or cylinder wear have produced ridges, they must be completely removed with a special tool **(see illustration)**. Follow the manufacturer's instructions provided with the tool. Failure to remove the ridges before attempting to remove the piston/connecting rod assemblies may result in piston breakage.
2 After the cylinder ridges have been removed, turn the engine upside-down so the crankshaft is facing up.
3 Before the connecting rods are removed, check the endplay with feeler gauges. Slide them between the first connecting rod and the crankshaft throw until the play is removed **(see illustration)**. The endplay is equal to the thickness of the feeler gauge(s). If the endplay exceeds the service limit, new connecting rods will be required. If new rods (or a new crankshaft) are installed, the endplay may fall under the minimum listed in this Chapter's Specifications (if it does, the rods will have to be machined to restore it – consult an automotive machine shop for advice if necessary). Repeat the procedure for the remaining connecting rods.
4 Check the connecting rods and caps for identification marks. If they aren't plainly marked, use a small centerpunch to make the appropriate number of indentations on each rod and cap (1, 2, 3, etc., depending on the cylinder they're associated with).
5 Loosen each of the connecting rod cap nuts 1/2-turn at a time until they can be removed by hand. Remove the number one connecting rod cap and bearing insert. Don't drop the bearing insert out of the cap.
6 Slip a short length of plastic or rubber hose over each connecting rod cap bolt to protect the crankshaft journal and cylinder wall as the piston is removed **(see illustration)**.
7 Remove the bearing insert and push the connecting rod/piston assembly out through the top of the engine. Use a wooden hammer handle to push on the upper bearing surface in the connecting rod. If resistance is felt, double-check to make sure that all of the ridge was removed from the cylinder.
8 Repeat the procedure for the remaining cylinders.
9 After removal, reassemble the connecting rod caps and bearing inserts in their respective connecting rods and install the cap nuts finger tight. Leaving the old bearing inserts in place until reassembly will help prevent the connecting rod bearing surfaces from being accidentally nicked or gouged.
10 Don't separate the pistons from the connecting rods (see Section 18 for additional information).

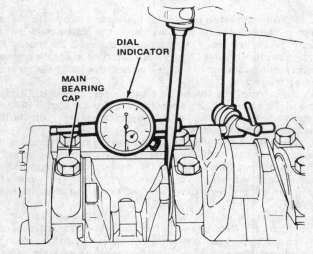

14.1 Checking crankshaft endplay with a dial indicator

14 Crankshaft – removal

Refer to illustrations 14.1 and 14.4
Note: *The crankshaft can be removed only after the engine has been removed from the vehicle. It's assumed that the flywheel or driveplate, crankshaft pulley, timing belt, oil pan, oil pump and piston/connecting rod assemblies have already been removed. The rear main oil seal housing must be unbolted and separated from the block before proceeding with crankshaft removal.*

1 Before the crankshaft is removed, check the endplay. Mount a dial indicator with the stem in line with the crankshaft and just touching one of the crank throws **(see illustration)**.

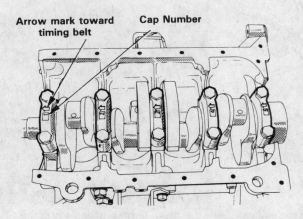

14.4 The arrow on the main bearing cap indicates the front (timing belt end) of the engine

2 Push the crankshaft all the way to the rear and zero the dial indicator. Next, pry the crankshaft to the front as far as possible and check the reading on the dial indicator. The distance that it moves is the endplay. If it's greater than listed in this Chapter's Specifications, check the crankshaft thrust surfaces for wear. If no wear is evident, new main bearings should correct the endplay.

3 If a dial indicator isn't available, feeler gauges can be used. Gently pry or push the crankshaft all the way to the front of the engine. Slip feeler gauges between the crankshaft and the front face of the thrust main bearing to determine the clearance.

4 Check the main bearing caps to see if they're marked to indicate their locations. They should be numbered consecutively from the front of the engine to the rear. If they aren't, mark them with number stamping dies or a center-punch. Main bearing caps generally have a cast-in arrow, which points to the front of the engine **(see illustration)**. Loosen the main bearing cap bolts 1/4-turn at a time each, until they can be removed by hand.

5 Gently tap the caps with a soft-face hammer, then separate them from the engine block. If necessary, use the bolts as levers to remove the caps. Try not to drop the bearing inserts if they come out with the caps.

6 Carefully lift the crankshaft out of the engine. It may be a good idea to have an assistant available, since the crankshaft is quite heavy. With the bearing inserts in place in the engine block and main bearing caps, return the caps to their respective locations on the engine block and tighten the bolts finger tight.

15 Engine block – cleaning

Refer to illustrations 15.1, 15.8 and 15.10
Caution: *The core plugs (also known as freeze or soft plugs) may be difficult or impossible to retrieve if they're driven into the block coolant passages.*

1 Drill a small hole in the center of each core plug and pull them out with an auto body type dent puller **(see illustration)**.

2 Using a gasket scraper, remove all traces of gasket material from the engine block. Be very careful not to nick or gouge the gasket sealing surfaces.

3 Remove the main bearing caps and separate the bearing inserts from the caps and the engine block. Tag the bearings, indicating which cylinder they were removed from and whether they were in the cap or the block, then set them aside.

4 Remove all of the threaded oil gallery plugs from the block. The plugs are usually very tight – they may have to be drilled out and the holes re-tapped. Use new plugs when the engine is reassembled.

5 If the engine is extremely dirty it should be taken to an automotive machine shop to be steam cleaned or hot tanked.

6 After the block is returned, clean all oil holes and oil galleries one more time. Brushes specifically designed for this purpose are available at most auto parts stores. Flush the passages with warm water until the water runs clear, dry the block thoroughly and wipe all machined surfaces with a light, rust preventive oil. If you have access to compressed air, use it to speed the drying process and to blow out all the oil holes and galleries. **Warning:** *Wear eye protection when using compressed air!*

7 If the block isn't extremely dirty or sludged up, you can do an adequate cleaning job with hot soapy water and a stiff brush. Take plenty of time and do a thorough job. Regardless of the cleaning method used, be sure to clean all oil holes and galleries very thoroughly, dry the block completely and coat all machined surfaces with light oil.

8 The threaded holes in the block must be clean to ensure accurate torque readings during reassembly. Run the proper size tap into each of the holes to remove rust, corrosion, thread sealant or sludge and restore damaged threads **(see illustration)**. If possible, use compressed air to clear the holes of debris produced by this operation. Now is a good time to clean the threads on the head bolts and the main bearing cap bolts as well.

9 Reinstall the main bearing caps and tighten the bolts finger tight.

10 After coating the sealing surfaces of the new core plugs with Permatex no. 2 sealant, install them in the engine block **(see illustration)**. Make sure they're driven in straight and seated properly or leakage could result. Special tools are available for this purpose, but a large socket, with an outside diameter that will just slip into the core plug, a 1/2-inch drive extension and a hammer will work just as well.

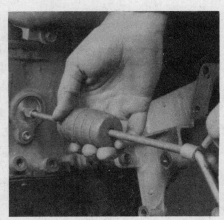

15.1 The core plugs should be removed with a puller – if they're driven into the block, they may be impossible to retrieve

15.8 All bolt holes in the block – particularly the main bearing cap and head bolt holes – should be cleaned and restored with a tap (be sure to remove debris from the holes after this is done)

15.10 A large socket on an extension can be used to drive the new core plugs into the bores

Chapter 2 Part B General engine overhaul procedures   2B – 21

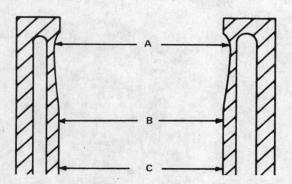

16.4a Measure the diameter of each cylinder just under the wear ridge (A), at the center (B) and at the bottom (C)

11 Apply non-hardening sealant (such as Permatex no. 2 or Teflon pipe sealant) to the new oil gallery plugs and thread them into the holes in the block. Make sure they're tightened securely.
12 If the engine isn't going to be reassembled right away, cover it with a large plastic trash bag to keep it clean.

16 Engine block – inspection

Refer to illustrations 16.4a, 16.4b and 16.4c

1 Before the block is inspected, it should be cleaned as described in Section 15.
2 Visually check the block for cracks, rust and corrosion. Look for stripped threads in the threaded holes. It's also a good idea to have the block checked for hidden cracks by an automotive machine shop that has the special equipment to do this type of work. If defects are found, have the block repaired, if possible, or replaced.
3 Check the cylinder bores for scuffing and scoring.
4 Measure the diameter of each cylinder at the top (just under the ridge area), center and bottom of the cylinder bore, parallel to the crankshaft axis **(see illustrations)**.
5 Next, measure each cylinder's diameter at the same three locations across the crankshaft axis. Compare the results to the Specifications.
6 If the required precision measuring tools aren't available, the piston-to-cylinder clearances can be obtained, though not quite as accurately, using feeler gauge stock. Feeler gauge stock comes in 12-inch lengths and various thicknesses and is generally available at auto parts stores.
7 To check the clearance, select a feeler gauge and slip it into the cylinder along with the matching piston. The piston must be positioned exactly as it normally would be. The feeler gauge must be between the piston and cylinder on one of the thrust faces (90-degrees to the piston pin bore).
8 The piston should slip through the cylinder (with the feeler gauge in place) with moderate pressure.
9 If it falls through or slides through easily, the clearance is excessive and a new piston will be required. If the piston binds at the lower end of the cylinder and is loose toward the top, the cylinder is tapered. If tight spots are encountered as the piston/feeler gauge is rotated in the cylinder, the cylinder is out-of-round.
10 Repeat the procedure for the remaining pistons and cylinders.
11 If the cylinder walls are badly scuffed or scored, or if they're out-of-round or tapered beyond the limits given in the Specifications, have the engine block rebored and honed at an automotive machine shop. If a rebore is done, oversize pistons and rings will be required.
12 If the cylinders are in reasonably good condition and not worn to the outside of the limits, and if the piston-to-cylinder clearances can be maintained properly, then they don't have to be rebored. Honing is all that's necessary (see Section 17).

17 Cylinder honing

Refer to illustrations 17.3a and 17.3b

1 Prior to engine reassembly, the cylinder bores must be honed so the new piston rings will seat correctly and provide the best possible combustion chamber seal. **Note:** *If you don't have the tools or don't want to tackle the honing operation, most automotive machine shops will do it for a reasonable fee.*
2 Before honing the cylinders, install the main bearing caps and tighten the bolts to the torque listed in this Chapter's Specifications.
3 Two types of cylinder hones are commonly available – the flex hone or "bottle brush" type and the more traditional surfacing hone with spring-loaded stones. Both will do the job, but for the less experienced mechanic the "bottle brush" hone will probably be easier to use. You'll also need some kerosene or honing oil, rags and an electric drill motor. Proceed as follows:

  a) Mount the hone in the drill motor, compress the stones and slip it into the first cylinder **(see illustration)**. Be sure to wear safety goggles or a face shield!
  b) Lubricate the cylinder with plenty of honing oil, turn on the drill and move the hone up-and-down in the cylinder at a pace that will produce a fine crosshatch pattern on the cylinder walls. Ideally, the crosshatch lines should intersect at approximately a 60-degree

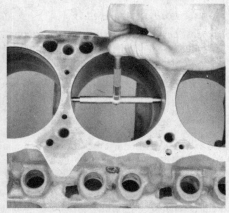

16.4b The ability to "feel" when the telescoping gauge is at the correct point will be developed over time, so work slowly and repeat the check until you're satisfied the bore measurement is accurate

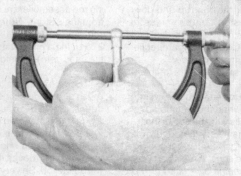

16.4c The gauge is then measured with a micrometer to determine the bore size

17.3a A "bottle brush" hone will produce better results if you've never honed cylinders before

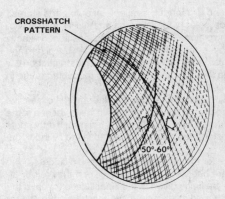

17.3b The cylinder hone should leave a smooth, crosshatch pattern with the lines intersecting at approximately a 60-degree angle

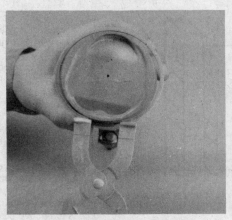

18.2 Use a special tool to remove the piston rings from the piston

18.4a The piston ring grooves can be cleaned with a special tool, as shown here, . . .

18.4b . . . or a section of a broken ring

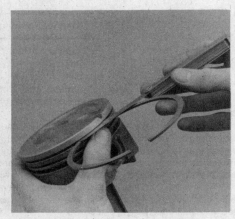

18.10 Check the ring side clearance with a feeler gauge at several points around the groove

18.11 Measure the piston diameter at a 90-degree angle to the piston pin and in line with it

angle **(see illustration)**. Be sure to use plenty of lubricant and don't take off any more material than is absolutely necessary to produce the desired finish. **Note:** *Piston ring manufacturers may specify a smaller crosshatch angle than the traditional 60-degrees – read and follow any instructions included with the new rings.*

c) Don't withdraw the hone from the cylinder while it's running. Instead, shut off the drill and continue moving the hone up-and-down in the cylinder until it comes to a complete stop, then compress the stones and withdraw the hone. If you're using a "bottle brush" type hone, stop the drill motor, then turn the chuck in the normal direction of rotation while withdrawing the hone from the cylinder.

d) Wipe the oil out of the cylinder and repeat the procedure for the remaining cylinders.

4 After the honing job is complete, chamfer the top edges of the cylinder bores with a small file so the rings won't catch when the pistons are installed. Be very careful not to nick the cylinder walls with the end of the file.

5 The entire engine block must be washed again very thoroughly with warm, soapy water to remove all traces of the abrasive grit produced during the honing operation. **Note:** *The bores can be considered clean when a lint-free white cloth – dampened with clean engine oil – used to wipe them out doesn't pick up any more honing residue, which will show up as gray areas on the cloth. Be sure to run a brush through all oil holes and galleries and flush them with running water.*

6 After rinsing, dry the block and apply a coat of light rust preventive oil to all machined surfaces. Wrap the block in a plastic trash bag to keep it clean and set it aside until reassembly.

18 Pistons/connecting rods – inspection

Refer to illustrations 18.2, 18.4a, 18.4b, 18.10 and 18.11

1 Before the inspection process can be carried out, the piston/connecting rod assemblies must be cleaned and the original piston rings removed from the pistons. **Note:** *Always use new piston rings when the engine is reassembled.*

2 Using a piston ring removal tool **(see illustration)**, carefully remove the rings from the pistons. Be careful not to nick or gouge the pistons in the process.

3 Scrape all traces of carbon from the top of the piston. A hand-held wire brush or a piece of fine emery cloth can be used once the majority of the deposits have been scraped away. Do not, under any circumstances, use a wire brush mounted in a drill motor to remove deposits from the pistons. The piston material is soft and may be eroded away by the wire brush.

4 Use a piston ring groove cleaning tool to remove carbon deposits from the ring grooves. If a tool isn't available, a piece broken off the old ring will do the job. Be very careful to remove only the carbon deposits – don't remove any metal and do not nick or scratch the sides of the ring grooves **(see illustrations)**.

5 Once the deposits have been removed, clean the piston/rod assemblies with solvent and dry them with compressed air (if available). Make sure the oil return holes in the back sides of the ring grooves are clear.

6 If the pistons and cylinder walls aren't damaged or worn excessively,

Chapter 2 Part B General engine overhaul procedures

19.3 Rubbing a penny lengthwise on each journal will reveal its condition – if copper rubs off and is embedded in the crankshaft, the journals should be reground

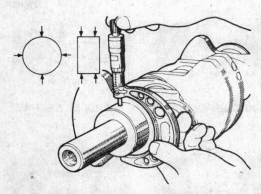

19.6 Measure the diameter of each crankshaft journal at several points to detect taper and out-of-round conditions

and if the engine block is not rebored, new pistons won't be necessary. Normal piston wear appears as even, vertical wear on the piston thrust surfaces and slight looseness of the top ring in its groove. New piston rings, however, should always be used when an engine is rebuilt.

7 Carefully inspect each piston for cracks around the skirt, at the pin bosses and at the ring lands.

8 Look for scoring and scuffing on the thrust faces of the skirt, holes in the piston crown and burned areas at the edge of the crown. If the skirt is scored or scuffed, the engine may have been suffering from overheating and/or abnormal combustion, which caused excessively high operating temperatures. The cooling and lubrication systems should be checked thoroughly. A hole in the piston crown is an indication that abnormal combustion (preignition) was occurring. Burned areas at the edge of the piston crown are usually evidence of spark knock (detonation). If any of the above problems exist, the causes must be corrected or the damage will occur again. The causes may include intake air leaks, incorrect fuel/air mixture, incorrect ignition timing and EGR system malfunctions.

9 Corrosion of the piston, in the form of small pits, indicates that coolant is leaking into the combustion chamber and/or the crankcase. Again, the cause must be corrected or the problem may persist in the rebuilt engine.

10 Measure the piston ring side clearance by laying a new piston ring in each ring groove and slipping a feeler gauge in beside it **(see illustration)**. Check the clearance at three or four locations around each groove. Be sure to use the correct ring for each groove – they are different. If the side clearance is greater than specified, new pistons will have to be used.

11 Check the piston-to-bore clearance by measuring the bore (see Section 16) and the piston diameter. Make sure the pistons and bores are correctly matched. Measure the piston across the skirt 5/64-inch above the bottom of the piston, at a 90-degree angle to and in line with the piston pin **(see illustration)**. Subtract the piston diameter from the bore diameter to obtain the clearance. If it's greater than specified, the block will have to be rebored and new pistons and rings installed.

12 Check the piston-to-rod clearance by twisting the piston and rod in opposite directions. Any noticeable play indicates excessive wear, which must be corrected. The piston/connecting rod assemblies should be taken to an automotive machine shop to have the pistons and rods resized and new pins installed.

13 If the pistons must be removed from the connecting rods for any reason, they should be taken to an automotive machine shop. While they are there have the connecting rods checked for bend and twist, since automotive machine shops have special equipment for this purpose. **Note:** *Unless new pistons and/or connecting rods must be installed, do not disassemble the pistons and connecting rods.*

14 Check the connecting rods for cracks and other damage. Temporarily remove the rod caps, lift out the old bearing inserts, wipe the rod and cap bearing surfaces clean and inspect them for nicks, gouges and scratches. After checking the rods, replace the old bearings, slip the caps into place

and tighten the nuts finger tight. **Note:** *If the engine is being rebuilt because of a connecting rod knock, be sure to install new rods.*

19 Crankshaft – inspection

Refer to illustration 19.3 and 19.6

1 Clean the crankshaft with solvent and dry it with compressed air (if available). Be sure to clean the oil holes with a stiff brush and flush them with solvent.

2 Check the main and connecting rod bearing journals for uneven wear, scoring, pits and cracks.

3 Rub a penny across each journal several times. If a journal picks up copper from the penny, it's too rough and must be reground **(see illustration)**.

4 Remove all burrs from the crankshaft oil holes with a stone, file or scraper.

5 Check the rest of the crankshaft for cracks and other damage. It should be magnafluxed to reveal hidden cracks – an automotive machine shop will handle the procedure.

6 Using a micrometer, measure the diameter of the main and connecting rod journals and compare the results to the Specifications **(see illustration)** listed in this Chapter. By measuring the diameter at a number of points around each journal's circumference, you'll be able to determine whether or not the journal is out-of-round. Take the measurement at each end of the journal, near the crank throws, to determine if the journal is tapered.

7 If the crankshaft journals are damaged, tapered, out-of-round or worn beyond the limits given in the Specifications, have the crankshaft reground by an automotive machine shop. Be sure to use the correct size bearing inserts if the crankshaft is reconditioned.

8 Check the oil seal journals at each end of the crankshaft for wear and damage. If the seal has worn a groove in the journal, or if it's nicked or scratched, the new seal may leak when the engine is reassembled. In some cases, an automotive machine shop may be able to repair the journal by pressing on a thin sleeve. If repair isn't feasible, a new or different crankshaft should be installed.

9 Refer to Section 20 and examine the main and rod bearing inserts.

20 Main and connecting rod bearings – inspection

Refer to illustration 20.1

1 Even though the main and connecting rod bearings should be replaced with new ones during the engine overhaul, the old bearings should

FATIGUE FAILURE

IMPROPER SEATING

SCRATCHED BY DIRT / LACK OF OIL

EXCESSIVE WEAR

TAPERED JOURNAL

20.1 Typical bearing failures

be retained for close examination, as they may reveal valuable information about the condition of the engine (see illustration).
2   Bearing failure occurs because of lack of lubrication, the presence of dirt or other foreign particles, overloading the engine and corrosion. Regardless of the cause of bearing failure, it must be corrected before the engine is reassembled to prevent it from happening again.
3   When examining the bearings, remove them from the engine block, the main bearing caps, the connecting rods and the rod caps and lay them out on a clean surface in the same general position as their location in the engine. This will enable you to match any bearing problems with the corresponding crankshaft journal.
4   Dirt and other foreign particles get into the engine in a variety of ways. It may be left in the engine during assembly, or it may pass through filters or the PCV system. It may get into the oil, and from there into the bearings. Metal chips from machining operations and normal engine wear are often present. Abrasives are sometimes left in engine components after reconditioning, especially when parts are not thoroughly cleaned using the proper cleaning methods. Whatever the source, these foreign objects often end up embedded in the soft bearing material and are easily recognized. Large particles will not embed in the bearing and will score or gouge the bearing and journal. The best prevention for this cause of bearing failure is to clean all parts thoroughly and keep everything spotlessly clean during engine assembly. Frequent and regular engine oil and filter changes are also recommended.
5   Lack of lubrication (or lubrication breakdown) has a number of interrelated causes. Excessive heat (which thins the oil), overloading (which squeezes the oil from the bearing face) and oil leakage or throw off (from excessive bearing clearances, worn oil pump or high engine speeds) all contribute to lubrication breakdown. Blocked oil passages, which usually are the result of misaligned oil holes in a bearing shell, will also oil starve a bearing and destroy it. When lack of lubrication is the cause of bearing failure, the bearing material is wiped or extruded from the steel backing of the bearing. Temperatures may increase to the point where the steel backing turns blue from overheating.
6   Driving habits can have a definite effect on bearing life. Full throttle, low speed operation (lugging the engine) puts very high loads on bearings, which tends to squeeze out the oil film. These loads cause the bearings to flex, which produces fine cracks in the bearing face (fatigue failure). Eventually the bearing material will loosen in pieces and tear away from the steel backing. Short trip driving leads to corrosion of bearings because insufficient engine heat is produced to drive off the condensed water and corrosive gases. These products collect in the engine oil, forming acid and sludge. As the oil is carried to the engine bearings, the acid attacks and corrodes the bearing material.
7   Incorrect bearing installation during engine assembly will lead to bearing failure as well. Tight fitting bearings leave insufficient bearing oil clearance and will result in oil starvation. Dirt or foreign particles trapped behind a bearing insert result in high spots on the bearing which lead to failure.

21 Engine overhaul – reassembly sequence

1   Before beginning engine reassembly, make sure you have all the necessary new parts, gaskets and seals as well as the following items on hand:

   Common hand tools
   A 1/2-inch drive torque wrench
   Piston ring installation tool
   Piston ring compressor
   Short lengths of rubber or plastic hose
   to fit over connecting rod bolts
   Plastigage
   Feeler gauges
   A fine-tooth file
   New engine oil
   Engine assembly lube or moly-base grease
   Gasket sealant
   Thread locking compound

2   In order to save time and avoid problems, engine reassembly must be done in the following general order:

   Piston rings
   Crankshaft and main bearings
   Piston/connecting rod assemblies
   Oil pump
   Rear main oil seal housing
   Oil pan
   Cylinder head, camshaft and rocker arm assembly
   Timing belt and sprockets
   Timing belt cover
   Intake and exhaust manifolds
   Rocker arm cover
   Engine rear plate
   Flywheel/driveplate

22 Piston rings – installation

Refer to illustrations 22.3, 22.4, 22.5, 22.9a, 22.9b, 22.11 and 22.12

1   Before installing the new piston rings, the ring end gaps must be checked. It's assumed that the piston ring side clearance has been checked and verified correct (see Section 18).
2   Lay out the piston/connecting rod assemblies and the new ring sets so the ring sets will be matched with the same piston and cylinder during the end gap measurement and engine assembly.
3   Insert the top (number one) ring into the first cylinder and square it up with the cylinder walls by pushing it in with the top of the piston (see illustration). The ring should be near the bottom of the cylinder, at the lower limit of ring travel.

Chapter 2 Part B General engine overhaul procedures

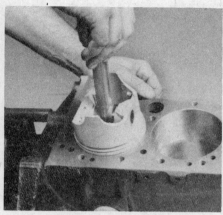

22.3 When checking piston ring end gap, the ring must be square in the cylinder bore (this is done by pushing the ring down with the top of a piston as shown)

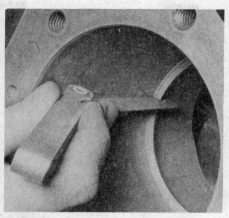

22.4 With the ring square in the cylinder, measure the end gap with a feeler gauge

22.5 If the end gap is too small, clamp a file in a vise and file the ring ends (from the outside in only) to enlarge the gap slightly

22.9a Installing the spacer/expander in the oil control ring groove

22.9b DO NOT use a piston ring installation tool when installing the oil ring side rails

4 To measure the end gap, slip feeler gauges between the ends of the ring until a gauge equal to the gap width is found (see illustration). The feeler gauge should slide between the ring ends with a slight amount of drag. Compare the measurement to the Specifications listed in this Chapter. If the gap is larger or smaller than specified, double-check to make sure you have the correct rings before proceeding.

5 If the gap is too small, it must be enlarged or the ring ends may come in contact with each other during engine operation, which can cause serious damage to the engine. The end gap can be increased by filing the ring ends very carefully with a fine file. Mount the file in a vise equipped with soft jaws, slip the ring over the file with the ends contacting the file face and slowly move the ring to remove material from the ends. When performing this operation, file only from the outside in (see illustration).

6 Excess end gap isn't critical unless it's greater than 0.039-inch. Again, double-check to make sure you have the correct rings for your engine.

7 Repeat the procedure for each ring that will be installed in the first cylinder and for each ring in the remaining cylinders. Remember to keep rings, pistons and cylinders matched up.

8 Once the ring end gaps have been checked/corrected, the rings can be installed on the pistons.

9 The oil control ring (lowest one on the piston) is usually installed first. It's composed of three separate components. Slip the spacer/expander into the groove (see illustration). If an anti-rotation tang is used, make sure it's inserted into the drilled hole in the ring groove. Next, install the lower side rail. Don't use a piston ring installation tool on the oil ring side rails, as they may be damaged. Instead, place one end of the side rail into the groove between the spacer/expander and the ring land, hold it firmly in place and slide a finger around the piston while pushing the rail into the groove (see illustration). Next, install the upper side rail in the same manner.

10 After the three oil ring components have been installed, check to make sure that both the upper and lower side rails can be turned smoothly in the ring groove.

11 The number two (middle) ring is installed next. It's usually stamped with a mark which must face up, toward the top of the piston (see illustration). Note: *On 1.8L Turbo and 2.0L engines (Cordia/Tredia only), the ring thickness varies with the engine. Follow the chart for the correct size.*

Number	Engine	Ring thickness	Identification mark
No.1	1.8L	0.059 in (1.5 mm)	T
	2.0L	0.079 in (2.0 mm)	N
No.2	1.8L	0.059 in (1.5 mm)	2T
	2.0L	0.079 in (2.0 mm)	N

Note: *Always follow the instructions printed on the ring package or box – different manufacturers may require different approaches. Do not mix up*

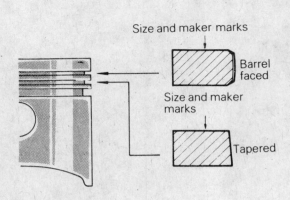

22.11 The no.1 (top) and no. 2 compression rings have different cross-sections – be sure to install them in the correct locations with the marks facing UP

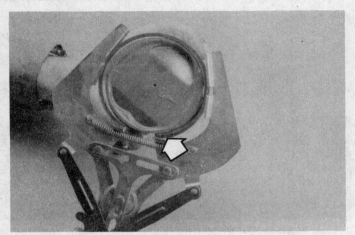

22.12 Installing the compression rings with a ring expander – the mark (arrow) must face up

23.11 Lay the Plastigage strips (arrow) on the main bearing journals, parallel to the crankshaft centerline

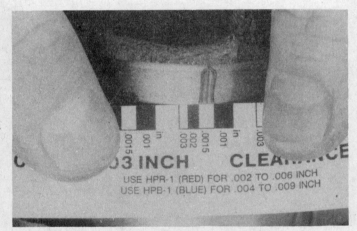

23.15 Measuring the width of the crushed Plastigage to determine the rod bearing oil clearance (be sure to use the correct scale – standard and metric ones are included)

the top and middle rings, as they have different cross-sections.

12 Use a piston ring installation tool and make sure the identification mark is facing the top of the piston, then slip the ring into the middle groove on the piston (see illustration). Don't expand the ring any more than necessary to slide it over the piston.

13 Install the number one (top) ring in the same manner. Make sure the mark is facing up. Be careful not to confuse the number one and number two rings.

14 Repeat the procedure for the remaining pistons and rings.

23 Crankshaft – installation and main bearing oil clearance check

Refer to illustrations 23.11 and 23.15

1 Crankshaft installation is the first step in engine reassembly. It's assumed at this point that the engine block and crankshaft have been cleaned, inspected and repaired or reconditioned.
2 Position the engine with the bottom facing up.
3 Remove the main bearing cap bolts and lift out the caps. Lay them out in the proper order to ensure correct installation.
4 If they're still in place, remove the original bearing inserts from the block and the main bearing caps. Wipe the bearing surfaces of the block and caps with a clean, lint-free cloth. They must be kept spotlessly clean.

Main bearing oil clearance check

5 Clean the back sides of the new main bearing inserts and lay one in each main bearing saddle in the block. If one of the bearing inserts from each set has a large groove in it, make sure the grooved insert is installed in the block. Lay the other bearing from each set in the corresponding main bearing cap. Make sure the tab on the bearing insert fits into the recess in the block or cap. **Caution:** *The oil holes in the block must line up with the oil holes in the bearing insert. Do not hammer the bearing into place and don't nick or gouge the bearing faces. No lubrication should be used at this time.*
6 The flanged thrust bearing must be installed in the center cap and saddle (see illustration 8.5).
7 Clean the faces of the bearings in the block and the crankshaft main bearing journals with a clean, lint-free cloth.
8 Check or clean the oil holes in the crankshaft, as any dirt here can go only one way – straight through the new bearings.
9 Once you're certain the crankshaft is clean, carefully lay it in position in the main bearings.
10 Before the crankshaft can be permanently installed, the main bearing oil clearance must be checked.
11 Cut several pieces of the appropriate size Plastigage (they must be slightly shorter than the width of the main bearings) and place one piece on each crankshaft main bearing journal, parallel with the journal axis (see illustration).

Chapter 2 Part B General engine overhaul procedures

24.3 Support the housing on a couple of wood blocks and drive out the old seal with a punch or screwdriver and hammer

24.5 Drive the new seal into the housing with a block of wood or a section of pipe, if you have one large enough – make sure that you don't cock the seal in the bore

12 Clean the faces of the bearings in the caps and install the caps in their respective positions (don't mix them up) with the arrows pointing toward the front of the engine. Don't disturb the Plastigage.
13 Starting with the center main and working out toward the ends, tighten the main bearing cap bolts, in three steps, to the torque listed in this Chapter's Specifications. Don't rotate the crankshaft at any time during this operation.
14 Remove the bolts and carefully lift off the main bearing caps. Keep them in order. Don't disturb the Plastigage or rotate the crankshaft. If any of the main bearing caps are difficult to remove, tap them gently from side-to-side with a soft-face hammer to loosen them.
15 Compare the width of the crushed Plastigage on each journal to the scale printed on the Plastigage envelope to obtain the main bearing oil clearance (see illustration). Check the Specifications listed in this Chapter to make sure it's correct.
16 If the clearance is not as specified, the bearing inserts may be the wrong size (which means different ones will be required). Before deciding that different inserts are needed, make sure that no dirt or oil was between the bearing inserts and the caps or block when the clearance was measured. If the Plastigage was wider at one end than the other, the journal may be tapered (refer to Section 19).
17 Carefully scrape all traces of the Plastigage material off the main bearing journals and/or the bearing faces. Use your fingernail or the edge of a credit card – don't nick or scratch the bearing faces.

Final crankshaft installation

18 Carefully lift the crankshaft out of the engine.
19 Clean the bearing faces in the block, then apply a thin, uniform layer of moly-base grease or engine assembly lube to each of the bearing surfaces. Be sure to coat the thrust faces as well as the journal face of the thrust bearing.
20 Make sure the crankshaft journals are clean, then lay the crankshaft back in place in the block.
21 Clean the faces of the bearings in the caps, then apply lubricant to them.
22 Install the caps in their respective positions with the arrows pointing toward the front of the engine.
23 Install the bolts.
24 Tighten all except the thrust bearing cap bolts to the torque listed in this Chapter's Specifications (work from the center out and approach the final torque in three steps).
25 Tighten the thrust bearing cap bolts to 10-to-12 ft-lbs.
26 Tap the ends of the crankshaft forward and backward with a lead or brass hammer to line up the main bearing and crankshaft thrust surfaces.
27 Retighten all main bearing cap bolts to the torque listed in this Chapter's Specifications, starting with the center main and working out toward the ends.
28 On manual transmission equipped models, install a new pilot bearing in the end of the crankshaft (see Chapter 8).
29 Rotate the crankshaft a number of times by hand to check for any obvious binding.
30 Recheck the crankshaft endplay with a feeler gauge or a dial indicator as described in Section 14. The endplay should be correct if the crankshaft thrust faces aren't worn or damaged and new bearings have been installed.
31 Refer to Section 24 and install the new rear main oil seal, then bolt the housing to the block.

24 Rear main oil seal installation

Refer to illustrations 24.3 and 24.5

1 The crankshaft must be installed first and the main bearing caps bolted in place, then the new seal should be installed in the housing and the housing bolted to the block.
2 Check the seal contact surface on the crankshaft very carefully for scratches and nicks that could damage the new seal lip and cause oil leaks. If the crankshaft is damaged, the only alternative is a new or different crankshaft.
3 The old seal can be removed from the housing by driving it out from the back side with a hammer and punch (see illustration). Be sure to note how far it's recessed into the bore before removing it; the new seal will have to be recessed an equal amount. Be very careful not to scratch or otherwise damage the bore in the housing or oil leaks could develop.
4 Make sure the housing is clean, then apply a thin coat of engine oil to the outer edge of the new seal. The seal must be pressed squarely into the bore, so hammering it into place isn't recommended. If you don't have access to a press, sandwich the housing and seal between two smooth pieces of wood and press the seal into place with the jaws of a large vise. The pieces of wood must be thick enough to distribute the force evenly around the entire circumference of the seal. Work slowly and make sure the seal enters the bore squarely.
5 As a last resort, the seal can be tapped into the housing with a hammer. Use a block of wood to distribute the force evenly and make sure the seal is driven in squarely (see illustration).
6 The seal lips must be lubricated with clean engine oil or moly-based grease before the seal/housing is slipped over the crankshaft and bolted to the block. Use a new gasket and sealant.
7 Tighten the bolts a little at a time until they're all secure.
8 If the engine is equipped with an oil separator, be sure to install it with the oil hole at the bottom of the oil seal housing.

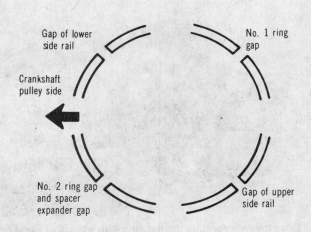

25.5 Position the ring gaps as shown here before installing the piston/connecting rod assemblies in the engine

25.9 The front mark must point towards the timing belt end of the engine when the pistons are installed

25.11 The piston can be driven (gently) into the cylinder bore with the end of a wooden or plastic hammer handle

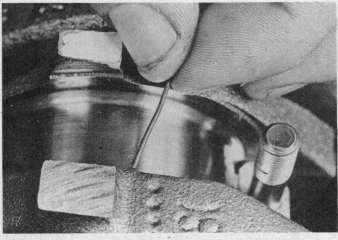

25.13 Lay the Plastigage strips on each rod bearing journal, parallel to the crankshaft centerline

Pistons/connecting rods – installation and rod bearing oil clearance check

Refer to illustrations 25.5, 25.9, 25.11, 25.13, 25.14 and 25.17

1 Before installing the piston/connecting rod assemblies, the cylinder walls must be perfectly clean, the top edge of each cylinder must be chamfered, and the crankshaft must be in place.

2 Remove the cap from the end of the number one connecting rod (refer to the marks made during removal). Remove the original bearing inserts and wipe the bearing surfaces of the connecting rod and cap with a clean, lint-free cloth. They must be kept spotlessly clean.

Connecting rod bearing oil clearance check

3 Clean the back side of the new upper bearing insert, then lay it in place in the connecting rod. Make sure the tab on the bearing fits into the recess in the rod. Don't hammer the bearing insert into place and be very careful not to nick or gouge the bearing face. Don't lubricate the bearing at this time.

4 Clean the back side of the other bearing insert and install it in the rod cap. Again, make sure the tab on the bearing fits into the recess in the cap, and don't apply any lubricant. It's critically important that the mating surfaces of the bearing and connecting rod are perfectly clean and oil free when they're assembled.

5 Position the piston ring gaps at 90-degree intervals around the piston **(see illustration)**.

6 Slip a section of plastic or rubber hose over each connecting rod cap bolt.

7 Lubricate the piston and rings with clean engine oil and attach a piston ring compressor to the piston. Leave the skirt protruding about 1/4-inch to guide the piston into the cylinder. The rings must be compressed until they're flush with the piston.

8 Rotate the crankshaft until the number one connecting rod journal is at BDC (bottom dead center) and apply a coat of engine oil to the cylinder walls.

9 With the mark on top of the piston **(see illustration)** facing the front (timing belt end) of the engine, gently insert the piston/connecting rod assembly into the number one cylinder bore and rest the bottom edge of the ring compressor on the engine block.

Chapter 2 Part B General engine overhaul procedures

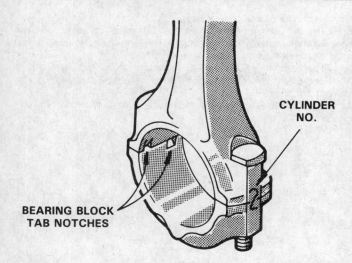

25.14 Match up the cylinder number marks and the bearing tab notches when installing the connecting rod caps

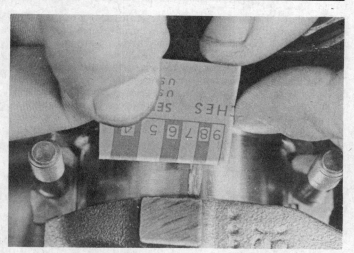

25.17 Measuring the width of the crushed Plastigage to determine the rod bearing oil clearance (be sure to use the correct scale – standard and metric ones are included)

10 Tap the top edge of the ring compressor to make sure it's contacting the block around its entire circumference.

11 Gently tap on the top of the piston with the end of a wooden or plastic hammer handle **(see illustration)** while guiding the end of the connecting rod into place on the crankshaft journal. The piston rings may try to pop out of the ring compressor just before entering the cylinder bore, so keep some downward pressure on the ring compressor. Work slowly, and if any resistance is felt as the piston enters the cylinder, stop immediately. Find out what's hanging up and fix it before proceeding. Do not, for any reason, force the piston into the cylinder – you might break a ring and/or the piston.

12 Once the piston/connecting rod assembly is installed, the connecting rod bearing oil clearance must be checked before the rod cap is permanently bolted in place.

13 Cut a piece of the appropriate size Plastigage slightly shorter than the width of the connecting rod bearing and lay it in place on the number one connecting rod journal, parallel with the journal axis **(see illustration)**.

14 Clean the connecting rod cap bearing face, remove the protective hoses from the connecting rod bolts and install the rod cap. Make sure the mating mark on the cap is on the same side as the mark on the connecting rod **(see illustration)**. **Note:** *Check to make sure the identification mark on the connecting rod faces toward the front (timing belt) end of the engine.*

15 Install the nuts and tighten them to the torque listed in this Chapter's Specifications, working up to it in three steps. **Note:** *Use a thin-wall socket to avoid erroneous torque readings that can result if the socket is wedged between the rod cap and nut. If the socket tends to wedge itself between the nut and the cap, lift up on it slightly until it no longer contacts the cap.* Do not rotate the crankshaft at any time during this operation.

16 Remove the nuts and detach the rod cap, being very careful not to disturb the Plastigage.

17 Compare the width of the crushed Plastigage to the scale printed on the Plastigage envelope to obtain the oil clearance **(see illustration)**. Compare it to the Specifications (listed in this Chapter) to make sure the clearance is correct.

18 If the clearance is not as specified, the bearing inserts may be the wrong size (which means different ones will be required). Before deciding that different inserts are needed, make sure that no dirt or oil was between the bearing inserts and the connecting rod or cap when the clearance was measured. Also, recheck the journal diameter. If the Plastigage was wider at one end than the other, the journal may be tapered (refer to Section 19).

Final connecting rod installation

19 Carefully scrape all traces of the Plastigage material off the rod journal and/or bearing face. Be very careful not to scratch the bearing – use your fingernail or the edge of a credit card.

20 Make sure the bearing faces are perfectly clean, then apply a uniform layer of clean moly-base grease or engine assembly lube to both of them. You'll have to push the piston into the cylinder to expose the face of the bearing insert in the connecting rod – be sure to slip the protective hoses over the rod bolts first.

21 Slide the connecting rod back into place on the journal, remove the protective hoses from the rod cap bolts, install the rod cap and tighten the nuts to the torque listed in this Chapter's Specifications. Again, work up to the torque in three steps.

22 Repeat the entire procedure for the remaining pistons/connecting rods.

23 The important points to remember are . . .
  a) Keep the back sides of the bearing inserts and the insides of the connecting rods and caps perfectly clean when assembling them.
  b) Make sure you have the correct piston/rod assembly for each cylinder.
  c) The mark on the piston must face the front of the engine.
  d) Lubricate the cylinder walls with clean oil.
  e) Lubricate the bearing faces when installing the rod caps after the oil clearance has been checked.

24 After all the piston/connecting rod assemblies have been properly installed, rotate the crankshaft a number of times by hand to check for any obvious binding.

25 As a final step, the connecting rod endplay must be checked. Refer to Section 13 for this procedure.

26 Compare the measured endplay to the Specifications to make sure it's correct. If it was correct before disassembly and the original crankshaft and rods were reinstalled, it should still be right. If new rods or a new crankshaft were installed, the endplay may be inadequate. If so, the rods will have to be removed and taken to an automotive machine shop for resizing.

26 Initial start-up and break-in after overhaul

Warning: *Have a fire extinguisher handy when starting the engine for the first time.*

1 Once the engine has been installed in the vehicle, double-check the engine oil and coolant levels. Add transaxle fluid as needed.

2 With the spark plugs out of the engine and the ignition system disabled (see Section 3), crank the engine until the oil pressure light goes out.

3 Install the spark plugs, hook up the plug wires and restore the ignition system functions (see Section 3).

4 Start the engine. It may take a few moments for the fuel system to build up pressure, but the engine should start without a great deal of effort. **Note:** *If backfiring occurs through the carburetor, recheck the valve timing and ignition timing.*

5   After the engine starts, it should be allowed to warm up to normal operating temperature. While the engine is warming up, make a thorough check for fuel, oil and coolant leaks. Check the automatic transaxle fluid level (if so equipped).

6   Shut the engine off and recheck the engine oil and coolant levels.

7   Drive the vehicle to an area with minimum traffic, accelerate at full throttle from 30 to 50 mph, then allow the vehicle to slow to 30 mph with the throttle closed. Repeat the procedure 10 or 12 times. This will load the piston rings and cause them to seat properly against the cylinder walls. Check again for oil and coolant leaks.

8   Drive the vehicle gently for the first 500 miles (no sustained high speeds) and keep a constant check on the oil level. It is not unusual for an engine to use oil during the break-in period.

9   At approximately 500 to 600 miles, change the oil and filter.

10   For the next few hundred miles, drive the vehicle normally. Do not pamper it or abuse it.

11   After 2000 miles, change the oil and filter again and consider the engine broken in.

Chapter 3 Cooling, heating and air conditioning systems

Contents

Air conditioning compressor – removal and installation 15	Engine cooling fan – check and replacement 5
Air conditioning condenser – removal and installation 16	General information 1
Air conditioning receiver/drier – removal and installation 14	Heater and air conditioner blower motor – removal
Air conditioning system – check and maintenance 13	and installation 10
Antifreeze – general information 2	Heater and air conditioner control assembly – removal,
Coolant level check .... See Chapter 1	installation and cable adjustment 12
Coolant reservoir – removal and installation 7	Heater core – replacement 11
Coolant temperature sending unit – check and replacement 6	Radiator – removal and installation 4
Cooling system check .... See Chapter 1	Thermostat – check and replacement 3
Cooling system servicing (draining, flushing	Underhood hose check and replacement .... See Chapter 1
and refilling) .... See Chapter 1	Water pump – check 8
Drivebelt check, adjustment and replacement ..... See Chapter 1	Water pump – removal and installation 9

Specifications

General

Thermostat rating (opening temperature)
   1993 Galant SOHC 16-valve engine .... 180-degrees F
   All others .... 190-degrees F
Cooling system capacity .... See Chapter 1
Refrigerant capacity .... 1.9 to 2.3 lbs (approximate)

Torque specifications   Ft-lbs (unless otherwise indicated)

Water pump-to-engine block
   Alternator brace bolt (head mark 7) .... 14 to 20
   All others (head mark 4) .... 9 to 11
Water pump pulley bolts .... 72 to 84 in-lbs
Thermostat housing cover bolts
   1990 and later Precis and Mirage .... 144 to 168 in-lbs
   All others .... 86 to 114 in-lbs

Chapter 3 Cooling, heating and air conditioning systems

3.10 Remove the two thermostat housing cover bolts (arrows) to gain access to the thermostat (Galant 2.0L engine shown)

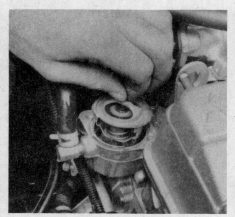

3.13 Make sure the thermostat is installed correctly – if it's installed wrong, the engine will overheat and damage will result (Galant 2.0L shown)

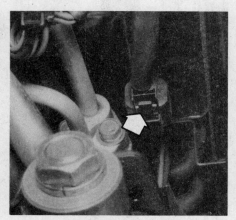

4.5 Press the tab (arrow) to release the connector for the fan motor

1 General information

Engine cooling system

All vehicles covered by this manual employ a pressurized engine cooling system with thermostatically controlled coolant circulation. An impeller type water pump mounted on the front (drivebelt end) of the block pumps coolant through the engine. The coolant flows around each cylinder and toward the rear of the engine. Cast-in coolant passages direct coolant around the intake and exhaust ports, near the spark plug areas and in close proximity to the exhaust valve guides.

A wax pellet type thermostat is located in a housing near the transaxle end of the engine. During warm-up, the closed thermostat prevents coolant from circulating through the radiator. As the engine nears normal operating temperature, the thermostat opens and allows hot coolant to travel through the radiator, where it's cooled before returning to the engine.

The cooling system is sealed by a pressure type radiator cap, which raises the boiling point of the coolant and increases the cooling efficiency of the radiator. If the system pressure exceeds the cap pressure relief valve, the excess pressure in the system forces the spring-loaded valve inside the cap off its seat and allows the coolant to escape through the overflow tube into a coolant reservoir. When the system cools, the excess coolant is automatically drawn from the reservoir back into the radiator.

The coolant reservoir does double duty as both the point at which fresh coolant is added to the cooling system to maintain the proper fluid level and as a holding tank for overheated coolant.

This type of cooling system is known as a closed design because coolant that escapes past the pressure cap is saved and reused.

Heating system

The heating system consists of a blower fan and heater core located in the heater box, the hoses connecting the heater core to the engine cooling system and the heater/air conditioning control head on the dashboard. Hot engine coolant is circulated through the heater core. When the heater mode is activated, a flap door opens to expose the heater box to the passenger compartment. A fan switch on the control head activates the blower motor, which forces air through the core, heating the air.

Air conditioning system

The air conditioning system consists of a condenser mounted in front of the radiator, an evaporator mounted adjacent to the heater core, a compressor mounted on the engine, a filter-drier which contains a high pressure relief valve and the plumbing connecting all of the above components.

A blower fan forces the warmer air of the passenger compartment through the evaporator core (sort of a radiator-in-reverse), transferring the heat from the air to the refrigerant. The liquid refrigerant boils off into low pressure vapor, taking the heat with it when it leaves the evaporator.

2 Antifreeze – general information

Warning: *Do not allow antifreeze to come in contact with your skin or painted surfaces of the vehicle. Rinse off spills immediately with plenty of water. Antifreeze is highly toxic if ingested. Never leave antifreeze lying around in an open container or in puddles on the floor; children and pets are attracted by it's sweet smell and may drink it. Check with local authorities about disposing of used antifreeze. Many communities have collection centers which will see that antifreeze is disposed of safely.*

The cooling system should be filled with a water/ethylene glycol based antifreeze solution, which will prevent freezing down to at least -20-degrees F, or lower if local climate requires it. It also provides protection against corrosion and increases the coolant boiling point.

The cooling system should be drained, flushed and refilled at the specified intervals (see Chapter 1). Old or contaminated antifreeze solutions are likely to cause damage and encourage the formation of corrosion and scale in the system. Use distilled water with the antifreeze.

Before adding antifreeze, check all hose connections, because antifreeze tends leak through very minute openings. Engines don't normally consume coolant, so if the level goes down, find the cause and correct it.

The exact mixture of antifreeze-to-water which you should use depends on the relative weather conditions. The mixture should contain at least 50-percent antifreeze, but should never contain more than 70-percent antifreeze. Consult the mixture ratio chart on the antifreeze container before adding coolant. Hydrometers are available at most auto parts stores to test the coolant. Use antifreeze which meets the vehicle manufacturer's specifications.

3 Thermostat – check and replacement

Warning: *Do not remove the radiator cap, drain the coolant or replace the thermostat until the engine has cooled completely.*

Check

1 Before assuming the thermostat is to blame for a cooling system problem, check the coolant level, drivebelt tension (see Chapter 1) and temperature gauge operation.

2 If the engine seems to be taking a long time to warm up (based on heater output or temperature gauge operation), the thermostat is probably stuck open. Replace the thermostat with a new one.

Chapter 3 Cooling, heating and air conditioning systems

4.9 Remove the bolts that retain the brackets and insulators to the radiator (2.0L Galant shown)

3   If the engine runs hot, use your hand to check the temperature of the upper radiator hose. If the hose isn't hot, but the engine is, the thermostat is probably stuck closed, preventing the coolant inside the engine from escaping to the radiator. Replace the thermostat. **Caution:** *Don't drive the vehicle without a thermostat. The computer may stay in open loop and emissions and fuel economy will suffer.*
4   If the upper radiator hose is hot, it means that the coolant is flowing and the thermostat is open. Consult the Troubleshooting Section at the front of this manual for cooling system diagnosis.

Replacement

Refer to illustrations 3.10 and 3.13

5   Disconnect the negative battery cable from the battery.
6   Drain the cooling system (see Chapter 1). If the coolant is relatively new or in good condition, save it and reuse it.
7   Follow the upper radiator hose to the engine to locate the thermostat housing.
8   Loosen the hose clamp, then detach the hose from the fitting. If it's stuck, grasp it near the end with a pair of adjustable pliers and twist it to break the seal, then pull it off. If the hose is old or deteriorated, cut it off and install a new one.
9   If the outer surface of the large fitting that mates with the hose is deteriorated (corroded, pitted, etc.) it may be damaged further by hose removal. If it is, the thermostat housing cover will have to be replaced.
10   Remove the bolts and detach the housing cover **(see illustrations)**. If the cover is stuck, tap it with a soft-face hammer to jar it loose. Be prepared for some coolant to spill as the gasket seal is broken.
11   Note how it's installed (which end is facing up), then remove the thermostat.
12   Stuff a rag into the engine opening, then remove all traces of old gasket material and sealant from the housing and cover with a gasket scraper. Remove the rag from the opening and clean the gasket mating surfaces with lacquer thinner or acetone.
13   Install the new thermostat in the housing. Make sure the correct end faces up – the spring end is normally directed into the engine **(see illustration)**.
14   Apply a thin, uniform layer of RTV sealant to both sides of the new gasket and position it on the housing.
15   Install the cover and bolts. Tighten the bolts to the torque listed in this Chapter's Specifications.
16   Reattach the hose to the fitting and tighten the hose clamp securely.
17   Refill the cooling system (Chapter 1).
18   Start the engine and allow it to reach normal operating temperature, then check for leaks and proper thermostat operation (as described in Steps 2 through 4).

4   Radiator – removal and installation

Refer to illustrations 4.5 and 4.9
Warning: *Wait until the engine is completely cool before beginning this procedure.*

Removal

1   Disconnect the negative battery cable from the battery.
2   Drain the cooling system (see Chapter 1). If the coolant is relatively new or in good condition, save it and reuse it.
3   Loosen the hose clamps and detach the radiator hoses from the fittings. If they're stuck, grasp each hose near the end with a pair of adjustable pliers and twist it to break the seal, then pull it off – be careful not to distort the radiator fittings! If the hoses are old or deteriorated, cut them off and install new ones.
4   Disconnect the reservoir hose from the radiator filler neck.
5   Unplug the fan electrical connector **(see illustration)**.
6   If the vehicle is equipped with an automatic transaxle, disconnect the cooler lines from the bottom of the radiator.
7   Plug the lines and fittings to prevent leakage.
8   If equipped, remove the air intake duct from the radiator and air filter housing.
9   Remove the radiator mounting bolts **(see illustration)**.
10   Carefully lift out the radiator. Don't spill coolant on the vehicle or scratch the paint.
11   With the radiator removed, it can be inspected for leaks and damage. If it needs repair, have a radiator shop or dealer service department perform the work as special techniques are required.
12   Bugs and dirt can be removed from the radiator with compressed air and a soft brush. Don't bend the cooling fins as this is done.

Installation

13   Installation is the reverse of the removal procedure.
14   After installation, fill the cooling system with the proper mixture of antifreeze and water. Refer to Chapter 1 if necessary.
15   Start the engine and check for leaks. Allow the engine to reach normal operating temperature, indicated by the upper radiator hose becoming hot. Recheck the coolant level and add more if required.
16   If you're working on an automatic transaxle equipped vehicle, check and add fluid as needed.

5   Engine cooling fan – check and replacement

Refer to illustrations 5.8 and 5.9

Check

1   The engine cooling fan is controlled by a temperature switch mounted in the bottom of the radiator. When the coolant reaches a predetermined temperature, the switch provides a ground return for the fan motor circuit.
2   First, check the fuses (see Chapter 12).
3   To test the fan motor, unplug the motor electrical connector and use jumper wires to connect the fan directly to the battery. If the fan still does not work, replace the motor.
4   If the motor tested OK, the fault lies in the coolant temperature switch or the wiring harness (see Chapter 12).
5   Test the temperature switch by unplugging the electrical connector and grounding the wire from the fan motor (located on the connector for the temperature switch) with a jumper wire.
6   If the fan does not operate, check the wiring (see Chapter 12).

Replacement

7   Remove the radiator from the vehicle (see Section 4).

Chapter 3 Cooling, heating and air conditioning systems

5.8 Unclip the wiring harness and remove the bolts (arrows)

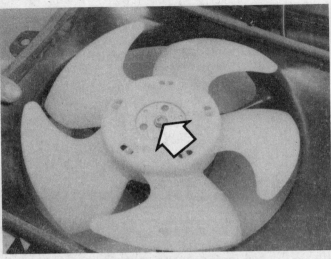

5.9 Hold the fan blades to keep them from turning and remove the nut (arrow)

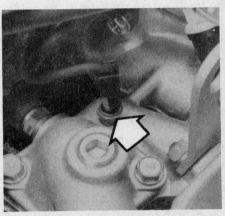

6.3a The coolant temperature sending unit (arrow) is usually located near the thermostat housing on the intake manifold

6.3b On some models, the temperature sending unit (arrow) is located on the underside of the intake manifold, behind the distributor (the distributor has been removed for clarity)

7.2 Pull out the cap and hose and remove the bolt (arrow), then lift out the coolant reservoir

8   Detach the wires and unbolt the fan shroud (see illustration).
9   Remove the nut and detach the fan blade assembly from the motor shaft (see illustration).
10 Remove the screws holding the fan motor to the bracket and detach the motor.
11 Installation is the reverse of removal. **Note:** *Air conditioned vehicles have an additional fan to cool the condenser, which is mounted to the left of the radiator. It can be checked by following Steps 2, 3 and 6.*

6   Coolant temperature sending unit – check and replacement

Refer to illustrations 6.3a and 6.3b
Warning: *The engine must be completely cool before removing the sending unit.*

Check

1   If the coolant temperature gauge is inoperative, check the fuses first (see Chapter 12).
2   If the temperature indicator shows excessive temperature after running a while, see the Troubleshooting Section in the front of the manual.
3   If the temperature gauge indicates Hot shortly after the engine is started cold, disconnect the wire(s) at the coolant temperature sending unit (see illustrations). If the gauge reading drops, replace the sending unit. If the reading remains high, the wire to the gauge or light may be shorted to ground or the gauge is faulty.
4   If the coolant temperature gauge fails to indicate after the engine has been warmed up (approximately 10 minutes) and the fuses checked out OK, shut off the engine. Disconnect the wire at the sending unit and using a jumper wire, connect it to a clean ground on the engine. Turn on the ignition without starting the engine. If the gauge now indicates Hot, replace the sending unit.
5   If the gauge still does not work, the circuit may be open or the gauge may be faulty. See Chapter 12 for additional information.

Replacement

6   With the engine completely cool, remove the cap from the radiator to release any pressure, squeeze the upper radiator hose and replace the cap. This reduces coolant loss during sending unit replacement.
7   Disconnect the electrical connector from the sending unit.
8   Prepare the new sending unit for installation by applying sealer to the threads.
9   Unscrew the sending unit from the engine and quickly install the new one to prevent coolant loss.

Chapter 3 Cooling, heating and air conditioning systems

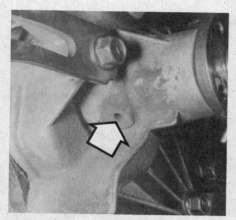

8.4 If coolant is leaking from the weep hole (arrow) the water pump must be replaced

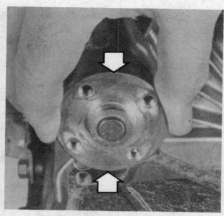

8.5 If there is play in the shaft, replace the water pump

9.3 Loosen the water pump pulley bolts before removing the drivebelts

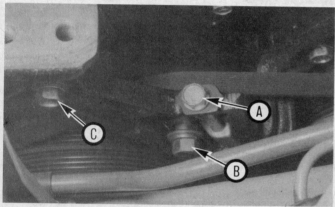

9.4 First loosen bolt B, then back off the adjusting bolt (A) and remove the air conditioning belt and the remaining A/C bracket bolt (C)

9.7a Remove the mounting bolts (arrows) (typical SOHC engine)

10 Tighten the sending unit securely and connect the wiring harness.
11 Refill the cooling system and run the engine. Check for leaks and proper gauge operation.

7 Coolant reservoir – removal and installation

Refer to illustration 7.2

1 Lift the cap off the coolant reservoir and pull out the overflow hose.
2 Remove the coolant reservoir-to-fender bolt (see illustration) and lift the coolant reservoir out. The windshield washer reservoir can remain in place.
3 Installation is the reverse of removal.

8 Water pump – check

Refer to illustrations 8.4 and 8.5

1 A failure in the water pump can cause serious engine damage due to overheating.
2 There are three ways to check the operation of the water pump while it's installed on the engine. If the pump is defective, it should be replaced with a new or rebuilt unit.
3 With the engine running at normal operating temperature, squeeze the upper radiator hose. If the water pump is working properly, a pressure surge should be felt as the hose is released. **Warning:** *Keep your hands away from the fan blades!*

4 Water pumps are equipped with weep or vent holes. If a failure occurs in the pump seal, coolant will leak from the hole. In most cases you'll need a flashlight to find the hole on the water pump from underneath to check for leaks **(see illustration)**.
5 If the water pump shaft bearings fail there may be a howling sound at the front of the engine while it's running. Shaft wear can be felt if the water pump pulley is rocked up and down **(see illustration)**. Don't mistake drivebelt slippage, which causes a squealing sound, for water pump bearing failure.

9 Water pump – removal and installation

Refer to illustrations 9.3, 9.4, 9.7a and 9.7b
Warning: *Wait until the engine is completely cool before beginning this procedure.*

Removal

1 Disconnect the negative battery cable from the battery.
2 Drain the cooling system (see Chapter 1). If the coolant is relatively new or in good condition, save it and reuse it.
3 Remove the left engine mount (see Chapter 2, Part A) and loosen the water pump pulley bolts **(see illustration)**.
4 Remove the drivebelts (see Chapter 1) and the pulley at the end of the water pump shaft. **Note:** *To remove the air conditioning belt, completely remove the adjustment bracket assembly* **(see illustration)**.
5 Remove the timing belt and tensioner (see Chapter 2, Part A).
6 Remove the alternator bracket from the water pump.
7 Remove the bolts **(see illustrations)** and detach the water pump

3-6 Chapter 3 Cooling, heating and air conditioning systems

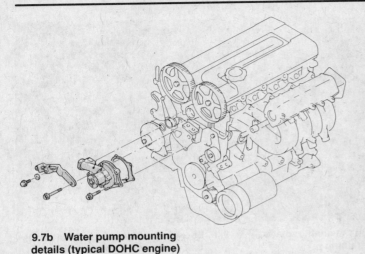

9.7b Water pump mounting details (typical DOHC engine)

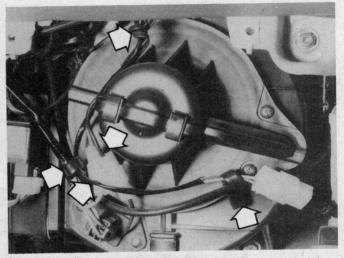

10.3 Typical blower motor and mounting details – unplug the electrical connectors and release the clips (arrows), then remove the mounting screws (Precis models shown, others similar)

from the engine. Note the locations of the various lengths and different types of bolts as they're removed to ensure correct installation.

Installation

8 Clean the bolt threads and the threaded holes in the engine to remove corrosion and sealant.
9 Compare the new pump to the old one to make sure they're identical.
10 Remove all traces of old gasket material from the engine with a gasket scraper.
11 Clean the engine and new water pump mating surfaces with lacquer thinner or acetone.
12 Install a new O-ring in the groove at the front end of the coolant pipe and lubricate the O-ring with coolant.
13 Apply a thin coat of RTV sealant to the engine side of the new gasket and to the gasket mating surface of the new pump, then carefully mate the gasket and the pump. Slip a couple of bolts through the pump mounting holes to hold the gasket in place.
14 Carefully attach the pump and gasket to the engine and thread the bolts into the holes finger tight. Note that the bolt on the left side of the pump (that attaches the alternator brace) is longer than the other three bolts. Be sure to install the bolt(s) with the correct length into the corresponding water pump holes.
15 Install the remaining bolts (be sure to reposition the alternator bracket at this time). Tighten them to the torque listed in this Chapter's Specifications in 1/4-turn increments. Don't overtighten them or the pump may be distorted.
16 Reinstall all parts removed for access to the pump.
17 Refill the cooling system and check the drivebelt tension (see Chapter 1). Run the engine and check for leaks.

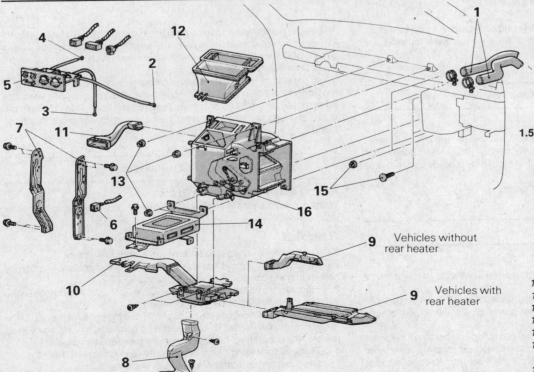

1.5 Typical heater unit details

1 Heater hoses
2 Air selection control cable
3 Temperature control cable
4 Mode selection control cable
5 Heater control assembly
6 Electrical connector
7 Instrument panel center support
8 Rear heater duct A
9 Duct
10 Duct
11 Duct
12 Center ventilation duct
13 Heater unit mounting nuts
14 Control unit
15 Evaporator mounting fasteners
16 Heater unit

Chapter 3 Cooling, heating and air conditioning systems

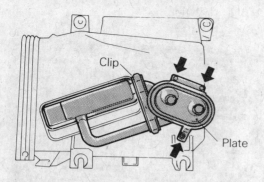

11.9 The retainer plate and clip are located on the backside of the heater unit

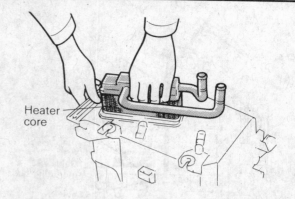

11.10 Carefully lift the heater core from the housing

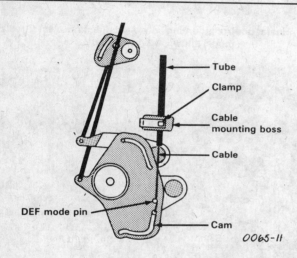

12.4a Mode control cable mounting details (Precis model shown, others similar)

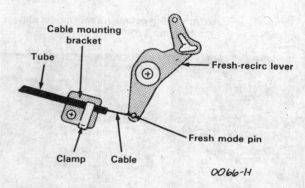

12.4b Fresh-recirc cable mounting details (Precis model shown, others similar)

10 Heater and air conditioner blower motor – removal and installation

Refer to illustration 10.3

1 Disconnect the negative cable from the battery.
2 Remove the lower dash panel below the glove compartment.
3 Disconnect the electrical connectors from the blower and detach the wiring harness from the clips **(see illustration)**.
4 Remove the screws holding the blower motor to the heater housing and lift the unit out.
5 If you are replacing the motor, detach the fan and transfer it to the new motor.
6 Installation is the reverse of removal. Run the blower and check for proper operation.

11 Heater core – replacement

Refer to illustrations 11.5, 11.9 and 11.10

1 Disconnect the negative cable from the battery.
2 Drain the cooling system (see Chapter 1).
3 Working in the engine compartment or under the dash, disconnect the heater hoses from the heater control valve.
4 Remove the instrument panel and the center console (if equipped).
5 Remove the heater controls (see Section 12) and disconnect the heater hoses from the heater core tubes **(see illustration)**. Be prepared for coolant spillage.
6 Remove the radio, then remove the bolts that retain the instrument panel center support.
7 Label and detach the air ducts, wiring and controls still attached to the heating unit.
8 Unbolt the heating unit and detach it from the vehicle.
9 Remove the screws and clip and separate the core retainer plate from the housing **(see illustration)**.
10 Take out the old heater core and install the new unit **(see illustration)**.
11 Reassemble the heater unit and check the operation of the air control flaps. If any parts bind, correct the problem before installation.
12 Reinstall the remaining parts in the reverse order of removal.

12 Heater and air conditioner control assembly – removal, installation and cable adjustment

Removal and installation

Refer to illustrations 12.4a, 12.4b, 12.4c and 12.5

1 Disconnect the negative cable from the battery.
2 Remove the trim panel which surrounds the control assembly.
3 Remove the glove compartment and the covers on the sides of the console under the dash.
4 Disconnect the control cables at the ends opposite from the control by

Chapter 3 Cooling, heating and air conditioning systems

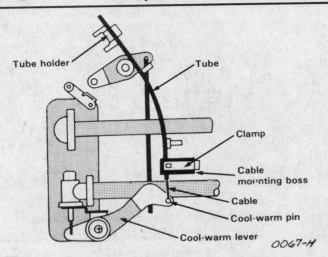

12.4c Temperature cable mounting details (Precis model shown, others similar)

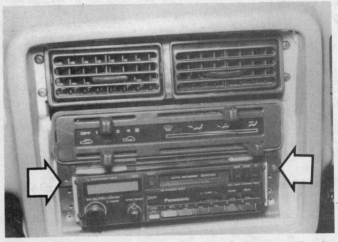

12.5 Heater control mounting screw locations (arrows) (Precis model shown, others similar)

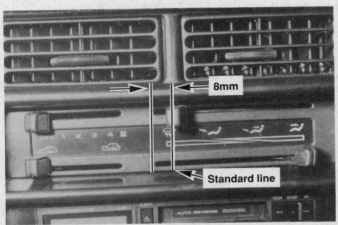

12.8 Mode control in the defrost position for cable adjustment (Precis model shown, others similar)

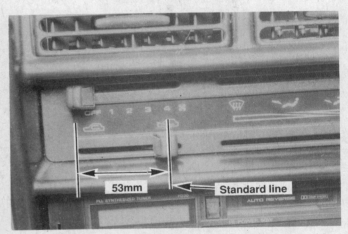

12.9 Fresh-recirc control in the outside air position for cable adjustment (Precis model shown, others similar)

6 Carefully slip the unit out of the dash.
7 Installation is the reverse of removal.

Cable adjustment
Refer to illustrations 12.8, 12.9 and 12.10

Defrost-vent cable

8 Set the control to defrost **(see illustration)**. Move the lever all the way back, connect the cable to the lever and reinstall the clamp.

Fresh-recirc cable

9 Set the control in the outside-air position **(see illustration)**. Move the lever all the way back, connect the cable to the lever and reinstall the clamp.

Temperature control cable

10 Set the control to the left position **(see illustration)**. Move the water valve lever to the highest position, connect the cable to the lever and reinstall the clamp.

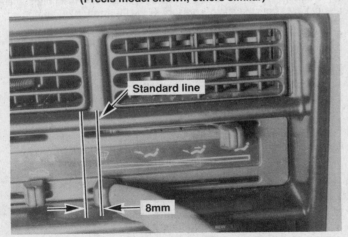

12.10 Place the temperature lever all the way to the left for cable adjustment (Precis model shown, others similar)

13 Air conditioning system – check and maintenance

Warning: *The air conditioning system is under high pressure. Do not loosen any hose fittings or remove any components until after the system has been discharged by a dealer service department or service station. Always wear eye protection when disconnecting air conditioning system fittings.*

detaching the cable clamps and separating the cables from the pins on the operating levers **(see illustrations)**.
5 Remove the control assembly mounting screws **(see illustration)**.

Chapter 3 Cooling, heating and air conditioning systems

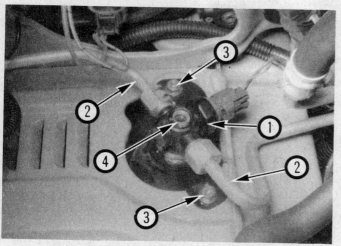

14.2 Receiver/drier and related components

1. Pressure switch
2. Refrigerant lines
3. Bolt
4. Sight glass

15.6 Air conditioner compressor viewed from above showing refrigerant line fittings and upper mounting bolts (arrows)

1   The following maintenance checks should be performed on a regular basis to ensure that the air conditioner continues to operate at peak efficiency.
  a) Check the compressor drivebelt. If it's worn or deteriorated, replace it (see Chapter 1).
  b) Check the drivebelt tension and, if necessary, adjust it (see Chapter 1).
  c) Check the system hoses. Look for cracks, bubbles, hard spots and deterioration. Inspect the hoses and all fittings for oil bubbles and seepage. If there's any evidence of wear, damage or leaks, replace the hose(s).
  d) Inspect the condenser fins for leaves, bugs and other debris. Use a "fin comb" or compressed air to clean the condenser.
  e) Make sure the system has the correct refrigerant charge.

2   It's a good idea to operate the system for about 10 minutes at least once a month, particularly during the winter. Long term non-use can cause hardening, and subsequent failure, of the seals.
3   Because of the complexity of the air conditioning system and the special equipment necessary to service it, in-depth troubleshooting and repairs are not included in this manual. However, simple checks and component replacement procedures are provided in this Chapter.
4   The most common cause of poor cooling is simply a low system refrigerant charge. If a noticeable drop in cool air output occurs, one of the following quick checks will help you determine if the refrigerant level is low.
5   Warm the engine up to normal operating temperature.
6   Place the air conditioning temperature selector at the coldest setting and put the blower at the highest setting. Open the doors (to make sure the air conditioning system doesn't cycle off as soon as it cools the passenger compartment).
7   With the compressor engaged – the clutch will make an audible click and the center of the clutch will rotate – inspect the sight glass on the receiver/drier (see Section 14). If the refrigerant looks foamy, it's low. Have a certified air conditioning shop charge the system. Due to environmental regulations, 14-ounce refrigerant cans are no longer available to the home mechanic. Bulk refrigerant drums are quite expensive, but, if you can obtain one (and a set of manifold gauges), recharging instructions are included in the Haynes Automotive Heating and Air Conditioning Manual.

14 Air conditioning receiver/drier – removal and installation

Refer to illustration 14.2
Warning: *The air conditioning system is under high pressure. Do not loosen any fittings or remove any components until after the system has been discharged by a dealer service department or service station. Always wear eye protection when disconnecting refrigerant fittings.*

Removal
1   The receiver/drier, which acts as a reservoir and filter for the refrigerant, is located in the front corner of the engine compartment.
2   Unplug the electrical connector from the pressure switch **(see illustration)**.
3   Detach the two refrigerant lines from the receiver/drier.
4   Immediately cap the open fittings to prevent the entry of dirt and moisture.
5   Unbolt the receiver/drier and lift it out of the engine compartment.

Installation
6   Install new O-rings on the lines and lubricate them with clean refrigerant oil.
7   Installation is the reverse of removal. **Note:** *Do not remove the sealing caps until you are ready to reconnect the lines. Do not mistake the inlet (marked IN) and the outlet (marked OUT) connections.*
8   If a new receiver/drier is installed, add 1.8 US fluid ounces (30 cc) of refrigerant oil to the system.
9   Have the system evacuated, charged and leak tested by the shop that discharged it.

15 Air conditioning compressor – removal and installation

Refer to illustrations 15.6 and 15.7
Warning: *The air conditioning system is under high pressure. Do not loosen any fittings or remove any components until after the system has been discharged by a dealer service department or service station. Always wear eye protection when disconnecting refrigerant fittings.*

Removal
1   Disconnect the negative cable from the battery.
2   Set the parking brake and block the rear tires.
3   Raise the front of the vehicle and support it securely on jackstands.
4   Unbolt the splash pan and remove the compressor drivebelt (see Chapter 1).
5   Disconnect the clutch electrical connector from the compressor.
6   Detach the refrigerant lines from the back of the compressor **(see illustration)** and immediately cap the open fittings to prevent the entry of dirt and moisture.

3–10 Chapter 3 Cooling, heating and air conditioning systems

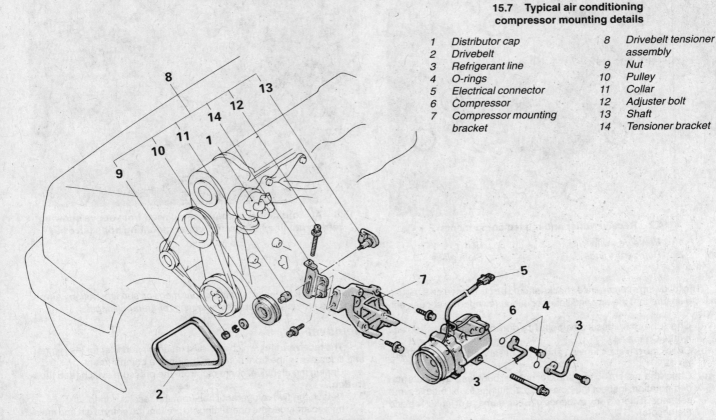

15.7 Typical air conditioning compressor mounting details

1. Distributor cap
2. Drivebelt
3. Refrigerant line
4. O-rings
5. Electrical connector
6. Compressor
7. Compressor mounting bracket
8. Drivebelt tensioner assembly
9. Nut
10. Pulley
11. Collar
12. Adjuster bolt
13. Shaft
14. Tensioner bracket

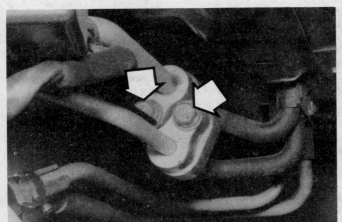

16.2a Remove the bolts that connect the refrigerant lines to the condenser

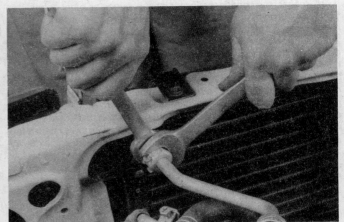

16.2b Use a back-up wrench to avoid twisting the lines

7 Remove the mounting bolts **(see illustration)** and lower the compressor from the engine compartment. Note the location and thickness of any shims and reinstall them in the same place. **Note:** *Keep the compressor level during handling and storage. If the compressor seized or you find metal particles in the refrigerant lines, the system must be flushed out by an air conditioning technician and the receiver/drier must be replaced.*

Installation

8 Prior to installation, turn the center of the clutch six times to disperse any oil that has collected in the head.
9 Install the compressor in the reverse order of removal.

10 If you are installing a new compressor, refer to the manufacturer's instructions for adding refrigerant oil to the system.
11 Have the system evacuated, charged and leak tested by the shop that discharged it.

16 Air conditioning condenser – removal and installation

Refer to illustrations 16.2a, 16.2b, 16.3 and 16.4
Warning: *The air conditioning system is under high pressure. Do not loosen any fittings or remove any components until after the system has been discharged by a dealer service department or service station. Always wear eye protection when disconnecting refrigerant fittings.*

Chapter 3 Cooling, heating and air conditioning systems

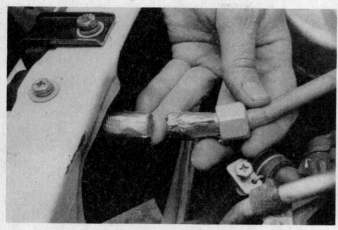

16.3 Immediately seal off the ends – tape works well

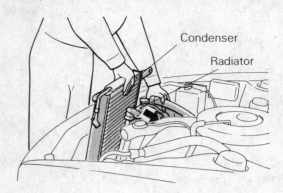

16.4 Push the radiator/cooling fan back as far as possible, then lift the condenser out

Removal

1 Remove the hood, grille and vertical brace (see Chapter 11).
2 Disconnect the refrigerant lines from the condenser. Be sure to use a back-up wrench to avoid twisting the lines **(see illustrations)**.
3 Immediately cap the open fittings to prevent the entry of dirt and moisture **(see illustration)**.
4 Unbolt the radiator upper mount and move the radiator/cooling fan assembly back as far as possible (don't disconnect the hoses). Unbolt the condenser and lift it out of the vehicle **(see illustration)**. Store it upright to prevent oil loss.

Installation

5 Installation is the reverse of removal.
6 If a new condenser was installed, add 1.8 US ounces (30 cc) of refrigerant oil to the system.
7 Have the system evacuated, charged and leak tested by the shop that discharged it.

Note:

Chapter 4 Fuel and exhaust systems

Contents

Air cleaner assembly – removal and installation	8
Air filter replacement	See Chapter 1
Carburetor – diagnosis and overhaul	10
Carburetor – on-vehicle checks and adjustments	11
Carburetor – removal and installation	9
Exhaust system check	See Chapter 1
Exhaust system servicing – general information	17
Fuel filter replacement	See Chapter 1
Fuel filter – general information	14
Fuel injection system – check	12
Fuel lines and fittings – inspection and replacement	4
Fuel pressure relief procedure (fuel-injected models)	2
Fuel pump/fuel pressure – testing	3
Fuel pump – removal and installation	5
Fuel system check	See Chapter 1
Fuel tank – cleaning and repair	7
Fuel tank – removal and installation	6
General information	1
Throttle cable – removal, installation and adjustment	16
Throttle Body Injection (TBI) assembly – component removal, overhaul and installation	13
Turbocharger – general information	18
Turbocharger – inspection	19
Turbocharger – removal and installation	20
Multi-Point Injection (MPI) – component removal and installation	15
Underhood hose check and replacement	See Chapter 1

Specifications

Choke heater resistance .... 6 ohms at room temperature

Dashpot adjustment engine speed .... 1800 rpm

Solenoid valve resistance
Bowl vent valve (BVV) .... 80 ohms
Deceleration solenoid valve (DSV) .... 49.7 ohms
Enrichment solenoid valve (ESV) .... 49.7 ohms
Idle-up control solenoid valve .... Approx. 40 ohms
Jet mixture solenoid valve (JSV) .... 49.7 ohms

Chapter 4 Fuel and exhaust systems

Throttle position sensor output voltage
Carbureted .... 250 mV at full closed throttle
Fuel injected .... must be taken to a dealer service department or other qualified shop

Fuel pressure (fuel-injected models)
MPI (Approximate regulated pressure)
  Galant
    1986 .... 36.26 psi
    1987 .... 28.4 psi
    1988 .... 37.7 psi
    1989 .... 38 psi
    1990 .... 38 psi
  Mirage
    1989 .... 36.3 psi
    1990 .... 38 psi
  Precis (1990) .... 48 psi
TBI (Approximate regulated pressure) .... 35.6 psi

Fuel system identification

	Cordia and Tredia	Galant	Mirage	Precis
1983	1.8L – 2bbl non-FBC			
1984	2.0L – 2bbl FBC 1.8L – turbo TBI			
1985	2.0L – 2bbl FBC 1.8L – turbo TBI	2.4L – TBI	1.5L – 2bbl FBC 1.6L – TBI 1.6L – turbo TBI	
1986	2.0L – 2bbl FBC 1.8L – turbo TBI	2.4L – MPI	1.5L – 2bbl FBC 1.6L – TBI 1.6L – turbo TBI	
1987	2.0L – 2bbl FBC 1.8L – turbo TBI	2.4L – MPI	1.5L – 2bbl FBC 1.6L – TBI 1.6L – turbo TBI	1.5L – 2bbl FBC
1988	2.0L – 2bbl FBC 1.8L – turbo TBI	3.0L – MPI (not covered in this manual)	1.5L – 2bbl FBC 1.6L – TBI 1.6L – turbo TBI	1.5L – 2bbl FBC
1989		2.0L – MPI	1.5L – MPI 1.6L – MPI	1.5L – 2bbl FBC
1990		2.0L – MPI	1.5L – MPI 1.6L – MPI	1.5L – MPI

1 General information

Note: *Refer to the accompanying chart to determine what type of fuel system (FBC or non-FBC carbureted or TBI/MPI fuel injection) your vehicle is equipped with.*

Carbureted

The fuel system consists of a fuel tank, two fuel filters (one in-tank and one in the engine compartment), a fuel pump, an air cleaner assembly and a two-barrel carburetor.

The fuel pump is a mechanical type mounted on the cylinder head. The pump is driven off the camshaft by a pushrod.

The Feedback Carburetor (FBC) system used on 1984 and later models is controlled by a computer which monitors changes in engine operation with various information sensors, compares this data to parameters stored in its memory and alters fuel delivery accordingly by means of actuators installed on the carburetor. Refer to Chapter 6 for more information.

Fuel injection

Multi Point Injection (MPI)

This fuel system consists of a fuel tank, two fuel filters (one in-tank and one in the fuel feed line), an electric fuel pump, a fuel pressure regulator, a fuel pressure solenoid (on some models), an injector delivery pipe, injectors and an air cleaner assembly.

Chapter 4 Fuel and exhaust systems

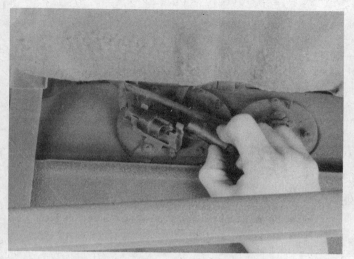

2.4 Disconnecting the fuel pump (Galant is shown, others are similar)

3.4 Check the fuel pump breather hole (arrow) – if any gas or oil is leaking out, replace the pump (intake manifold removed for clarity)

The fuel pump supplies pressurized fuel to the injectors. The regulator returns excess fuel to the tank to maintain the pressure at a set limit.

This computerized fuel control system uses various input sensors to feed information to the Electronic Control Unit (ECU). The ECU uses this input to determine the amount of time to hold the injectors (placed at each intake valve) open. Refer to Chapter 6 for more information.

Throttle Body Injection (TBI)

This fuel system consists of a fuel tank, two fuel filters (one in-tank and one in the fuel feed line), an electric fuel pump, a fuel pressure regulator, two injectors and an air cleaner assembly.

The fuel pump supplies pressurized fuel to the injectors. The regulator returns excess fuel to the tank to maintain the pressure at a set limit.

This computerized fuel control system uses various input sensors to feed information to the Electronic Control Unit (ECU). The ECU uses this input to determine the amount of time to hold the injectors (placed at the throttle body) open.

Exhaust system

All vehicles are equipped with an exhaust manifold, a front catalytic converter, a rear converter, the connecting pipe between the two converters and a main muffler assembly (the rear part of the exhaust pipe and the muffler itself). The exhaust system is suspended from the underside of the vehicle and insulated from vibration by a series of rubber hangers.

2 Fuel pressure relief procedure (fuel-injected models)

Refer to illustration 2.4

1 Locate the fuel pump electrical connector – see Section 5, if necessary – (some have an access cover inside the luggage compartment or under the rear seat).
2 Remove the fuel filler cap.
3 Set the parking brake, start the engine and let it idle.
4 Disconnect the fuel pump **(see illustration)**.
5 Let the engine run to use the remaining fuel in the lines. The engine will stall.
6 To be sure all fuel is removed from the lines, crank the engine over for about five seconds.
7 After repairs are completed and all lines and fittings are connected securely, re-connect the fuel pump and start the engine – it may take unusually long to start, since fuel must be pumped to the injectors.
8 Inspect all connections for leaks.

3 Fuel pump/fuel pressure – testing

Fuel pump – check

Carbureted models

Refer to illustration 3.4

Warning: *Gasoline is extremely flammable, so take extra precautions when you work on any part of the fuel system. Don't smoke or allow open flames or bare light bulbs near the work area, and don't work in a garage where a natural gas-type appliance (such as a water heater or clothes dryer) with a pilot light is present. If you spill any fuel on your skin, rinse it off immediately with soap and water. When you perform any kind of work on the fuel tank, wear safety glasses and have a Class B type fire extinguisher on hand.*

1 If you suspect insufficient fuel delivery, first inspect all fuel lines to ensure that the problem is not simply a leak in a line.
2 If there are no leaks evident in the fuel lines, inspect the fuel pump itself. The following checks will tell you if the fuel pump is leaking and whether it is pumping fuel.
3 Before performing the following checks, detach the cable from the negative terminal of the battery and remove the air cleaner, if necessary (see Section 8).

Breather hole check

4 Locate the fuel pump, on the side of the cylinder head. Note whether there is any fuel or oil leaking from the breather hole **(see illustration)**. If there is, either the oil seal or the diaphragm in the fuel pump is defective. Replace the fuel pump if leakage is noted (see Section 5).

Fuel pump output check

5 Hook up a remote starter switch in accordance with the manufacturer's instructions. If you don't have a remote starter switch, you will need an assistant to help you with this and the following procedure.
6 Trace the fuel outlet hose from the pump to the carburetor and detach it at the carburetor.
7 Attach the cable to the negative terminal of the battery.
8 Detach the wires from the ignition coil primary terminals (see Chapter 5).
9 Place a metal or approved gasoline container under the open end of the fuel pump outlet hose.
10 Direct the fuel pump outlet hose into the container while cranking the engine for about ten seconds with the remote starter (or while an assistant cranks the engine with the ignition key).

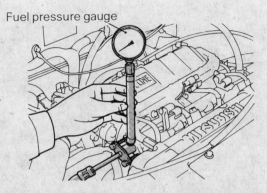

3.20 On MPI systems, you must use a special adapter (MD 998742 or equivalent) along with a pressure gauge to tee into the fuel pressure line

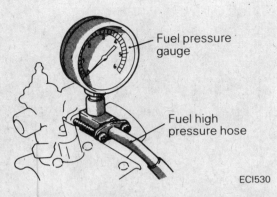

3.33 TBI systems use the same special adapter as MPI systems (MD 998742 or equivalent) to tee into the fuel pressure line

11 If fuel is emitted in well-defined spurts, the pump is operating satisfactorily. If fuel dribbles or trickles out the hose, the pump is defective. Replace it.

Inlet valve check
12 Detach the inlet hose from the fuel pump.
13 Attach a vacuum gauge to the inlet fitting.
14 Crank the engine with a remote starter switch (or have an assistant crank it with the ignition key).
15 A fairly steady vacuum, uninterrupted by alternating blowback pulses (sudden pulses of pressure), should be evident.
16 If blowback is evident, the fuel pump inlet valve is not seating properly. Replace the pump.
17 Replace the wires on the ignition coil primary terminals and reconnect the battery terminal.

MPI models
Refer to illustration 3.20
18 Relieve the fuel pressure (see Section 2).
19 At the fuel delivery pipe, detach the high-pressure line.
20 Using a special adapter (MD 998742 or equivalent) and a fuel pressure gauge, tee into the fuel delivery circuit **(see illustration)**.
21 With the electrical harness for the fuel pump connected, have an assistant start the engine while watching for leaks. **Warning:** *If fuel leaks are present, have the assistant immediately turn the engine off and repair any leaks.*
22 Allow the engine to idle.
23 With the engine speed at idle, note the pressure on the gauge and compare the reading to the specifications at the front of this Chapter.
24 Detach the vacuum hose from the pressure regulator and plug it.
25 Note the reading on the gauge with the engine speed at idle.
26 Rev the engine two or three times.
27 With the engine speed at idle, note the reading on the gauge.
28 At the return hose to the pressure regulator, lightly squeeze the hose with your hand to feel for pressure in the hose. To be sure that pressure remains in the return hose, rev the engine a few times.
29 If the fuel pressure was low in both tests, either the fuel filter is plugged, the feed line is plugged or the pump is faulty.
30 If the reading was high with the regulator connected, either the regulator is faulty or no vacuum is getting to the regulator.
31 If the reading was low with the regulator connected but okay with the regulator disconnected, the regulator is faulty.

TBI models
32 Relieve the fuel pressure.
33 Using a special adapter (MD 998742 or equivalent) and a fuel pressure gauge, tee into the fuel delivery circuit to the throttle body **(see illustration)**.

34 With the fuel pump electrical wiring connected, have an assistant start the engine while watching for leaks. **Warning:** *If fuel leaks are present, have the assistant immediately turn the engine off and repair any leaks.*
35 With the engine running at idle, compare the reading on the gauge with the specification at the front of this Chapter.
36 If the reading is low, check for a plugged fuel filter, a defective regulator, pinched feed line or a defective pump.
37 Any further diagnosis should be left to a dealer service department or other qualified repair facility.

4 Fuel lines and fittings – inspection and replacement

Warning: *Gasoline is extremely flammable, so take extra precautions when you work on any part of the fuel system. Don't smoke or allow open flames or bare light bulbs near the work area, and don't work in a garage where a natural gas-type appliance (such as a water heater or clothes dryer) with a pilot light is present. If you spill any fuel on your skin, rinse it off immediately with soap and water. When you perform any kind of work on the fuel tank, wear safety glasses and have a Class B type fire extinguisher on hand.*

Inspection
1 Once in a while, you will have to raise the vehicle to service or replace some component (an exhaust pipe hanger, for example). Whenever you work under the vehicle, always inspect the fuel lines and fittings for possible damage or deterioration.
2 Check all hoses and pipes for cracks, kinks, deformation or obstructions.
3 Make sure all hose and pipe clips attach their associated hoses or pipes securely to the underside of the vehicle.
4 Verify all hose clamps attaching rubber hoses to metal fuel lines or pipes are snug enough to assure a tight fit between the hoses and pipes.

Replacement
5 If you must replace any damaged sections, use original equipment replacement hoses or pipes constructed from exactly the same material as the section you are replacing. Do not install substitutes constructed from inferior or inappropriate material or you could cause a fuel leak or a fire.
6 Always, before detaching or disassembling any part of the fuel line system, note the routing of all hoses and pipes and the orientation of all clamps and clips to assure that replacement sections are installed in exactly the same manner. On fuel-injected models, relieve the fuel system pressure (see Section 2) before disconnecting any fuel lines or fittings.
7 Before detaching any part of the fuel system on carbureted models, be sure to relieve the fuel tank pressure by removing the fuel filler cap.

Chapter 4 Fuel and exhaust systems

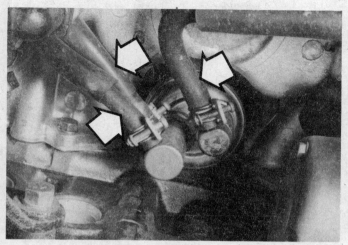

5.5 Remove the fuel inlet, outlet and return hoses from the fuel pump

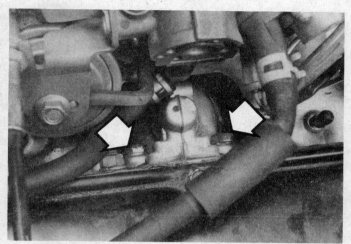

5.6 Remove the fuel pump bolts (arrows), then carefully break the pump loose with your hand.

5 Fuel pump – removal and installation

Warning: *Gasoline is extremely flammable, so take extra precautions when you work on any part of the fuel system. Don't smoke or allow open flames or bare light bulbs near the work area, and don't work in a garage where a natural gas-type appliance (such as a water heater or clothes dryer) with a pilot light is present. If you spill any fuel on your skin, rinse it off immediately with soap and water. When you perform any kind of work on the fuel tank, wear safety glasses and have a Class B type fire extinguisher on hand.*

Mechanical pump (carbureted models)
Refer to illustrations 5.5 and 5.6

1 Detach the cable from the negative battery terminal.
2 Remove the fuel tank filler cap to relieve fuel tank pressure.
3 Remove the air cleaner assembly (see Section 8).
4 Apply the parking brake and place blocks behind the rear wheels. Raise the front of the vehicle and support it with jackstands.
5 Detach the fuel inlet, outlet and return hoses **(see illustration)** from the pump.
6 Remove the fuel pump mounting bolts **(see illustration)**.
7 Carefully break the fuel pump loose with your hand (do not use a prybar) and remove the pump, gaskets and insulator. Note that the insulator is sandwiched between the two gaskets.
8 Using a scraper, remove the old gasket material from the insulator, the pump (if it will be reused), and the pump mating surface on the cylinder head.

9 Inspect the condition of the fuel inlet, outlet and return hoses. If they're damaged or worn, replace them (see Section 2).
10 Installation is the reverse of removal. Be sure to use new gaskets.

Electric pump (fuel-injected models)
Chassis-mounted (Cordia and Tredia)
Refer to illustration 5.14

11 Relieve the fuel pressure (see Section 2).
12 Remove the left rear wheel.
13 Loosen the fuel tank mounting band and lower the fuel tank a little.
14 Remove the bolts mounting the pump **(see illustration)**.
15 Detach the fuel lines from the pump.
16 Detach the electrical connector from the pump.
17 Remove the fuel pump and bracket.
18 Installation is the reverse of removal.

In-tank (all others)
Refer to illustrations 5.21a and 5.21b

19 Relieve the fuel pressure (see Section 2).
20 Remove the cable from the negative battery terminal.
21 Remove the fuel tank, if necessary, or disconnect the fuel lines and electrical connector at the fuel pump mounting point **(see illustrations)**.

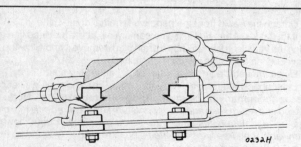

5.14 On the fuel-injected Cordia or Tredia, remove the left rear wheel and then remove the bolts mounting the pump (arrows)

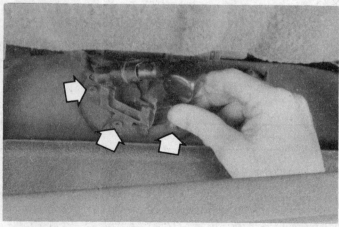

5.21a On some models, such as this Galant, the fuel pump is removed through the side of the fuel tank, so the fuel tank doesn't have to be removed – just disconnect the electrical connector and lines and remove the mounting screws, as shown

Chapter 4 Fuel and exhaust systems

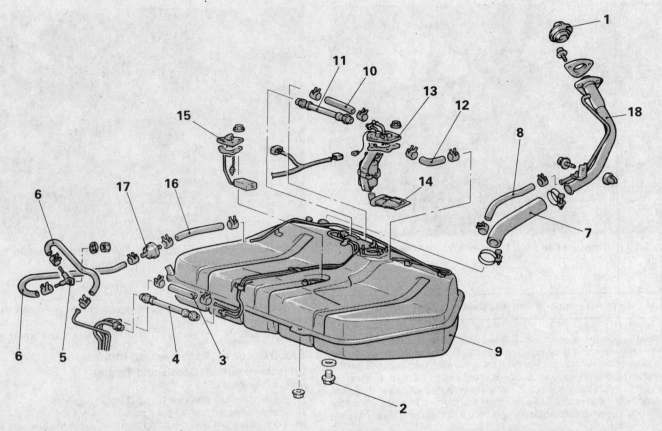

5.21b On other models, such as this Mirage, the fuel pump is removed through the top of the tank, so the tank must be removed first (see Section 6)

1 Tank cap	7 Filler hose	13 Fuel pump
2 Drain plug	8 Breather hose	14 In-tank fuel filter
3 Return hose	9 Fuel tank	15 Fuel gauge unit
4 High pressure fuel hose	10 Return hose	16 Vapor hose
5 Check valve	11 High pressure fuel hose	17 Two-way valve
6 Vapor hose	12 Vapor hose	18 Filler neck

22 Remove the mounting screws.
23 Pull the fuel pump out of the tank.
24 Check the condition of the fuel pump gasket. If it's dried out, cracked or deteriorated, replace it.
25 Inspect the filter on the lower end of the fuel pump (see illustration 5.21b). If it's dirty, remove it, clean it with solvent and blow it out with compressed air. If it's too dirty to be cleaned, replace it.
26 Installation is the reverse of removal.

6 Fuel tank – removal and installation

Refer to illustration 6.6

Warning: *Gasoline is extremely flammable, so take extra precautions when you work on any part of the fuel system. Don't smoke or allow open flames or bare light bulbs near the work area, and don't work in a garage where a natural gas-type appliance (such as a water heater or clothes dryer) with a pilot light is present. If you spill any fuel on your skin, rinse it off immediately with soap and water. When you perform any kind of work on the fuel tank, wear safety glasses and have a Class B type fire extinguisher on hand.*

Note: *The following procedure is much easier to perform if the fuel tank is empty. Some tanks have a drain plug for this purpose. If the tank does not have a drain plug, schedule fuel tank removal when the tank is nearly empty.*

1 Remove the fuel tank filler cap to relieve fuel tank pressure.
2 If the vehicle is fuel-injected, relieve the fuel system pressure (see Section 2).
3 Detach the cable from the negative terminal of the battery.
4 If the tank still has fuel in it, you can drain it at the fuel feed line after raising the vehicle. If the tank has a drain plug, remove it and allow the fuel to collect in an approved gasoline container.
5 Raise the vehicle and place it securely on jackstands.
6 Disconnect the fuel lines, the vapor return line and the fuel filler neck (see the accompanying illustration and illustration 5.21b). **Note:** *The fuel feed and return lines and the vapor return line are three different diameters, so reattachment is simplified. If you have any doubts, however, clearly label the three lines and the fittings. Be sure to plug the hoses to prevent leakage and contamination of the fuel system.*
7 Siphon the fuel from the tank at the fuel feed – not the return line.
8 Support the fuel tank with a floor jack or jackstands. Position a piece of wood between the jack head and the fuel tank to protect the tank.
9 Disconnect both fuel tank retaining straps and pivot them down until they are hanging out of the way.

Chapter 4 Fuel and exhaust systems

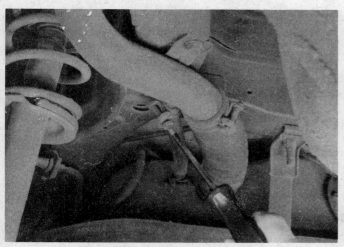

6.6 Detach the lines from the fuel tank

8.3a The mounting bolt locations (arrows) for the air cleaner assembly on carbureted engines. Don't forget to remove the wing nut and label and detach any hoses or wiring interfering with removal

8 Air cleaner assembly - removal and installation

Refer to illustrations 8.3a and 8.3b

Warning: *Gasoline is extremely flammable, so take extra precautions when you work on any part of the fuel system. Don't smoke or allow open flames or bare light bulbs near the work area, and don't work in a garage where a natural gas-type appliance (such as a water heater or clothes dryer) with a pilot light is present. If you spill any fuel on your skin, rinse it off immediately with soap and water. When you perform any kind of work on the fuel tank, wear safety glasses and have a Class B type fire extinguisher on hand.*

1 Remove the air cleaner cover and filter element (see Chapter 1). On later models with MPI, remove the battery, intake hoses and air flow sensor.

Note: *Always inspect the filter element for contamination or moisture when you remove it.*

2 Label and detach any wiring or hoses interfering with the air cleaner assembly removal.
3 Remove the nuts or bolts retaining the air cleaner assembly **(see illustrations)**.
4 Remove the air cleaner assembly.
5 Installation is the reverse of removal.

8.3b Use a ratchet, socket and proper-length extension to remove the mounting bolts for the air cleaner assembly (fuel-injected engines)

10 Lower the tank enough to disconnect the electrical wires and ground strap from the fuel pump/fuel gauge sending unit, if you have not already done so.
11 Remove the tank from the vehicle.
12 Installation is the reverse of removal.

7 Fuel tank – cleaning and repair

1 All repairs to the fuel tank or filler neck should be carried out by a professional who has experience in this critical and potentially dangerous work. Even after cleaning and flushing of the fuel system, explosive fumes can remain and ignite during repair of the tank.
2 If the fuel tank is removed from the vehicle, it should not be placed in an area where sparks or open flames could ignite the fumes coming out of the tank. Be especially careful inside garages where a natural gas-type appliance is located, because the pilot light could cause an explosion.

9 Carburetor – removal and installation

Warning: *Gasoline is extremely flammable, so take extra precautions when you work on any part of the fuel system. Don't smoke or allow open flames or bare light bulbs near the work area, and don't work in a garage where a natural gas-type appliance (such as a water heater or clothes dryer) with a pilot light is present. If you spill any fuel on your skin, rinse it off immediately with soap and water. When you perform any kind of work on the fuel tank, wear safety glasses and have a Class B type fire extinguisher on hand.*

Removal

1 Remove the fuel filler cap to relieve fuel tank pressure.
2 Remove the air cleaner assembly (see Section 8).
3 Disconnect the throttle cable from the throttle lever (see Section 16).
4 If the vehicle is equipped with an automatic transmission, disconnect the transmission kick-down cable (if equipped) from the throttle lever (see Chapter 7B).
5 Clearly label all vacuum hoses and fittings, then disconnect the hoses.

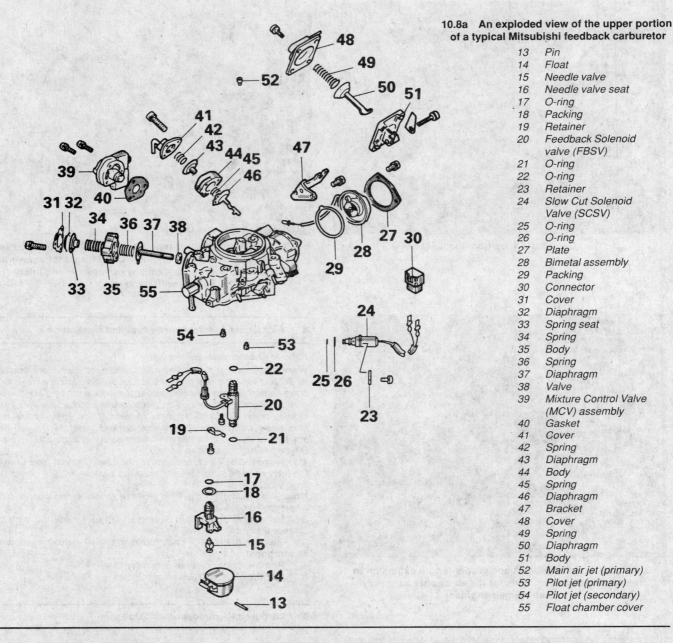

10.8a An exploded view of the upper portion of a typical Mitsubishi feedback carburetor

13	Pin
14	Float
15	Needle valve
16	Needle valve seat
17	O-ring
18	Packing
19	Retainer
20	Feedback Solenoid valve (FBSV)
21	O-ring
22	O-ring
23	Retainer
24	Slow Cut Solenoid Valve (SCSV)
25	O-ring
26	O-ring
27	Plate
28	Bimetal assembly
29	Packing
30	Connector
31	Cover
32	Diaphragm
33	Spring seat
34	Spring
35	Body
36	Spring
37	Diaphragm
38	Valve
39	Mixture Control Valve (MCV) assembly
40	Gasket
41	Cover
42	Spring
43	Diaphragm
44	Body
45	Spring
46	Diaphragm
47	Bracket
48	Cover
49	Spring
50	Diaphragm
51	Body
52	Main air jet (primary)
53	Pilot jet (primary)
54	Pilot jet (secondary)
55	Float chamber cover

6   Disconnect the fuel line from the carburetor.
7   Label the wires and terminals, then unplug all wire harness connectors.
8   Remove the mounting bolts and detach the carburetor from the intake manifold. Remove the carburetor mounting gasket. Stuff a shop rag into the intake manifold opening.

Installation

9   Use a gasket scraper to remove all traces of gasket material and sealant from the intake manifold (and the carburetor, if it's being reinstalled), then remove the shop rag from the manifold openings. Clean the mating surfaces with lacquer thinner or acetone.
10 Place a new gasket on the intake manifold.
11 Position the carburetor on the gasket and install the mounting fasteners.
12 To prevent carburetor distortion or damage, tighten the fasteners in a criss-cross pattern, 1/4-turn at a time.
13 The remaining installation steps are the reverse of removal.

14 Check and, if necessary, adjust the idle speed (see Chapter 1).
15 If the vehicle is equipped with an automatic transmission, refer to Chapter 7B for the kickdown cable adjustment procedure.
16 Start the engine and check carefully for fuel leaks.

10 Carburetor – diagnosis and overhaul

Refer to illustrations 10.8a and 10.8b

Warning: *Gasoline is extremely flammable, so take extra precautions when you work on any part of the fuel system. Don't smoke or allow open flames or bare light bulbs near the work area, and don't work in a garage where a natural gas-type appliance (such as a water heater or clothes dryer) with a pilot light is present. If you spill any fuel on your skin, rinse it off immediately with soap and water. When you perform any kind of work on the fuel tank, wear safety glasses and have a Class B type fire extinguisher on hand.*

Chapter 4 Fuel and exhaust systems

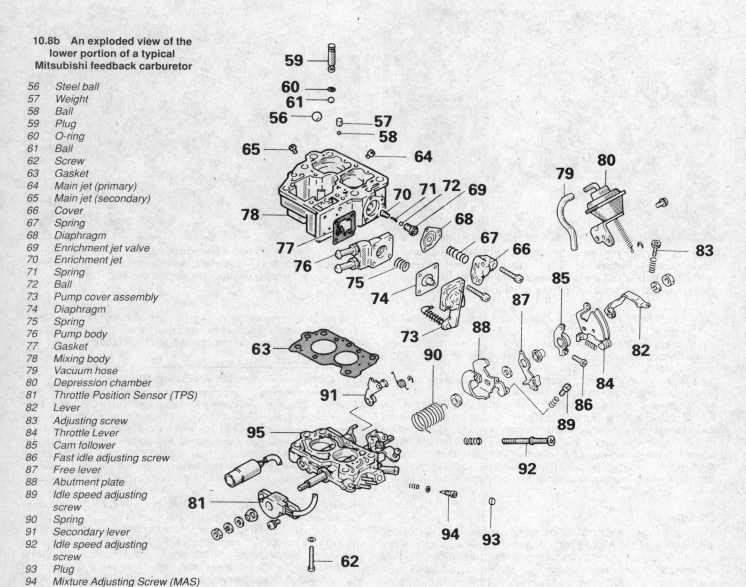

10.8b An exploded view of the lower portion of a typical Mitsubishi feedback carburetor

56	Steel ball
57	Weight
58	Ball
59	Plug
60	O-ring
61	Ball
62	Screw
63	Gasket
64	Main jet (primary)
65	Main jet (secondary)
66	Cover
67	Spring
68	Diaphragm
69	Enrichment jet valve
70	Enrichment jet
71	Spring
72	Ball
73	Pump cover assembly
74	Diaphragm
75	Spring
76	Pump body
77	Gasket
78	Mixing body
79	Vacuum hose
80	Depression chamber
81	Throttle Position Sensor (TPS)
82	Lever
83	Adjusting screw
84	Throttle Lever
85	Cam follower
86	Fast idle adjusting screw
87	Free lever
88	Abutment plate
89	Idle speed adjusting screw
90	Spring
91	Secondary lever
92	Idle speed adjusting screw
93	Plug
94	Mixture Adjusting Screw (MAS)
95	Throttle body

Diagnosis

1 A thorough road test and check of carburetor adjustments should be done before any major carburetor service work. Specifications for some adjustments are listed on the Vehicle Emissions Control Information (VECI) label found in the engine compartment.

2 Carburetor problems usually show up as flooding, hard starting, stalling, severe backfiring and poor acceleration. A carburetor that's leaking fuel and/or covered with wet looking deposits definitely needs attention.

3 Some performance complaints directed at the carburetor are actually a result of loose, out-of-adjustment or malfunctioning engine or electrical components. Others develop when vacuum hoses leak, are disconnected or are incorrectly routed. The proper approach to analyzing carburetor problems should include the following items:

a) Inspect all vacuum hoses and actuators for leaks and correct installation (see Chapters 1 and 6).
b) Tighten the intake manifold and carburetor mounting nuts/bolts evenly and securely.
c) Perform a cylinder compression test (see Chapter 2).
d) Clean or replace the spark plugs as necessary (see Chapter 1).
e) Check the spark plug wires (see Chapter 1).
f) Inspect the ignition primary wires.
g) Check the ignition timing (follow the instructions printed on the Vehicle Emissions Control Information label).
h) Check the fuel pump (see Section 3).
i) Check the heat control valve in the air cleaner for proper operation (see Chapter 1).
j) Check/replace the air filter element (see Chapter 1).
k) Check the PCV system (see Chapter 6).
l) Check/replace the fuel filter (see Chapter 1). Also, the strainer in the tank could be restricted.
m) Check for a plugged exhaust system.
n) Check EGR valve operation (see Chapter 6).
o) Check the choke – it should be completely open at normal engine operating temperature (see Chapter 1).
p) Check for fuel leaks and kinked or dented fuel lines (see Chapters 1 and 4).
q) Check accelerator pump operation with the engine off (remove the air cleaner cover and operate the throttle as you look into the carbu-

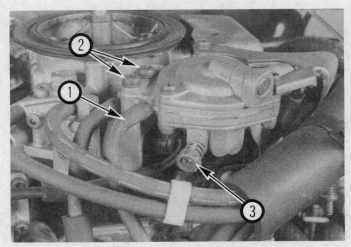

11.2a Inspect the idle-up actuator hose (1) for cracks and deterioration – remove it from the vacuum fitting before testing the actuator; take out the two screws (2) to remove the actuator; adjust the actuator with the adjustment screw (3)

11.2b Inspect the hoses and wiring harness connected to the idle-up actuator (1) – to detach the solenoid mounting bracket from the engine, take out the two bolts (2)

retor throat – you should see a steady stream of gasoline enter the carburetor).
r) Check for incorrect fuel or bad gasoline.
s) Check the valve clearances and camshaft lobe lift (see Chapters 1 and 2).
t) Have a dealer service department or repair shop check the electronic engine and carburetor controls.

4   Diagnosing carburetor problems may require that the engine be started and run with the air cleaner off. While running the engine without the air cleaner, backfires are possible. This situation is likely to occur if the carburetor is malfunctioning, but just the removal of the air cleaner can lean the fuel/air mixture enough to produce an engine backfire. **Warning:** *Do not position any part of your body, especially your face, directly over the carburetor during inspection and servicing procedures. Wear eye protection!*

Overhaul

5   Once it's determined that the carburetor needs an overhaul, several options are available. If you're going to attempt to overhaul the carburetor yourself, first obtain a good quality carburetor rebuild kit (which will include all necessary gaskets, internal parts, instructions and a parts list). You'll also need some special solvent and a means of blowing out the internal passages of the carburetor with air.

6   An alternative is to obtain a new or rebuilt carburetor. They are readily available from dealers and auto parts stores. Make absolutely sure the exchange carburetor is identical to the original. A tag is usually attached to the top of the carburetor or a number is stamped on the float bowl. It will help determine the exact type of carburetor you have. When obtaining a rebuilt carburetor or a rebuild kit, make sure the kit or carburetor matches your application exactly. Seemingly insignificant differences can make a large difference in engine performance.

7   If you choose to overhaul your own carburetor, allow enough time to disassemble it carefully, soak the necessary parts in the cleaning solvent (usually for at least one-half day or according to the instructions listed on the carburetor cleaner) and reassemble it, which will usually take much longer than disassembly. When disassembling the carburetor, match each part with the illustration in the carburetor kit and lay the parts out in order on a clean work surface. Overhauls by inexperienced mechanics can result in an engine which runs poorly or not at all. To avoid this, use care and patience when disassembling the carburetor so you can reassemble it correctly.

8   Because carburetor designs are constantly modified by the manufacturer in order to meet increasingly more stringent emissions regulations, it isn't feasible to include a step-by-step overhaul of each type. You'll receive a tailed, well illustrated set of instructions with any carburetor overhaul kit;

they will apply in a more specific manner to the carburetor on your vehicle. Exploded views of the carburetors are included here to help you if you decide to overhaul the carburetor yourself **(see illustrations)**.

11   Carburetor – on-vehicle checks and adjustments

Warning: *Gasoline is extremely flammable, so take extra precautions when you work on any part of the fuel system. Don't smoke or allow open flames or bare light bulbs near the work area, and don't work in a garage where a natural gas-type appliance (such as a water heater or clothes dryer) with a pilot light is present. If you spill any fuel on your skin, rinse it off immediately with soap and water. When you perform any kind of work on the fuel tank, wear safety glasses and have a Class B type fire extinguisher on hand.*

Note: *If your vehicle's engine is hard to start or does not start at all, has an unstable idle or poor driveability and you suspect the carburetor is malfunctioning, it's best to first take the vehicle to a dealer service department that has the Feedback Carburetor equipment necessary to diagnose this highly complicated system. The following procedures are intended to help the home mechanic verify proper operation of components, make minor adjustments, and replace some components. They are not intended as troubleshooting procedures.*

Idle speed and mixture adjustment

1   Idle speed adjustment is covered in Chapter 1. Idle mixture adjustment cannot be performed by the home mechanic because an infrared gas analyzer is required. A dealer service department must perform this procedure.

Electrical and power steering load idle-up system check

Refer to illustrations 11.2a and 11.2b

2   Inspect the vacuum hoses and the idle-up control solenoid harness – make sure they're properly connected **(see illustrations)**.
3   Detach the vacuum hose from the fitting on the idle-up actuator/dashpot.
4   Attach a vacuum pump to the idle-up actuator/dashpot vacuum fitting.
5   Hook up a tachometer in accordance with the manufacturer's instructions.
6   Start the engine and allow it to idle.
7   Apply 11.8 in-Hg vacuum with the vacuum pump. Engine speed should increase.
8   If engine speed does not increase, replace the idle-up actuator/dashpot, as discussed below. Stop the engine and remove the vacuum pump.

Chapter 4 Fuel and exhaust systems

4–11

11.11 To detach the idle-up actuator rod from the free lever, pull the rod in the direction shown by the arrow

11.23 To adjust the idle-up actuator for proper idle speed under an air conditioning load, turn the throttle adjusting screw (arrow) until the tachometer indicates the specified speed

Idle-up actuator/dashpot replacement and adjustment

Refer to illustration 11.11

Replacement

9 Remove the throttle return spring from the throttle lever.
10 Remove the two idle-up actuator/dashpot attaching screws (see illustration 11.2a).
11 Detach the idle-up actuator rod from the free lever (see illustration) and remove the idle-up actuator.
12 Install the new idle-up actuator/dashpot and reattach the vacuum hose.
13 Adjust the idle-up actuator/dashpot as discussed below.

Adjustment

Note: *The following procedure adjusts the idle-up actuator's control of the idle speed when electric or power steering loads are applied.*

14 Make sure that the curb idle speed is set as specified in Chapter 1. If it's not, re-adjust the idle speed to the specified speed before proceeding.
15 Unbolt the solenoid mounting bracket (see illustration 11.2b). Remove the electrical connector from the bottom of the idle-up control solenoid. Using jumper wires, connect one of the solenoid's terminals to the positive terminal of the battery and the other solenoid terminal to the negative terminal of the battery. This applies intake manifold vacuum to the idle-up actuator, which activates the actuator.
16 Open the throttle slightly – until engine speed reaches about 2000 rpm – then slowly close it.
17 Note the indicated engine speed. Adjust it (if necessary) to the specified rpm with the throttle opener adjustment screw (see illustration 11.2a).
18 Repeat Step 16 above and check the engine speed again.
19 Remove the jumper wire attached in Step 15 and reattach the wire harness.

Idle-up air conditioning load actuator adjustment

Refer to illustration 11.23

Note: *The following procedure adjusts the idle-up actuator's control of the idle speed when an air conditioning load is applied.*

20 Hook up a tachometer in accordance with the manufacturer's instructions.
21 Start the engine.
22 Turn on the air conditioner switch. This opens the solenoid valve, which allows the intake manifold vacuum to move the actuator to its full open position. Note the indicated engine speed and compare your reading to the specified rpm.
23 If the engine speed is out of specification, adjust it with the throttle adjusting screw (see illustration).

Idle-up control solenoid valve check

24 Set the ignition switch in the "OFF" position.
25 Remove the solenoid (follow the vacuum line from the dashpot to locate it) and detach the electrical connector from the idle-up control solenoid (see illustration 11.2b).
26 Check the solenoid valve coil with an ohmmeter. Compare your reading with the specified resistance. If the indicated resistance is not as specified, there is an open or short in the solenoid coil. Replace the valve.

Deceleration, enrichment, jet mixture solenoid valves and bowl vent valve check

Refer to illustrations 11.28, 11.30a, 11.30b, 11.30c and 11.30d

27 Unplug the solenoid valve connector.
28 Check each solenoid valve coil with an ohmmeter between the indicated connector terminals. For the bowl vent valve, measure between the connector terminal and the negative post of the battery (see illustration).

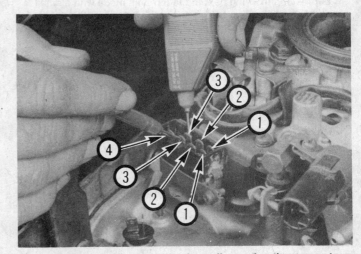

11.28 To check the solenoid valve coils, unplug the connector and measure the resistance across the indicated pairs of terminals with an ohmmeter (for the bowl vent valve, measure between the indicated terminal and the negative post of the battery), then compare your readings with the specified resistance for each coil:

1 Jet mixture solenoid valve
2 Enrichment solenoid valve
3 Deceleration solenoid valve
4 Bowl vent valve

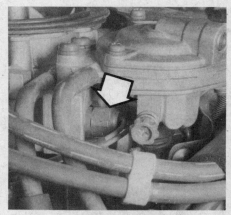
11.30a Location of the deceleration solenoid valve

11.30b Location of the enrichment solenoid valve

11.30c Location of the jet mixture solenoid valve

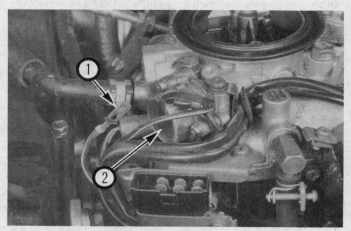
11.30d If replacing one of the carburetor solenoids, detach the terminal blade (1) from the solenoid control valve connector by depressing the tang next to the terminal blade with a small screwdriver and pulling the blade out the bottom of the connector – to remove the bowl vent solenoid and valve assembly (2), take out the three mounting screws

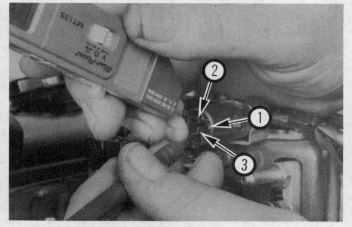
11.32 To check the Throttle Position Sensor (TPS), unplug the connector and measure the resistance between terminals 2 and 3

29 Compare your readings with the specified resistance.
30 If the indicated resistance for any solenoid valve is not as specified, it should be replaced (see illustrations).

Throttle Position Sensor (TPS) check and adjustment

Refer to illustrations 11.32, 11.35, 11.37, 11.40a and 11.40b

Check
31 Unplug the TPS connector.
32 Check the resistance between terminals 2 and 3 with an ohmmeter (see illustration).
33 Verify that the resistance changes smoothly as the throttle valve is slowly turned from a closed position to wide open. Note your readings and compare them to the specified resistance. If the resistance does not change smoothly or is not in specification, replace the sensor and adjust it.

Adjustment
34 Warm up the engine.
35 Loosen the throttle cable and verify the fast idle cam is released (see illustration).
36 Stop the engine.
37 While counting and recording the number of turns, back off (turn counterclockwise) the speed adjusting screws (SAS-1 and SAS-2) (see illustration) until the throttle valve is fully closed.

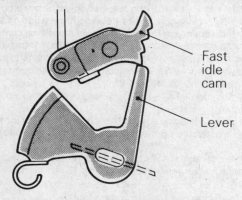
11.35 Verify the fast idle cam is released (the lever must not be resting on the cam)

38 Attach a digital voltmeter between terminals 2 and 3 of the TPS connector (see illustration 11.32). Note: *Don't disconnect the TPS connector from the main wire harness.*
39 Turn the ignition on (don't start the engine), measure the TPS output voltage and compare your reading to the specified output voltage.
40 If the output voltage is incorrect, adjust the output voltage. Depending on the type of TPS on the vehicle, either turn the adjusting screw or loosen the TPS mounting screws and rotate it (see illustrations).
41 Turn the ignition switch to Off.

Chapter 4 Fuel and exhaust systems

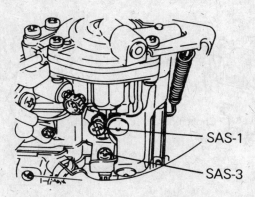
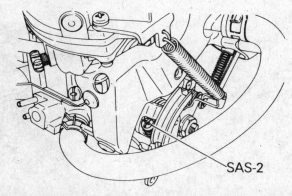

11.37 Back off the speed adjusting screws (SAS-1 and SAS-2) until the throttle valve is fully closed – be sure to count the number of turns (down to 1/4 -turn) so you can later return the screws to the same positions

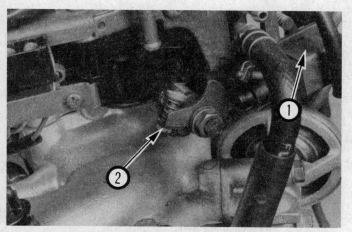

11.40a If your carburetor has this type of TPS (1), use the screw (2) to adjust it.

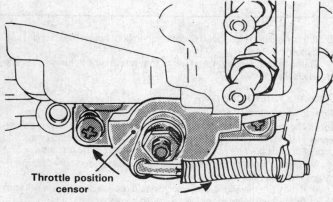

11.40b If your carburetor has this type of TPS, loosen the mounting screws and adjust it by rotating the sensor.

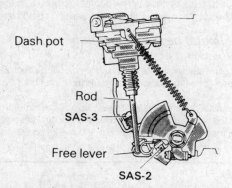

11.46 With the engine idling, open the throttle valve until the free lever contacts SAS-3

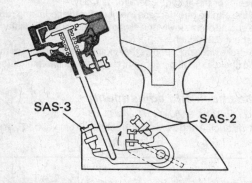

11.47 Close the throttle valve until SAS-2 contacts the free lever and note the indicated idle speed at that point

42 Retighten (turn clockwise) the SAS-1 and SAS-2 screws the same number of turns that you backed them off in Step 37 above.
43 Adjust the throttle cable freeplay (see Section 16).
44 Start the engine and verify the idle speed is within specifications (see Chapter 1).

Dashpot check and adjustment

Refer to illustrations 11.46 and 11.47

45 Before checking the dashpot:
  a) Start the engine and warm it up (coolant temperature must be between 176 and 203-degrees F).
  b) Turn off all lights and electrical accessories. Make sure that the cooling fan is off.
  c) Place the transmission in Neutral (manual transaxle) or Park (automatic transaxle).
  d) If the vehicle has power steering, make sure that the wheel are pointed straight ahead.
46 With the engine idling, open the throttle valve the full stroke of the rod until the free lever contacts SAS-3 **(see illustration)**.
47 Close the throttle valve until the SAS-2 contacts the free lever **(see illustration)** and note the indicated engine idle speed at that moment.

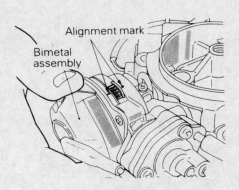

11.50 Verify the alignment marks on the electric choke and bimetal assembly line up properly

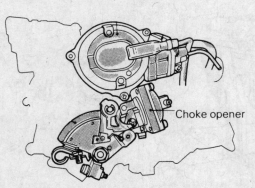

11.60 Detach the vacuum hose from the choke opener

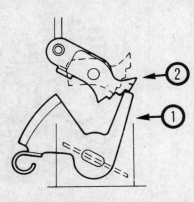

11.61 Set the lever (1) on the second highest step of the fast idle cam (2)

48 If the indicated idle speed is not as specified, adjust the dashpot setting by turning SAS-3.
49 Release the free lever and verify that the engine returns to its idle speed slowly.

Choke system check

Refer to illustration 11.50
Note: *The choke related parts are factory adjusted, so no further adjustments should be necessary unless you rebuild the carburetor, or a smog inspection indicates that choke related parts need to be adjusted.*

50 Verify that the alignment marks on the electric choke/bimetal assembly are lined up **(see illustration)**.
  a) If they are misaligned in a clockwise direction, the engine will start better but the spark plugs are probably sooty.
  b) If they are misaligned in a counterclockwise direction, the engine will be hard to start and will be more likely to stall.
51 With the engine coolant temperature below 50-degrees F, depress the accelerator pedal. The choke should be fully closed. If it isn't, either the bimetal heater or the linkage operation is faulty.
52 Start the engine and depress the accelerator pedal and run it at idle; the choke should open slowly and slightly (immediately after starting), with a gap of about 0.059-inch.
53 If the choke does not open as the engine warms up, check the bimetal heater, the choke pull-off diaphragm(s) and the choke pull-off(s) vacuum circuits.

Fast idle check and adjustment

Refer to illustrations 11.60, 11.61 and 11.63

54 Start the engine and warm up the coolant to between 176 and 203-degrees F.
55 Make sure that all lights, the cooling fan and all electrical accessories are off.
56 Place the transaxle in Neutral (manual transaxle) or Park (automatic transaxle).
57 If the vehicle is equipped with power steering, make sure that the wheels are pointed straight ahead.
58 Remove the air cleaner assembly (see Section 8).
59 Hook up a tachometer in accordance with the manufacturer's instructions.
60 Detach the vacuum hose from the choke opener **(see illustration)**.
61 Set the lever on the second highest step of the fast idle cam **(see illustration)**.
62 Start the engine and check the fast idle speed. Compare your reading with the specified value.
63 If the fast idle speed is out of specification, adjust it with the fast idle adjusting screw **(see illustration)**.
  a) If you turn the fast idle screw in a clockwise direction, the valve opening will be larger and the fast idle speed should increase.
  b) If you turn the fast idle screw counterclockwise, the valve opening will be smaller and the the fast idle speed should decrease.

Choke bimetal heater inspection

Refer to illustration 11.64

64 Unplug the electric choke heater connector and check the heater with an ohmmeter **(see illustration)**. It should indicate about 6 ohms at room temperature.
65 If the resistance is not as specified, replace the bimetal assembly (by drilling out the housing rivets). When installing a new assembly, be sure to line up the marks on the assembly with the marks on the housing. Install new rivets.

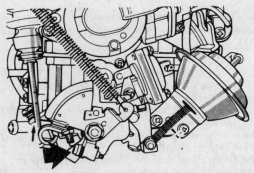

11.63 If the fast idle speed is out of specification, adjust it with the fast idle adjusting screw (arrow)

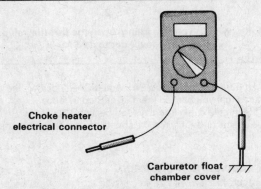

11.64 To check the electric choke heater, unplug the connector and check its resistance with an ohmmeter

Chapter 4 Fuel and exhaust systems

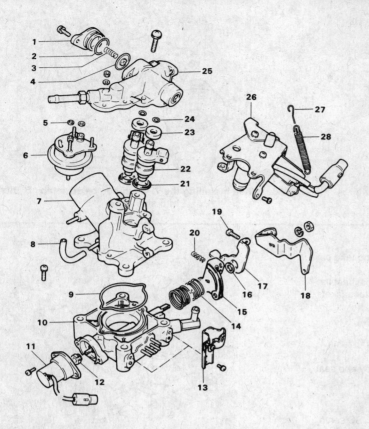

13.13 An exploded view of the injection mixer assembly

1. Pulsation damper cover
2. O-ring
3. Spring
4. Pulsation damper
5. O-rings (2)
6. Fuel pressure regulator
7. Mixing body
8. Hose
9. Seal ring
10. Throttle body assembly
11. Throttle position sensor
12. Joint
13. Connector bracket
14. Return spring
15. Throttle lever
16. Ring
17. Free lever
18. Kickdown lever
19. Adjusting screw
20. Spring
21. Seal rings (2)
22. Injectors
23. Collars (2)
24. O-rings (2)
25. Injector retainer
26. ISC servo assembly
27. Damper spring
28. Return spring

12 Fuel injection system – check

Warning: *Gasoline is extremely flammable, so take extra precautions when you work on any part of the fuel system. Don't smoke or allow open flames or bare light bulbs near the work area, and don't work in a garage where a natural gas-type appliance (such as a water heater or clothes dryer) with a pilot light is present. If you spill any fuel on your skin, rinse it off immediately with soap and water. When you perform any kind of work on the fuel tank, wear safety glasses and have a Class B type fire extinguisher on hand.*

Note: *the following procedure is based on the assumption that the fuel pump and fuel pressure are normal (see Section 3).*

1. Check all wiring harness connectors that are related to the system. Loose connectors and poor grounds can cause many problems that resemble more serious malfunctions.
2. Check to see that the battery is fully charged, as the control unit and sensors depend on an accurate supply voltage in order to properly meter the fuel.
3. Check the air filter element – a dirty or partially blocked filter will severely impede performance and economy (see Chapter 1).
4. If a blown fuse is found, replace it and see if it blows again. If it does, search for a grounded wire in the harness to the fuel pump.
5. Check the air intake duct from the airflow sensor/air filter to the intake manifold for leaks, which will result in an excessively lean mixture. Also check the condition of the vacuum hoses connected to the intake manifold.
6. Remove the air intake duct from the throttle body and check for dirt, carbon or other residue build-up in the throttle body. If it's dirty, clean with carburetor cleaner and a toothbrush.
7. On Multi Point Injection (MPI), with the engine running, place a screwdriver against each injector, one at a time, and listen through the handle for a clicking sound, indicating operation.
8. On all models, the remainder of the system checks should be left to a qualified repair shop, as there is a chance that the control unit may be damaged if not performed properly.

13 Throttle Body Injection (TBI) assembly – component removal, overhaul and installation

Refer to illustrations 13.13, 13.27, 13.35 and 13.48

Removal

1. Relieve the fuel pressure (see Section 2).
2. Disconnect the cable from the negative terminal of the battery.
3. Drain the coolant down to the intake manifold level or below.
4. Remove the water hose (connected between the injection mixer and the intake manifold) at the injection mixer nipple.
5. Detach the air intake pipe from the injection mixer.
6. Detach the throttle cable from the throttle lever of the injection mixer.
7. Carefully label, then detach the vacuum hoses from the injection mixer nipples.
8. Unplug the wire harness connectors at the injectors.
9. Unplug the wire harness connectors from the ISC servo and the TPS.
10. Detach the fuel inlet pipe and the fuel return hose from the injection mixer. **Warning:** *Even though the fuel pressure has been relieved, remove the two fuel inlet pipe bolts carefully and pull the pipe loose slowly to allow any remaining pressure to escape. If the fuel pipe is abruptly removed, residual pressure could cause gasoline to spray.*
11. Reference mark the position of the throttle position sensor to it's mounting. Remove the throttle position sensor.
12. Disconnect the rubber hose from the fuel pressure regulator and the mixing body.
13. Remove the injector retainer screws and remove the retainer (**see illustration**).

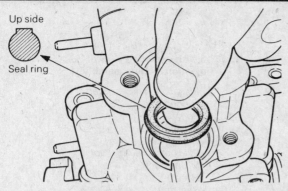

13.27 Be sure the injector seal ring is installed correctly

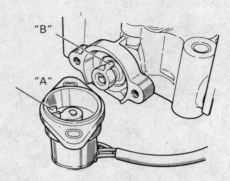

13.35 When installing the TPS sensor, be sure tab "B" fits into slot "A"

14 Remove the fuel pressure regulator from the retainer.
15 Remove the pulsation damper cover from the retainer and take out the spring and diaphragm.
16 Pull the injectors from the mixing body. Do not attempt to extract the injectors with pliers — you will damage them. Remove the injector gaskets from the body.
17 Remove the throttle return spring and the damper spring.
18 Remove the connector bracket.
19 Remove the ISC servo mounting bracket retaining bolts and remove the ISC servo and bracket.
20 Remove the two screws and then remove the mixing body and seal ring from the throttle body.

Overhaul

21 With the exception of the throttle position sensor and the ISC servo, clean all parts in solvent. Immersion of the TPS or the ISC servo in solvent will damage their insulation. Wipe them with a clean rag instead. Before cleaning the injectors and fuel pressure regulator with solvent, seal up the fuel inlet and outlet with Teflon tape. Check all vacuum ports and fuel passage for clogging. Clean them out with compressed air.

Installation

22 Install the new seal ring into the groove of the throttle body.
23 Install the mixing body onto the throttle body and tighten the screws firmly.
24 Install the ISC servo and bracket and tighten the screws firmly.
25 Install the connector bracket to the throttle body and clamp the ISC connector into the bracket.
26 Install the throttle return spring and the damper spring.
27 Insert the new seal rings into the mixing body. When installing the seal ring, be sure to install it with the flat face side up (see illustration).
28 Install the new O-rings onto the injectors.
29 Install the injectors into the mixing body. Push the injector down firmly by hand. If the injector(s) are being replaced, each injector has an identification mark stamped on the body. Make sure that the mark on the new injector matches the mark on the old injector.
30 Insert the pulsation damper diaphragm into the injector retainer.
31 Install the pulsation damper spring and cover, then tighten the screws.
32 Install the new O-rings onto the fuel pressure regulator, then install the regulator to the injector retainer.
33 Check the filters in the retainer for clogging or damage. Replace as necessary.
34 Install the injector retainer and push down firmly. Tighten the screws alternately a little at a time.
35 Using the reference marks made during disassembly, install the throttle position sensor onto the throttle body (see illustration).
36 Clean both gasket surfaces of the injection mixer and the intake manifold.
37 Install the new injection mixer gasket on the intake manifold.
38 Install the injection mixer assembly on the intake manifold and tighten

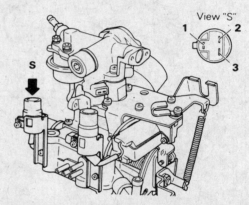

13.48 Connect a circuit tester across terminals 1 and 2 or 2 and 3 of the TPS connector to verify the resistance changes when the throttle valve is opened

the mounting bolts.
39 Install a new O-ring into the groove of the high-pressure hose and connect the high-pressure hose to the injection mixer. Tighten the hose mounting bolts firmly.
40 Connect the fuel return hose to the nipple of the fuel pressure regulator.
41 Plug in all harness electrical connectors to the injectors, ISC servo and sensors.
42 Clamp the connector for the throttle position sensor into its bracket.
43 Connect the vacuum hoses to their respective nipples on the injection mixer.
44 Connect the water hose to the nipple of the injection mixer.
45 Install the air intake pipe.
46 Refill the cooling system.
47 Connect the cable to the negative terminal of the battery.
48 Check for proper installation of the throttle position sensor. Measure the resistance value between terminals 1 and 2 or 2 and 3 while moving the throttle lever from open to closed (see illustration). If the resistance value changes, the throttle position sensor is properly installed.
49 Start the engine and warm it to normal operating temperature.
50 Run the engine and check for fuel and water leaks.
51 The throttle position sensor adjustment should be checked, take the vehicle to the dealer service department to have it checked and adjusted if necessary.

14 Fuel filter – general information

The models covered by this manual are equipped with two fuel filters – one inside the fuel tank and one either in the fuel feed line (fuel-injected

Chapter 4 Fuel and exhaust systems

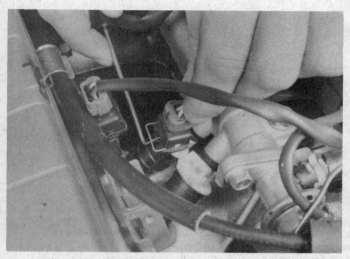

15.7 To disconnect the wiring to the injectors, first remove the metal retaining clips with a scribe or similar pointed tool

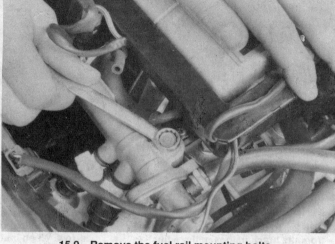

15.9 Remove the fuel rail mounting bolts

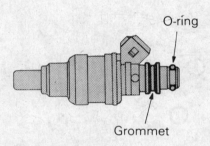

15.15 Always replace the injector O-rings. Use a light coating of fuel to ease installation and prevent damage

15.18 To remove a bolted-on fuel pressure regulator, simply remove the mounting bolts.

models) or in the engine compartment (carbureted models). See Chapter 1 for information on replacing the in-line or engine compartment fuel filter. The in-tank fuel filter seldom requires servicing.

15 Multi-Point Injection (MPI) – component removal and installation

Refer to illustrations 15.7, 15.9, 15.15 and 15.18

Throttle body

1 Remove the throttle return spring and disconnect the throttle cable from the throttle body (see Section 16).
2 Label and detach all wiring and hoses connected to the throttle body.
3 Remove the four nuts mounting the throttle body to the intake plenum.
4 Remove the throttle body.
5 Before installing the throttle body, scrape away all traces of old gasket material with a gasket scraper or putty knife. **Warning:** *Be careful not to gouge or otherwise damage the delicate aluminum surfaces.* Clean the gasket mating surfaces with lacquer thinner or acetone. The remainder of installation is the reverse of removal, being sure to install a new throttle body gasket.

Fuel rail

6 Relieve the fuel pressure (see Section 2)
7 Label and detach the wiring connected to the injectors **(see illustration)**.
8 Label and detach all hoses connected to the fuel rail.

9 Remove the two bolts retaining the rail assembly to the intake plenum **(see illustration)**.
10 Remove the fuel rail, the injectors, pulsation damper and the fuel regulator as an assembly.
11 Installation is the reverse of removal. **Note:** *Using a light coating of fuel on the injector O-rings will ease installation and reduce the possibility of damage* **(see illustration 15.15b)**. Use new O-rings whenever the injectors are removed or installed, regardless of whether the injector itself is being replaced.
12 Crank the engine to restore fuel pressure, then check for fuel leaks.

Fuel injector(s)

13 Remove the fuel rail (see procedure above).
14 Pull the injector(s) from the fuel rail.
15 Installation is the reverse of removal. **Note:** *Using a light coating of fuel on the injector O-rings will ease installation and reduce the possibility of damage* **(see illustration)**. Be sure to use new O-rings. The O-rings should be installed in the cylinder head.
16 Crank the engine to restore fuel pressure, then check for fuel leaks.

Fuel pressure regulator

Bolted-on type

17 Detach the vacuum line to the regulator.
18 Remove the mounting bolts and the regulator **(see illustration)**.
19 Installation is the reverse of removal. **Note:** *Use a light coating of fuel*

4-18 Chapter 4 Fuel and exhaust systems

16.2 To detach the throttle cable from the bracket, loosen the adjusting nut (arrow)

16.4 Remove the trim panel under the dash (if necessary) and detach the throttle cable at the point shown by the arrow. When installing a cable apply a dab of grease to the cable at the pedal arm contact area

16.5 To detach the throttle cable from the firewall, remove the two mounting bolts and separate the guide from the firewall

on the new regulator O-ring to ease installation.
20 Crank the engine to restore fuel pressure, then check for fuel leaks.

Screwed-in type
21 Detach the vacuum line.
22 Loosen the locknut on the regulator.
23 Unscrew the regulator.
24 To install, reverse the removal procedure.
25 Crank the engine to restore fuel pressure, then check for fuel leaks.

16 Throttle cable – removal, installation and adjustment

Models without cruise control
Cordia, Tredia, Mirage (except MPI) and Precis (except MPI)
Refer to illustrations 16.2, 16.4 and 16.5
1 On carbureted models, remove the air cleaner assembly (see Section 8).
2 On all models, loosen the throttle cable adjusting nut **(see illustration)** and lift the cable housing off the bracket.
3 Detach the throttle cable from the throttle lever.
4 Detach the throttle cable from the accelerator pedal **(see illustration)**.
5 Detach the throttle cable guide from the firewall **(see illustration)**.
6 From outside the vehicle, pull the throttle cable through the firewall.
7 Installation is the reverse of removal. Don't leave any sharp bends in the cable.
8 After installing the cable, adjust the freeplay, as follows:
  a) Run the engine until it reaches normal operating temperature. Verify the idle speed is correct and adjust it if necessary (see Chapter 1).
  b) Verify that the throttle cable has no slack in it.
  c) If the cable is slack, adjust it as follows:
  d) Turn the adjusting nut counterclockwise until the throttle lever is free.
  e) Remove any sharp bends from the accelerator cable.
  f) Loosen the locknut and turn the throttle cable adjusting nut clockwise to the point at which the throttle lever just begins to move, then back off the adjusting nut one turn and tighten the locknut securely.

Galant, Mirage and Precis (MPI)
Refer to illustration 16.12

Removal and installation
9 Detach the cable from the engine and firewall mounting points.

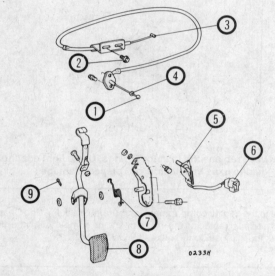

16.12 Galant, Mirage and Precis throttle cable assembly (without cruise control)

1 Accelerator side inner cable	6 Accelerator switch connector (A/T)
2 Adjusting bolt	7 Return spring
3 Throttle body side inner cable	8 Accelerator pedal
4 Bushing	9 Split pin
5 Accelerator arm bracket	

10 Detach the cable from the pedal **(see illustration 16.4)**.
11 Detach the throttle cable guide from the firewall **(see illustration 16.5)**. From outside the vehicle, pull the throttle cable through the firewall.
12 Remove the adjusting bolts near the throttle body **(see illustration)**, detach the cable from the throttle lever and remove the cable.
13 Installation is the reverse of removal. Don't leave any sharp bends in the cable.

Adjustment
14 Run the engine until it reaches normal operating temperature. Verify the idle speed is correct and adjust it if necessary (see Chapter 1). Turn off the engine.
15 Depress the accelerator pedal and check to be sure that the throttle lever moves smoothly from fully open to fully closed.

Chapter 4 Fuel and exhaust systems

16 Remove any sharp bends from the accelerator cable.
17 Check the arrangement and placement of the cable.
18 Turn the ignition key to the "ON" position for fifteen seconds (to actuate the ISC motor).
19 Loosen the two bolts retaining the cable near the throttle body.
20 Adjust the cable to remove the remainder of the slack from it.
21 Tighten the two bolts retaining the cable near the throttle body.
22 After adjusting the cable, be sure the throttle lever contacts the ISC motor.

Models with cruise control
Cordia and Tredia
Refer to illustration 16.23

Accelerator pedal cable
23 Back off the cable adjusting nuts and detach the cable from the throttle lever and bracket **(see illustration)**.
24 Detach the cable from the pedal **(see illustration 16.4)**.
25 Detach the cable from the firewall **(see illustration 16.5)**.
26 From the engine compartment, pull the cable through the firewall.
27 Installation is the reverse of removal. Be sure there are no sharp bends in the cable.

28 With the engine idling, use the adjusting nuts to remove the freeplay from the cable.
29 Tighten the adjusting nuts against each other.
30 Rev the engine to be sure a binding condition is not present.

Actuator cable
31 Detach the cable from the throttle lever.
32 Detach the cable from the actuator and remove the cable. Installation is the reverse of removal. After installation, adjust the cable freeplay as described beginning with the next Step.
33 With the engine idling, use the adjusting nuts to remove the freeplay from the cable.
34 Tighten the adjusting nuts against each other.
35 Rev the engine to be sure a binding condition is not present.

Galant
Refer to illustration 16.37

Removal and installation
Note: *This procedure applies to both accelerator cables A and B in illustration 16.37.*
36 Remove the cover from the speed control actuator.
37 Loosen the throttle cable adjusting nut (of the cable being replaced) and lift the cable housing off the bracket **(see illustration)**.

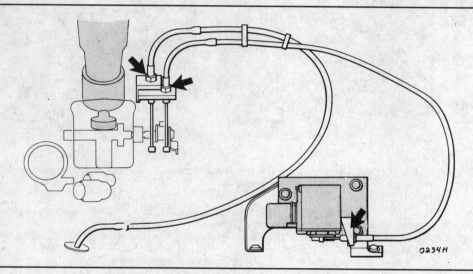

16.23 Throttle cable details for Cordia and Tredia with cruise control. Adjustment nuts are shown by the arrows

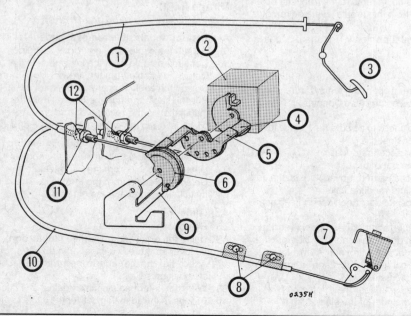

16.37 Throttle cable details for Galant with cruise control

1 Accelerator cable A
2 Cruise control actuator
3 Accelerator pedal
4 Selector
5 Intermediate link B
6 Intermediate link A
7 Throttle lever
8 Adjusting bolts
9 Stopper
10 Accelerator cable B
11 Adjusting nut A
12 Lock nut

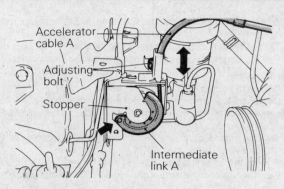

16.56 Throttle cable intermediate link details for Mirage MPI models.

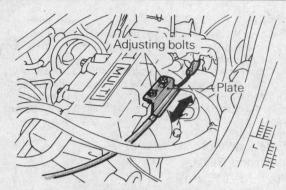

16.66 Use the adjusting bolts to adjust the freeplay for the throttle lever cable on Mirage MPI models

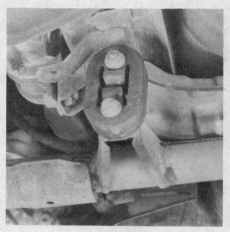

Front hanger

Middle hanger

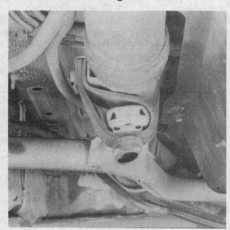

Rear hanger

17.1 Anytime you raise the vehicle – even if you're not planning to service the exhaust system – inspect and, if warranted, replace the rubber hangers (Precis shown)

38 Detach the throttle cable from the intermediate link.
39 If replacing the cable to the accelerator pedal, detach the cable from the pedal. Detach the throttle cable guide from the firewall (see illustration 16.4). From outside the vehicle, pull the throttle cable through the firewall.
40 If replacing the cable to the throttle body, remove the retaining bolts near the throttle body, detach the cable from the throttle lever and remove the cable.
41 Installation is the reverse of removal. Don't leave any sharp bends in the cable.

Adjustment

42 Run the engine until it reaches normal operating temperature. Verify the idle speed is correct and adjust it if necessary (see Chapter 1). Turn off the engine.
43 Depress the accelerator pedal and check to be sure the throttle lever moves smoothly from fully open to fully closed.
44 Remove any sharp bends from the accelerator cable.
45 Check the arrangement and placement of the cable.
46 With the cover of the speed control actuator removed, loosen the adjustment nuts and locknuts to free the intermediate links.
47 Turn the ignition key to the "ON" position for fifteen seconds (to actuate the ISC motor).
48 With intermediate link "A" (pedal cable) against the stopper (see illustration 16.37), use adjusting nut "A" to adjust the cable so intermediate link "A" just begins to move then, back off adjusting nut "A" two turns.
49 Using adjusting nut "B", tighten the cable so the lever on intermediate link "B" just contacts the lever on intermediate link "A".
50 Loosen the two bolts retaining the cable near the throttle body.
51 Adjust the cable to remove the remainder of the slack from it.
52 Tighten the intermediate link locknuts and the two bolts retaining the cable near the throttle body.
53 After adjusting, be sure the throttle lever contacts the ISC motor.
54 Install the cover of the speed control actuator.

Mirage (MPI)

Refer to illustrations 16.56 and 16.66

Accelerator pedal cable removal and installation

55 Detach the cable housing from the engine.
56 Detach the cable from the intermediate link (see illustration).
57 Detach the cable from the firewall (see illustration 16.5).
58 Detach the cable from the accelerator pedal (see illustration 16.4).
59 Pull the cable through the firewall from the engine compartment.
60 Installation is the reverse of removal. After installation, be sure no sharp bends are in the cable and that a binding condition is not present.

Throttle lever cable removal and installation

61 Detach the cable housing from the engine.
62 Detach the cable from the intermediate link
63 Detach the cable from the throttle lever and remove it.
64 Installation is the reverse of removal.

Adjustment

65 Remove the air cleaner.
66 Using the adjusting bolts (see illustration), adjust the cable to the throttle lever so there is a slight amount of freeplay at the lever.
67 After adjusting the cable to the throttle lever, be sure the lever touches the idle position switch.
68 With the intermediate link against the stopper (see illustration 16.56), loosen the adjusting bolt for the cable to the accelerator pedal and

Chapter 4 Fuel and exhaust systems

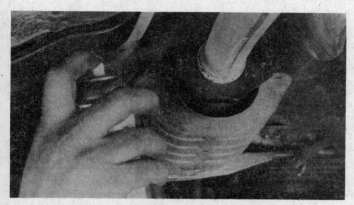

17.4 Exhaust system bolts and nuts, particularly those on the exhaust manifold and catalytic converters, can be very difficult to loosen – spraying them with a penetrant will free up the threads

remove all the freeplay from the cable.
69 Be sure to check that the throttle valve fully opens and closes with the use of the accelerator pedal.
70 Install the air cleaner.

17 Exhaust system servicing – general information

Refer to illustrations 17.1 and 17.4
Warning: *Inspection and repair of exhaust system components should be done only after enough time has elapsed after driving the vehicle to allow the system components to cool completely. Also, when working under the vehicle, make sure it is securely supported on jackstands.*

1 The exhaust system consists of the exhaust manifold(s), the catalytic converter, the muffler, the tailpipe and all connecting pipes, brackets, hangers and clamps. The exhaust system is attached to the body with mounting brackets and rubber hangers **(see illustration)**. If any of the parts are improperly installed, excessive noise and vibration will be transmitted to the body.
2 Conduct regular inspections of the exhaust system to keep it safe and quiet. Look for any damaged or bent parts, open seams, holes, loose connections, excessive corrosion or other defects which could allow exhaust fumes to enter the vehicle. Deteriorated exhaust system components should not be repaired; they should be replaced with new parts.
3 If the exhaust system components are extremely corroded or rusted together, welding equipment will probably be required to remove them. The convenient way to accomplish this is to have a muffler repair shop remove the corroded sections with a cutting torch. If, however, you want to save money by doing it yourself (and you don't have a welding outfit with a cutting torch), simply cut off the old components with a hacksaw. If you have compressed air, special pneumatic cutting chisels can also be used. If you do decide to tackle the job at home, be sure to wear safety goggles to protect your eyes from metal chips and work gloves to protect your hands.
4 Here are some simple guidelines to follow when repairing the exhaust system:
   a) Work from the back to the front when removing exhaust system components.
   b) Apply penetrating oil to the exhaust system component fasteners **(see illustration)** to make them easier to remove.
   c) Use new gaskets, hangers and clamps when installing exhaust systems components.
   d) Apply anti-seize compound to the threads of all exhaust system fasteners during reassembly.
   e) Be sure to allow sufficient clearance between newly installed parts and all points on the underbody to avoid overheating the floor pan and possibly damaging the interior carpet and insulation. Pay particularly close attention to the catalytic converter and heat shield.
5 To replace the front catalytic converter refer to Chapter 2A and remove the exhaust manifold and converter as an assembly. Then unbolt the converter from the manifold. Installation is the reverse of removal.

18 Turbocharger – general information

Refer to illustration 18.2
A turbocharger system is used to increase power. As increased power is required and the throttle is opened, more air/fuel mixture is forced into the combustion chambers by the turbocharger.
A turbine wheel in the turbocharger is driven by the exhaust stream and is connected by a shaft to drive a compressor wheel (fan) to pressurize the intake air stream **(see illustration)**. The amount of pressurized air al-

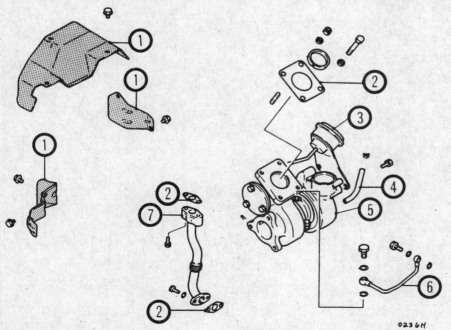

18.2 Details of the turbocharger system

1 Heat shields
2 Gasket
3 Wastegate
4 Vacuum line
5 Turbocharger
6 Oil-feed line
7 Oil-return line

lowed into the engine is controlled by an exhaust by-pass valve (wastegate).

19 Turbocharger – inspection

Caution: *Operation of the turbocharger without all the ducts and filters installed can result in personal injury and/or allow foreign objects to damage the wheel blades.*

Note: *The turbocharger is not serviceable and must be replaced as a unit.*

1 Every turbocharger has its own noise level when operating. If the noise level changes, suspect a problem.
   If the sound of the turbocharger goes up and down in pitch:
   a) Check for heavy dirt build up in the compressor housing and on the compressor wheel.
   b) Check for an air inlet restriction.
   If the noise level is a high pitch or whistling:
   c) Look for an inlet air or exhaust gas leak.
2 With the engine off and the turbocharger stopped, make a visual inspection of the turbocharger and components.
3 Check for loose duct connections from the air cleaner to the turbocharger.
4 Be sure the duct from the turbocharger-to-intake system is not loose.
5 Visually check the wheels of the turbocharger for damage from foreign objects.
6 Look for evidence of wheel-to-housing contact.
7 Be sure the shaft rotates freely. Rotating stiffness could indicate the presence of sludged oil or coking (hardened oil deposits) from overheating.
8 Push in on one of the shaft wheels while turning it. Be sure the wheels turn freely without contacting the housings.
9 Be sure the exhaust manifold has no loose connections or cracks.
10 Check the oil drain line for any restrictions.
11 Visually inspect the actuator and wastegate linkage for damage.
12 Check the hose to the wastegate.

20 Turbocharger – removal and installation

1 Remove the heat protectors **(see illustration 18.2)**.
2 Disconnect the oil return pipe from the oil pan.
3 Remove the oil pipe from the turbocharger and the oil filter bracket.
4 Remove the air intake pipe connecting bolt.
5 Disconnect the remaining hoses and electrical connectors from the turbocharger.
6 Remove the turbocharger mounting nuts and remove the turbocharger assembly from the exhaust manifold.
7 Pour clean engine oil into the turbocharger.
8 Installation is the reverse of removal. Be sure to use a new gasket.

Chapter 5 Engine electrical systems

Contents

Alternator – removal and installation	16
Battery cables – check and replacement	4
Battery check and maintenance	See Chapter 1
Battery – emergency jump starting	2
Battery – removal and installation	3
Centrifugal advance mechanism – check and component replacement	12
Charging system – check	15
Charging system – general information and precautions	14
Distributor pickup air gap – check and adjustment	13
Distributor – removal and installation	8
Drivebelt check, adjustment and replacement	See Chapter 1
General information	1
Igniter – replacement	10
Ignition coil – check and replacement	7
Ignition system – check	6
Ignition system – general information and precautions	5
Ignition timing check and adjustment	See Chapter 1
Pickup coil check and replacement	11
Spark plug replacement	See Chapter 1
Spark plug wire, distributor cap and rotor check and replacement	See Chapter 1
Starter motor – testing in vehicle	19
Starter motor – removal and installation	20
Starter solenoid – removal and installation	21
Starting system – general information and precautions	18
Vacuum advance unit – check and replacement	9
Voltage regulator/alternator brushes – replacement	17

Specifications

Distributor pickup air gap

Note: *The air gap check and adjustment procedure is applicable only to the models listed below.*

Cordia and Tredia	
1983	0.008 to 0.015 in
1984 and 1985	
1.8L	0.008 to 0.015 in
2.0L	0.031 in
1986	0.008 to 0.015 in
1987 and 1988	
1.8L	Not available
2.0L	0.008 to 0.015 in
Galant	
1985 and 1986	0.031 in
1987 on	Not available
Mirage (1985 through 1988 1.5L)	0.008 to 0.015 in
Precis (1987 through 1989)	0.031 in

Ignition coil

Primary resistance
  Cordia/Tredia
    1983 .... 1.04 to 1.27 ohms
    1984 through 1986
      1.8L .... 1.15 ohms
      2.0L .... 1.2 ohms
    1987
      1.8L .... 1.25 ohms
      2.0L .... 1.2 ohms
    1988
      1.8L .... 1.13 to 1.37 ohms
      2.0L .... 1.08 to 1.32 ohms
  Galant
    1985 and 1986 .... 1.2 ohms
    1987 .... 0.72 to 0.88 ohms
    1988 .... Not available
    1989 on
      SOHC .... 0.72 to 0.88 ohms
      DOHC .... 0.77 to 0.95 ohms
  Mirage
    1985 through 1988 (all) .... 1.2 ohms
    1989 and later 1.6L engine .... 0.77 to 0.95 ohms
    1.5L engine
      1989 .... 0.72 to 0.88 ohms
      1990 and later .... 0.9 to 1.2 ohms
  Precis
    1987 .... Not available
    1988 .... 1.1 to 1.3 ohms
    1989 .... Not available
    1990 .... 0.72 to 0.88 ohms

Secondary resistance
  Cordia/Tredia
    1983 .... 7.10 to 9.60k ohms
    1984 through 1986
      1.8L .... 8.35k ohms
      2.0L .... 13.7k ohms
    1987
      1.8L .... 11.0k ohms
      2.0L .... 13.7k ohms
    1988
      1.8L .... 9.4 to 12.6k ohms
      2.0L .... 11.7 to 15.7k ohms
  Galant
    1985 .... 13.7k ohms
    1986 .... 26.0k ohms
    1987 .... 10.8 to 13.2k ohms
    1988 .... Not available
    1989 on
      SOHC .... 10.3 to 13.9k ohms
      DOHC .... 10.3 to 13.9k ohms
  Mirage
    1985 through 1988 .... 13.7k ohms
    1989 and 1990 1.5L engine; 1989 and later 1.6L engine .... 10.3 to 13.9K ohms
    1991 and later 1.5L engine .... 20 to 29k ohms
  Precis
    1987 through 1989 .... 11.6 to 15.8k ohms
    1990 and later .... 11.3 to 13.9k ohms

Pickup coil resistance

Note: *The pickup coil resistance check is applicable only to the models listed below*

Cordia and Tredia
  1.8L Turbo (1984 through 1988) .... 920 to 1120 ohms
  2.0L (1986 through 1988) .... 130 to 190 ohms
Mirage 1.6L (1985 through 1988) .... 920 to 1120 ohms

Chapter 5 Engine electrical systems

1 General information

The engine electrical systems include all ignition, charging and starting components. Because of their engine-related functions, these components are discussed separately from chassis electrical devices such as the lights, the instruments, etc. (which are included in Chapter 12).

Always observe the following precautions when working on the electrical systems:

a) Be extremely careful when servicing engine electrical components. They are easily damaged if checked, connected or handled improperly.
b) Never leave the ignition switch on for long periods of time with the engine off.
c) Don't disconnect the battery cables while the engine is running.
d) Maintain correct polarity when connecting a battery cable from another vehicle during jump starting.
e) Always disconnect the negative cable first and hook it up last or the battery may be shorted by the tool being used to loosen the cable clamps.

It's also a good idea to review the safety-related information regarding the engine electrical systems located in the Safety first section near the front of this manual before beginning any operation included in this Chapter.

2 Battery – emergency jump starting

Refer to the Booster battery (jump) starting procedure at the front of this manual.

3 Battery – removal and installation

Refer to illustration 3.1

1 **Caution:** *Always disconnect the negative cable first and hook it up last or the battery may be shorted by the tool being used to loosen the cable clamps.* Disconnect both cables from the battery terminals **(see illustration)**.
2 Remove the battery hold down clamp.
3 Lift out the battery. Be careful – it's heavy.
4 While the battery is out, inspect the carrier (tray) for corrosion (see Chapter 1).
5 If you are replacing the battery, make sure that you get one that's identical, with the same dimensions, amperage rating, cold cranking rating, etc.
6 Installation is the reverse of removal.

4 Battery cables – check and replacement

1 Periodically inspect the entire length of each battery cable for damage, cracked or burned insulation and corrosion. Poor battery cable connections can cause starting problems and decreased engine performance.
2 Check the cable-to-terminal connections at the ends of the cables for cracks, loose wire strands and corrosion. The presence of white, fluffy deposits under the insulation at the cable terminal connection is a sign that the cable is corroded and should be replaced. Check the terminals for distortion, missing mounting bolts and corrosion.
3 When removing the cables, always disconnect the negative cable first and hook it up last or the battery may be shorted by the tool used to loosen the cable clamps. Even if only the positive cable is being replaced, be sure to disconnect the negative cable from the battery first (see Chapter 1 for further information regarding battery cable removal).
4 Disconnect the old cables from the battery, then trace each of them to their opposite ends and detach them from the starter solenoid and ground terminals. Note the routing of each cable to ensure correct installation.

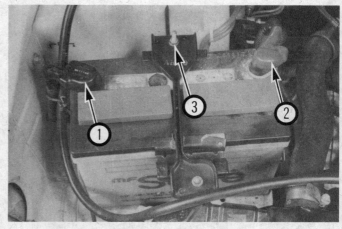

3.1 To remove the battery . . .
1 Detach the cable from the negative terminal.
2 Detach the cable for the positive terminal.
3 Remove the nuts and detach the clamp

5 If you are replacing either one or both of the cables, take them with you when buying new cables. It is vitally important that you replace the cables with identical parts. Cables have characteristics that make them easy to identify: positive cables are usually red, larger in cross-section and have a larger diameter battery post clamp; ground cables are usually black, smaller in cross-section and have a slightly smaller diameter clamp for the negative post.
6 Clean the threads of the solenoid or ground connection with a wire brush to remove rust and corrosion. Apply a light coat of battery terminal corrosion inhibitor, or petroleum jelly, to the threads to prevent future corrosion.
7 Attach the cable to the solenoid or ground connection and tighten the mounting nut/bolt securely.
8 Before connecting a new cable to the battery, make sure that it reaches the battery post without having to be stretched.
9 Connect the positive cable first, followed by the negative cable.

5 Ignition system – general information and precautions

When working on the ignition system, take the following precautions:

a) Do not keep the ignition switch on for more than 10 seconds if the engine will not start.
b) Always connect a tachometer in accordance with the manufacturer's instructions. Some tachometers may be incompatible with this ignition system. Consult a dealer service department before buying a tachometer for use with this vehicle.
c) Never allow the primary terminals of the ignition coil to touch ground.
d) Do not disconnect the battery when the engine is running.

Distributor equipped models – SOHC engines

The ignition system includes the ignition switch, the battery, the coil, the primary (low voltage) and secondary (high voltage) wiring circuits, the distributor and the spark plugs. On later distributors not using a vacuum or centrifugal advance, the ignition timing is controlled by the Electronic Control Unit (ECU).

Distributorless Ignition System (DIS) – DOHC engines

The DIS system consists of spark plugs, spark plug wires, ignition coils, crank angle sensor, power transistor, Electronic Control Unit (ECU)

Chapter 5 Engine electrical systems

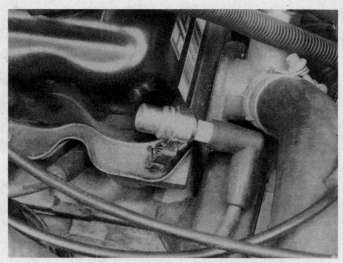

6.2 To use a calibrated ignition tester (available at most auto parts stores), simply disconnect a spark plug wire, attach the wire to the tester and clip the tester to a good ground – if there is enough power to fire the plug, sparks will be clearly visible between the electrode tip and the tester body as the engine is turned over

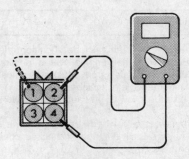

7.2 Terminal identification for the primary circuit of the 1989 DIS coil

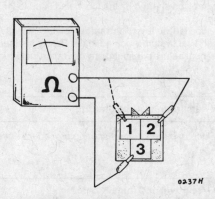

7.5 Terminal identification for the primary circuit of the 1990 DIS coil

and related input devices which monitor engine functions (such as rpm, intake air volume, engine temperature, etc.); the ECU ensures a perfectly timed spark under all conditions.

6 Ignition system – check

Refer to illustration 6.2

Warning: *Because of the very high voltage generated by the ignition system, extreme care should be taken when this check is performed.*

1 If the engine turns over but won't start, disconnect the spark plug wire from any spark plug and attach it to a calibrated ignition tester (available at most auto parts stores).
2 Connect the clip on the tester to a bolt or metal bracket on the engine **(see illustration).** If you're unable to obtain a calibrated ignition tester, remove the wire from one of the spark plugs and, using an insulated tool, hold the end of the wire about 1/4-inch from a good ground.
3 Crank the engine and watch the end of the tester or spark plug wire to see if bright blue, well-defined sparks occur. If you're not using a calibrated tester, have an assistant crank the engine for you. **Warning:** *Keep clear of drivebelts and other moving engine components that could injure you.*
4 If sparks occur, sufficient voltage is reaching the plug to fire it (repeat the check at the remaining plug wires to verify the wires, distributor cap and rotor [if equipped] and other coils [DIS systems] are OK). However, the plugs themselves may be fouled, so remove them and check them as described in Chapter 1.
5 On non-DIS models, if no sparks or intermittent sparks occur, remove the distributor cap and check the cap and rotor as described in Chapter 1. If moisture is present, dry out the cap and rotor, then reinstall the cap.
6 On non-DIS models, if there's still no spark, detach the coil secondary wire from the distributor cap and hook it up to the tester (reattach the plug wire to the spark plug), then repeat the spark check. Again, if you don't have a tester, hold the end of the wire about 1/4-inch from a good ground. If sparks occur now, the distributor cap, rotor or plug wire(s) may be defective.
7 If no sparks occur, check the wire connections at the coil to make sure they're clean and tight. Check for voltage to the coil. Make any necessary repairs, then repeat the check again.

8 On non-DIS models, if there's still no spark, the coil-to-cap wire may be bad (check the resistance with an ohmmeter – it should be 7000 ohms per foot or less). If a known good wire doesn't make any difference in the test results, the ignition module may be defective.

7 Ignition coil – check and replacement

Check

Distributorless Ignition System (DIS)
Refer to illustrations 7.2, 7.5 and 7.7

Primary circuit (1989)
1 Disconnect the primary wiring connector for the coils.
2 Using an ohmmeter, measure the resistance between the coil connector terminals **(see illustration)** No. 4 and 2 (No. 1 and 4 cylinders). Measure the resistance between terminals No. 4 and 1 (No. 2 and 3 cylinders).
3 Compare the readings to the Specifications at the beginning of this Chapter. Replace any coil if not to specification.

Primary circuit (1990)
4 Disconnect the primary wiring connector for the coils.
5 Using an ohmmeter, measure the resistance between the coil connector terminals **(see illustration)** No. 3 and 2 (No. 1 and 4 cylinders). Measure the resistance between terminals No. 3 and 1 (No. 2 and 3 cylinders).
6 Compare the readings to the Specifications at the beginning of this Chapter. Replace any coil if not to specification.

Chapter 5 Engine electrical systems

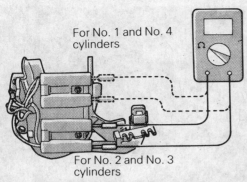

7.7 On the DIS coil, check between the two secondary terminals to check the resistance of the secondary circuit

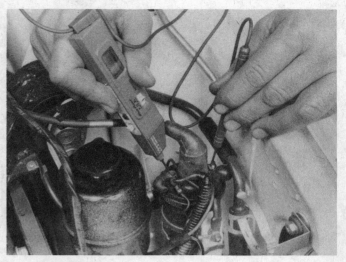

7.11 Checking the coil resistance of the primary circuit (typical coil is shown)

Secondary circuit

7   Using an ohmmeter, measure the resistance between the high voltage (secondary) terminals for cylinders the No. 1 and 4 **(see illustration)**. Measure the resistance between the high voltage terminals for cylinders No. 2 and 3.

8   Compare the readings to the Specifications at the beginning of this Chapter. Replace any coil if not to specification.

All others

Refer to illustration 7.11

9   Detach the cable from the negative terminal of the battery.
10 Locate the coil.
11 Using an ohmmeter, check the coil:
   a) Measure the resistance between the positive and negative terminals **(see illustration)**. Compare your reading with the primary coil resistance listed in this Chapter's Specifications.
   b) Detach the high tension lead from the coil. Measure the resistance between the positive terminal and high tension terminal. Compare your reading with the secondary coil resistance listed in this Chapter's Specifications.
12 If either of the above tests yield resistance values outside the specified resistance, replace the coil.

Replacement

13 Detach the cable from the negative terminal of the battery.
14 Detach the wiring to the coil assembly.
15 Detach the coil assembly from it's mounting.
16 Remove the coil assembly.
17 Installation is the reverse of removal.

8   Distributor – removal and installation

Refer to illustration 8.6

Removal

1   Detach the cable from the negative battery terminal.
2   Unplug the coil high tension lead from the distributor cap.
3   Detach the vacuum hose(s) from the advance unit (if equipped).
4   Look for a raised "1" on the distributor cap. This marks the location for the number one cylinder spark plug wire terminal. If the cap does not have a mark for the number one terminal, locate the number one spark plug and trace the wire back to the terminal on the cap.
5   Remove the distributor cap (see Chapter 1) and turn the engine over until the rotor is pointing toward the number one spark plug terminal (see locating TDC procedure in Chapter 2).
6   Make a mark on the edge of the distributor base directly below the rotor tip and in line with it **(see illustration)**. Also, mark the distributor base and the engine block to ensure that the distributor is installed correctly.
7   Remove the distributor hold-down nut and washer, then pull the distributor straight out to remove it. **Caution:** *DO NOT turn the crankshaft*

8.6 Before removing the distributor, paint or scribe an alignment mark on the edge of the distributor base directly beneath the rotor tip – DO NOT use a lead pencil

while the distributor is out of the engine, or the alignment marks will be useless.

8   Detach the primary wiring from the distributor.

Installation

Note: *If the crankshaft has been moved while the distributor is out, the number one piston must be repositioned at TDC. This can be done by feeling for compression pressure at the number one plug hole as the crankshaft is turned. Once compression is felt, align the TDC marks on the drivebelt pulley and the timing cover.*

9   Insert the distributor into the engine in exactly the same relationship to the block that it was in when removed.
10 To mesh the helical gears on the camshaft and the distributor, it may be necessary to turn the rotor slightly. Recheck the alignment marks between the distributor base and the block to verify that the distributor is in the same position it was in before removal. Also check the rotor to see if it's aligned with the mark you made on the edge of the distributor base.
11 Loosely install the nut.
12 Reconnect the electrical leads.
13 Install the distributor cap.
14 Reattach the vacuum line(s) to the advance unit (if equipped).

9.7 To detach the vacuum unit from the distributor remove the two screws (arrows) . . .

9.8 . . . then tilt it down, detach it from the pin (arrow) on the breaker base and pull it out

Replacement

6 Pull the rotor off the shaft (see Chapter 1).
7 Remove the vacuum unit mounting screws **(see illustration)**.
8 Remove the vacuum unit link from the pin on the breaker base, then detach the vacuum unit **(see illustration)**.
9 Installation is the reverse of removal.

10 Igniter – replacement

1 Refer to the exploded views for the procedure **(see illustrations 12.7a, 12.7b, 12.7c and 12.7d)**.
2 When installing the igniter, do not wipe away the conductive grease on it.
3 Adjust the air gap (see Section 13) before reinstalling the distributor.

11 Pickup coil check and replacement

Refer to illustration 11.3
Note: *This procedure applies only to the models listed below.*

Cordia/Tredia 1.8L Turbo (1984 and 1985); Mirage 1.6L (1985)

1 Remove the distributor cap.
2 Remove the rotor from the distributor.
3 Using an ohmmeter, check the resistance of the pickup coil by probing it's measuring terminals **(see illustration)**. Compare the results to this Chapter's Specifications.
4 Replace the pickup coil if not within the specifications **(see illustration 12.7d)**.

Cordia/Tredia 1.8L Turbo (1986 through 1988); Mirage 1.6L (1986 through 1988)

5 To check the pickup coil resistance, detach the distributor primary wires and use an ohmmeter to check the resistance between the two leads **(see illustration 12.7c)**.
6 Compare the reading to the specifications at the front of this chapter.
7 Replace the pickup coil if not within the specifications.

Cordia/Tredia 2.0L (1986 through 1988)

8 Disconnect the pickup coil leads **(see illustration 12.7b)**.
9 Using and ohmmeter, check the resistance between the terminals of the pickup coil.
10 Compare the reading to the specification at the beginning of

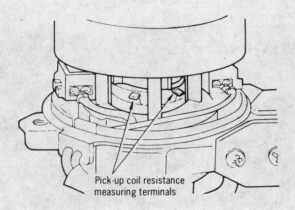

11.3 On Cordia/Tredia 1.8L Turbo (1984 and 1985) and Mirage 1.6L (1985), use the measuring terminals shown to check the resistance of the pickup coil

15 Reattach the spark plug wires to the plugs (if removed).
16 Connect the cable to the negative terminal of the battery.
17 Check the ignition timing (see Chapter 1) and tighten the distributor hold-down nut securely.

9 Vacuum advance unit – check and replacement

Refer to illustrations 9.7 and 9.8

Check

1 Remove the vacuum line from the vacuum advance unit and plug the line. If there are two lines, disconnect and plug the outer (furthest from the distributor) line.
2 Connect a timing light to the vehicle.
3 Attach a vacuum pump to the advance unit.
4 With the engine running, gradually apply vacuum to the advance unit while watching the timing marks with the timing light.
5 The timing should gradually advance. If it doesn't advance, but the unit holds vacuum, the advance plate in the distributor is binding. If it doesn't advance, and the unit does not hold vacuum, replace the unit.

Chapter 5 Engine electrical systems

this Chapter.
11 Replace the pickup coil if not within the specifications.

12 Centrifugal advance mechanism – check and component replacement

Refer to illustrations 12.7a, 12.7b, 12.7c and 12.7d
1 Detach the vacuum line(s) from vacuum advance unit on the distributor.
2 Connect a timing light according to it's manufacturer's instructions.
3 Start the engine.
4 While watching the timing marks with the use of the timing light, raise the engine speed while watching the timing marks advance.
5 The timing should advance smoothly.
6 If the timing does not advance smoothly, check the advance weights, springs and plate for damage or binding.
7 Replace parts as needed **(see illustrations)**.

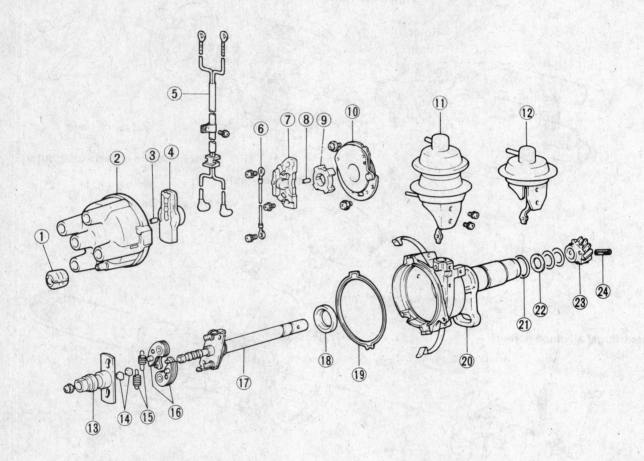

12.7a Exploded view of the distributor for Cordia and Tredia 2.0L (1984 and 1985), Galant (1985 and 1986), Mirage 1.5L (1985 through 1988) and Precis (1987 through 1989)

    1   Breather
    2   Cap
    3   Contact carbon
    4   Rotor
    5   Cable assembly
    6   Ground wire
    7   Igniter
    8   Dowel pin
    9   Signal rotor
    10 Advance plate
    11 Vacuum advance unit (dual-diaphragm type)
    12 Vacuum advance unit (single-diaphragm type)
    13 Rotor shaft
    14 Spring retainers
    15 Centrifugal advance springs
    16 Centrifugal advance weights
    17 Distributor shaft
    18 Oil seal
    19 Packing
    20 Housing
    21 O-ring
    22 Washer
    23 Driven gear
    24 Pin

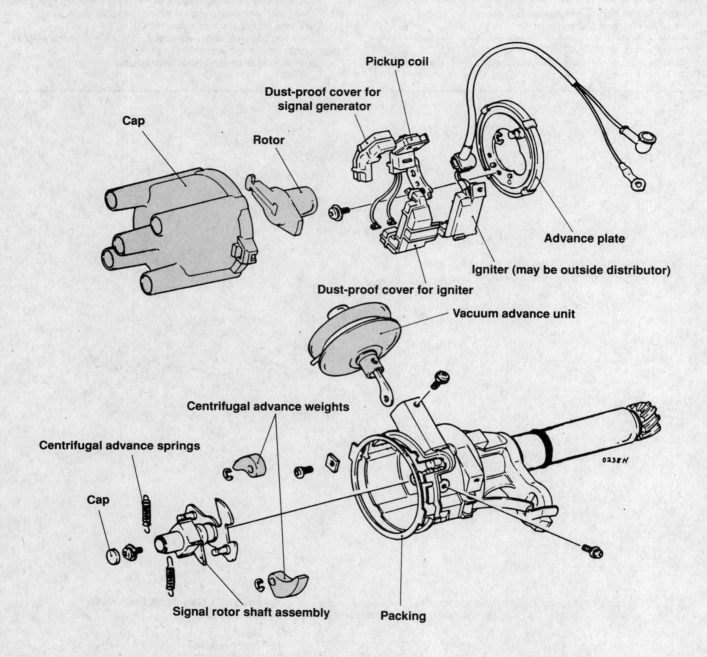

12.7b Exploded view of the distributor for Cordia and Tredia 2.0L (1986 through 1988)

Chapter 5 Engine electrical systems

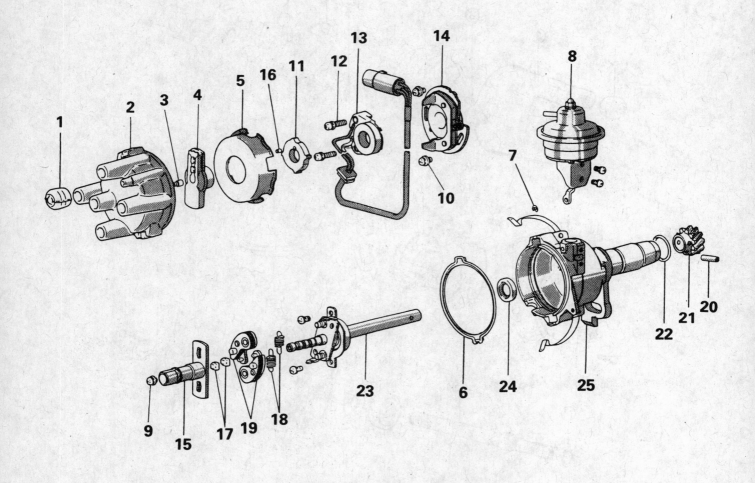

12.7c Exploded view of the distributor for Cordia and Tredia 1.8L Turbo (1986 through 1988) and Mirage 1.6L (1986 through 1988)

1	Breather	10	Screw	18	Centrifugal advance spring
2	Cap	11	Signal rotor	19	Centrifugal advance weight
3	Contact carbon	12	Screw	20	Spring pin
4	Rotor	13	Pick-up coil	21	Driven gear
5	Cover	14	Stator	22	O-ring
6	Seal ring	15	Rotor shaft	23	Shaft
7	Snap-ring	16	Pin	24	Oil seal
8	Vacuum advance unit	17	Spring retainer	25	Housing
9	Screw				

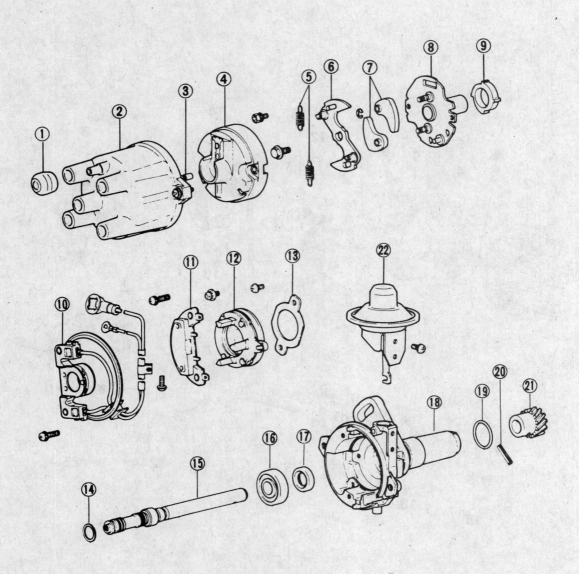

12.7d Exploded view of the distributor for Cordia and Tredia 1.8L Turbo (1984 and 1985) and Mirage 1.6L (1985)

1 Breather
2 Cap
3 Contact carbon
4 Rotor
5 Centrifugal advance springs
6 Centrifugal advance plate
7 Centrifugal advance weights
8 Centrifugal advance base
9 Signal rotor
10 Pickup coil
11 Ignitor
12 Breaker
13 Plate
14 Washer
15 Shaft
16 Bearing
17 Oil seal
18 Housing
19 O-ring
20 Spring pin
21 Gear
22 Vacuum control

Chapter 5 Engine electrical systems

13.2 To adjust the air gap between the igniter and the signal rotor, loosen the igniter mounting screws, insert a feeler gauge of the specified thickness between the igniter and one of the signal rotor projections, push the igniter against the gauge and tighten the mounting screws

13 Distributor pickup air gap – check and adjustment

Refer to illustration 13.2

Note: *This procedure applies to Cordia and Tredia 2.0L (1983 through 1988), Galant (all models), Mirage 1.5L (1985 through 1988) and Precis (1987 through 1989)*

1 Anytime you disturb the pickup air gap in the distributor it must be properly adjusted.
2 Loosen the pickup mounting screws. Place a feeler gauge of the thickness listed in this Chapter's Specifications between one of the projections on the signal rotor and the pickup **(see illustration)**.
3 Gently push the pickup toward the signal rotor until it's a snug – not tight – fit against the feeler gauge.
4 Tighten the pickup mounting screws.
5 Check the adjustment by noting the amount of drag on the feeler gauge when you pull it out of the gap between the signal rotor and the pickup. You should feel a slight amount of drag. If you feel excessive drag on the gauge, the gap is probably too small. If you don't feel any drag on the gauge when you pull it out, the air gap is too large.

14 Charging system – general information and precautions

The charging system includes the alternator, with an integral voltage regulator, the battery, the fusible link(s) and wiring between all the components. The charging system supplies electrical power for the ignition system, the lights, the radio, etc. The alternator is driven by a drivebelt at the front of the engine.

The purpose of the voltage regulator is to limit the alternator's voltage to a preset value. This prevents power surges, circuit overloads, etc., during peak voltage output.

The fusible link can be either a short length of insulated wire integral with the engine compartment wiring harness or a fuse-like device installed in the underhood electrical panel. See Chapter 12 for additional information regarding fusible links.

The charging system doesn't ordinarily require periodic maintenance. However, the drivebelt, battery and wires and connections should be inspected at the intervals outlined in Chapter 1.

The dashboard warning light should come on when the ignition key is turned to Start, then go off immediately. If it remains on, there is a malfunction in the charging system (see Section 14).

Be very careful when making electrical circuit connections to a vehicle equipped with an alternator and note the following:
  a) When reconnecting wires to the alternator from the battery, be sure to note the polarity.
  b) Before using arc welding equipment to repair any part of the vehicle, disconnect the wires from the alternator and the battery terminals.
  c) Never start the engine with a battery charger connected.
  d) Always disconnect both battery leads before using a battery charger.
  e) The alternator is turned by an engine drivebelt which could cause serious injury if your hands, hair or clothes become entangled in it with the engine running.
  f) Because the alternator is connected directly to the battery, it could arc or cause a fire if overloaded or shorted out.
  g) Wrap a plastic bag over the alternator and secure it with rubberbands before steam cleaning the engine.

15 Charging system – check

1 If a malfunction occurs in the charging circuit, don't automatically assume that the alternator is causing the problem. First check the following items:
  a) Check the drivebelt tension and condition (Chapter 1). Replace it if it's worn or deteriorated.
  b) Make sure the alternator mounting and adjustment bolts are tight.
  c) Inspect the alternator wiring harness and the connectors at the alternator. They must be in good condition and tight.
  d) Check the fusible link(s). If burned, determine the cause, repair the circuit and replace the link (the vehicle won't start and/or the accessories won't work if the fusible link blows). Sometimes a fusible link may look good, but still be bad. If in doubt, remove it and check for continuity.
  e) Start the engine and check the alternator for abnormal noises (a shrieking or squealing sound indicates a bad bearing).
  f) Check the specific gravity of the battery electrolyte. If it's low, charge the battery (doesn't apply to maintenance free batteries).
  g) Make sure the battery is fully charged (one bad cell in a battery can cause overcharging by the alternator).
  h) Disconnect the battery cables (negative first, then positive). Inspect the battery posts and the cable clamps for corrosion. Clean them thoroughly if necessary (see Chapter 1). Reconnect the cable to the positive terminal.
  i) With the key off, connect a test light between the negative battery post and the disconnected negative cable clamp.
    1) If the test light does not come on, reattach the clamp and proceed to the next step.
    2) If the test light comes on, there is a short (drain) in the electrical system of the vehicle. The short must be repaired before the charging system can be checked.
    3) Disconnect the alternator wiring harness.
      (a) If the light goes out, the alternator is bad.
      (b) If the light stays on, pull each fuse until the light goes out (this will tell you which component is shorted).
2 Using a voltmeter, check the battery voltage with the engine off. It should be approximately 12-volts.
3 Start the engine and check the battery voltage again. It should now be approximately 14-to-15 volts.
4 Turn on the headlights. The voltage should drop, and then come back up, if the charging system is working properly.
5 If the voltage reading is more than the specified charging voltage, replace the voltage regulator (refer to Section 16). If the voltage is less, the alternator diode(s), stator or rectifier may be bad or the voltage regulator may be malfunctioning.

5–12   Chapter 5   Engine electrical systems

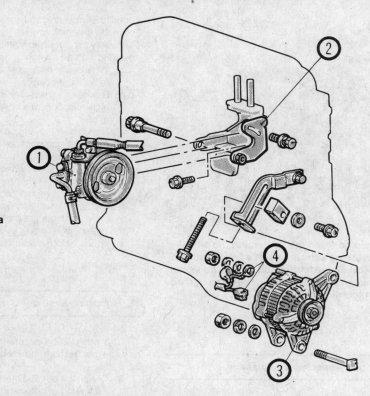

16.2 Typical Cordia and Tredia alternator mounting details (others similar)

1. Power steering oil pump assembly
2. Bracket
3. Alternator
4. Electrical connection

16.25 Mirage 1.5L (1989 on) alternator mounting details

1. Air conditioner drivebelt
2. Power steering drivebelt
3. Drivebelt
4. Water pump pulley
5. Water pump pulley
6. Alternator brace
7. Electrical connection
8. Alternator

16   Alternator – removal and installation

Cordia, Precis and Tredia

Refer to illustration 16.2

1. Remove the drivebelts for the power steering oil pump and the alternator (see Chapter 1).
2. Remove the power steering oil pump (**see illustration**).
3. Remove the adjusting bracket for the alternator.
4. Remove the lower mounting bolt for the alternator.
5. Detach the electrical connectors from the alternator.
6. Remove the alternator.
7. Installation is the reverse of removal.
8. Adjust the belts tension (see Chapter 1).

Chapter 5 Engine electrical systems

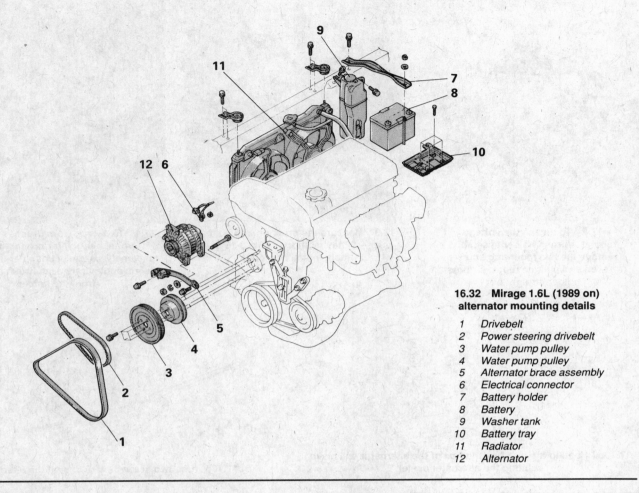

16.32 Mirage 1.6L (1989 on) alternator mounting details

1   Drivebelt
2   Power steering drivebelt
3   Water pump pulley
4   Water pump pulley
5   Alternator brace assembly
6   Electrical connector
7   Battery holder
8   Battery
9   Washer tank
10   Battery tray
11   Radiator
12   Alternator

Galant

9   Remove the left fan motor.
10   Remove the drivebelts for the power steering oil pump and the alternator (see Chapter 1).
11   Remove the alternator brace.
12   Label and detach the electrical connectors from the alternator.
13   Remove the lower mounting bolt from the alternator.
14   Remove the alternator.
15   Installation is the reverse of removal.
16   Adjust the belts tension (see Chapter 1).

Mirage (1985 through 1988)

17   Remove the alternator drivebelt and any other belt interfering with its removal (see Chapter 1).
18   Remove the alternator mounting bolts.
19   Label and detach the alternator electrical connectors.
20   Remove any component interfering with alternator removal.
21   Remove the alternator.
22   Installation is the reverse of removal.
23   Adjust the belt(s) tension (see Chapter 1).

Mirage (1989 on)

Refer to illustrations 16.25 and 16.32

1.5L

24   Remove the drivebelts for the power steering, air conditioning and the alternator (see Chapter 1).
25   Remove the alternator brace **(see illustration)**.
26   Label and detach the electrical connectors from the alternator.
27   Remove the lower mounting bolt from the alternator.
28   Remove the alternator.
29   Installation is the reverse of removal.
30   Adjust the belts tension (see Chapter 1).

1.6L

31   Remove the drivebelts for the power steering pump and the alternator (see Chapter 1).
32   Remove the water pump pulleys **(see illustration)**.
33   Remove the alternator brace.
34   Remove the radiator (see Chapter 3).
35   Label and detach the electrical connectors from the alternator.
36   Remove the lower mounting bolt from the alternator.
37   Remove the alternator.
38   Installation is the reverse of removal
39   Adjust the belts tension (see Chapter 1).

17   Voltage regulator/alternator brushes – replacement

1988 Precis (Bosch type)

Refer to illustrations 17.2, 17.3 and 17.4

Note: *If you are replacing the brushes (but not the regulator itself), the following procedure requires that you unsolder the old brush leads from the regulator and solder the new ones into place. Unless you are skilled with a soldering gun, have this procedure performed by someone who is. If you overheat and damage the regulator, you could end up spending a lot more money than necessary.*

1   Remove the alternator.

17.2 To detach the voltage regulator/brush holder assembly, remove the two mounting screws (arrows) and pull the regulator/brush holder straight out

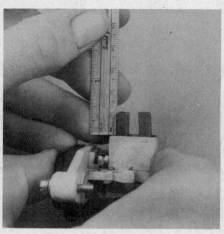

17.3 Measure the brushes with a small ruler – if they're shorter than 3/16 of an inch, install new ones

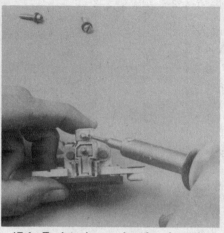

17.4 To detach worn brushes from the voltage regulator/brush holder assembly, carefully unsolder the brush leads and extract each brush and lead from the holder

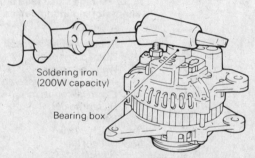

17.10a Heating the rear bearing box of the alternator will make splitting the alternator easier

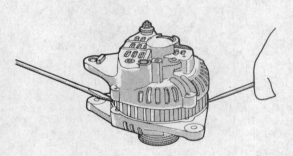

17.10b Use two screwdrivers to split the alternator

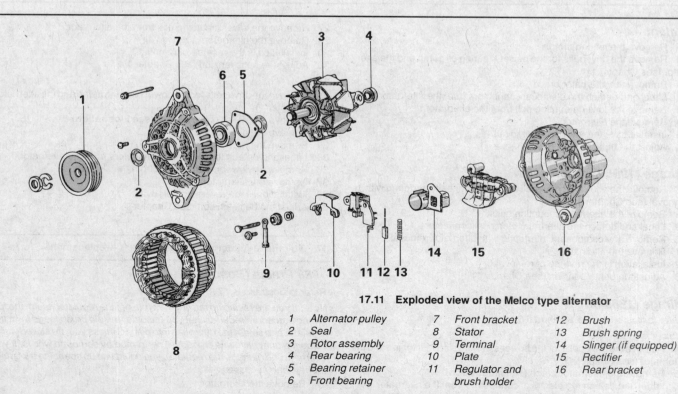

17.11 Exploded view of the Melco type alternator

1 Alternator pulley	7 Front bracket	12 Brush
2 Seal	8 Stator	13 Brush spring
3 Rotor assembly	9 Terminal	14 Slinger (if equipped)
4 Rear bearing	10 Plate	15 Rectifier
5 Bearing retainer	11 Regulator and brush holder	16 Rear bracket
6 Front bearing		

Chapter 5 Engine electrical systems

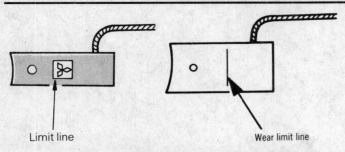

17.12a If the brushes are worn past the wear limit line, they should be replaced

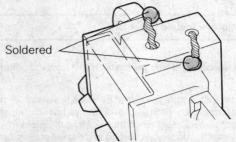

17.12b If the brushes are being replaced, unsolder and solder the pigtails at the area shown

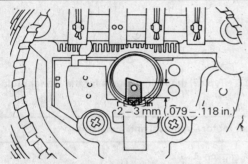

17.13 When installing new brushes, they should extend out of the holder the proper amount

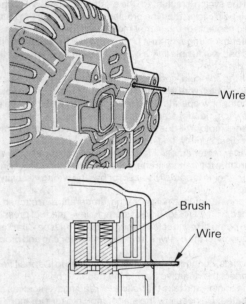

17.14 When placing the two covers together, use a piece of wire through the rear case and into the brush holder to retain the brushes in the holder

2 Remove the voltage regulator mounting screws **(see illustration)**.
3 Remove the regulator/brush holder assembly and measure the length of the brushes **(see illustration)**. If they are less than 3/16-inch long, replace them with new ones. **Note:** *If you're simply replacing the voltage regulator, skip the next step – the new regulator assembly includes a new set of brushes so the following step is unnecessary.*
4 Unsolder the brush wiring connections **(see illustration)** and remove the brushes and springs.
5 Installation is the reverse of removal. Be sure to solder the new brush leads properly.

All others (Melco type)

Refer to illustrations 17.10a, 17.10b, 17.11, 17.12a, 17.12b, 17.13 and 17.14

6 Remove the alternator (see Section 16).
7 Remove the bolts retaining the two halves of the alternator together.
8 Mount the front of the alternator face down in a vise. Using rags as a cushion, clamp to the front case portion of the alternator.
9 Remove all nuts from the back of the alternator.
10 Using a 200-watt soldering iron, heat the rear bearing area (bearing box) of the rear case **(see illustration)**. Insert two standard screwdrivers carefully between the two halves of the alternator (not too deep or you will damage the stator) and pry the rear case off the alternator **(see illustration)**. **Caution:** *Pry gently or you'll break the delicate aluminum case.*
11 Unsolder the regulator/brush holder **(see illustration)**. **Note:** *While applying heat to electrical components, it's a good idea to use a pair of needle-nose pliers as a heat sink. Don't apply heat for more than about five seconds.*
12 Inspect the brushes for excessive wear **(see illustration)** and replace them if necessary by unsoldering **(see illustration)**.
13 When installing new brushes, solder the pigtails so the brush limit line will be about 0.079 to 0.118 in. above the end of the brush holder **(see illustration)**.
14 To reassemble, compress the brushes into their holder and retain them with a straight piece of wire that can be pulled from the back of the alternator when reassembled **(see illustration)**.
15 To reassemble, reverse disassembly procedure.

18 Starting system – general information and precautions

The sole function of the starting system is to turn over the engine quickly enough to allow it to start.

The starting system consists of the battery, the starter motor, the starter solenoid and the wires connecting them. The solenoid is mounted directly on the starter motor.

The solenoid/starter motor assembly is installed on the transaxle bellhousing.

When the ignition key is turned to the Start position, the starter solenoid is actuated through the starter control circuit. The starter solenoid then connects the battery to the starter. The battery supplies the electrical energy to the starter motor, which does the actual work of cranking the engine.

Always observe the following precautions when working on the starting system:

a) Excessive cranking of the starter motor can overheat it and cause serious damage. Never operate the starter motor for more than 30 seconds at a time without pausing to allow it to cool for at least two minutes.
b) The starter is connected directly to the battery and could arc or cause a fire if mishandled, overloaded or shorted out.
c) Always detach the cable from the negative terminal of the battery before working on the starting system.

Chapter 5 Engine electrical systems

19 Starter motor – testing in vehicle

Note: *Before diagnosing starter problems, make sure the battery is fully charged.*

1 If the starter motor does not turn at all when the switch is operated, make sure that the shift lever is in Neutral or Park (automatic transmission) or that the clutch pedal is depressed (manual transmission).
2 Make sure that the battery is charged and that all cables, both at the battery and starter solenoid terminals, are clean and secure.
3 If the starter motor spins but the engine is not cranking, the overrunning clutch in the starter motor is slipping and the starter motor must be replaced.
4 If, when the switch is actuated, the starter motor does not operate at all but the solenoid clicks, then the problem lies with either the battery, the main solenoid contacts or the starter motor itself (or the engine is seized).
5 If the solenoid plunger cannot be heard when the switch is actuated, the battery is bad, the fusible link is burned (the circuit is open) or the solenoid itself is defective.
6 To check the solenoid, connect a jumper lead between the battery (+) and the ignition switch wire terminal (the small terminal) on the solenoid. If the starter motor now operates, the solenoid is OK and the problem is in the ignition switch, neutral start switch or the wiring.
7 If the starter motor still does not operate, remove the starter/solenoid assembly for disassembly, testing and repair.
8 If the starter motor cranks the engine at an abnormally slow speed, first make sure that the battery is charged and that all terminal connections are tight. If the engine is partially seized, or has the wrong viscosity oil in it, it will crank slowly.
9 Run the engine until normal operating temperature is reached. On engines equipped with a DIS ignition system, relieve the fuel pressure and leave the fuel pump disconnected throughout testing (see Chapter 4). On all others, disconnect the coil wire from the distributor cap and ground it on the engine.
10 On all models, connect a voltmeter positive lead to the positive battery post and connect the negative lead to the negative post.
11 Crank the engine and take the voltmeter readings as soon as a steady figure is indicated. Do not allow the starter motor to turn for more than 30 seconds at a time. A reading of 9 volts or more, with the starter motor turning at normal cranking speed, is normal. If the reading is 9 volts or more but the cranking speed is slow, the motor is faulty. If the reading is less than 9 volts and the cranking speed is slow, the solenoid contacts are probably burned, the starter motor is bad, the battery is discharged or there is a bad connection.

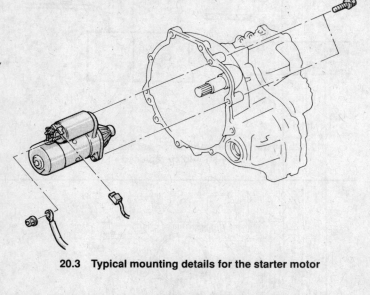

20.3 Typical mounting details for the starter motor

20 Starter motor – removal and installation

Refer to illustration 20.3
1 Detach the cable from the negative terminal of the battery.
2 Clearly label, then disconnect the wires from the terminals on the starter solenoid.
3 Remove the mounting bolts **(see illustration)** and remove the starter.
4 Installation is the reverse of removal.

21 Starter solenoid – removal and installation

Refer to illustration 21.3 and 21.4
1 Disconnect the cable from the negative terminal of the battery.
2 Remove the starter motor (see Section 19).
3 Disconnect the large motor lead from the solenoid terminal **(see illustration)**.
4 Remove the screws which secure the solenoid to the starter motor **(see illustration)**.
5 Pull the solenoid from the starter body flange.
6 Installation is the reverse of removal.

21.3 To separate the solenoid from the starter motor, remove the nut and detach the lead (arrow) . . .

21.4 . . . then remove the solenoid mounting screws (arrows) and pull the solenoid straight off the starter flange

Chapter 6 Emissions control systems

Contents

Catalytic converter	8
Computerized Engine Controls (CEC)	13
Deceleration devices (carbureted models)	9
Evaporative emission control system	3
Evaporative emission control system check and canister replacement	See Chapter 1
Exhaust Gas Recirculation (EGR) system	6
General information	1
Heated Air Intake (HAI) system (carbureted models)	4
High altitude compensation system	12
Idle-up system (carbureted models)	11
Jet air system (1988 and earlier models)	5
Mixture Control Valves (carbureted models)	10
Oxygen sensor replacement	See Chapter 1
Positive Crankcase Ventilation (PCV) system	2
Positive Crankcase Ventilation (PCV) valve check and replacement	See Chapter 1
Secondary air supply system	7
Thermostatically controlled air cleaner check	See Chapter 1

1 General information

To prevent pollution of the atmosphere from incompletely burned and evaporating gases, and to maintain good driveability and fuel economy, a number of emission control systems and devices are incorporated. They include the:

Positive Crankcase Ventilation (PCV) system
Evaporative emission control system
Heated Air Intake (HAI) system (carbureted models)
Jet air system (except 1989 and on models)
Exhaust Gas Recirculation (EGR) system
Secondary air supply system (except MPI models)
Catalytic converter
Deceleration devices (carbureted models)

Before assuming that an emissions control system is malfunctioning, check the fuel and ignition systems carefully. The diagnosis of emission control devices requires specialized tools, equipment and training.

Remember, the most frequent cause of emissions control problems is simply a loose or broken vacuum hose or wire, so always check the hose and wiring connections if a problem is suspected. **Note:** *Because of a Federally mandated extended warranty which covers the emission control system components, check with your dealer about warranty coverage before working on any emissions-related systems.*

Pay close attention to any special precautions outlined in this Chapter. It should be noted that the illustrations of the various systems may not exactly match the system installed on your vehicle because of changes made by the manufacture during production or from year-to-year.

A Vehicle Emissions Control Information (VECI) label is located in the engine compartment. This label contains important emissions specifications and adjustment information, as well as a vacuum hose schematic with emissions components identified. When servicing the engine or emissions systems, the VECI label in your particular vehicle should always be checked for up-to-date information.

2 Positive Crankcase Ventilation (PCV) system

1 The Positive Crankcase Ventilation (PCV) system reduces hydrocarbon emissions by scavenging crankcase vapors. It does this by circulating fresh air from the air cleaner through the crankcase, where it mixes with blow-by gases and is then rerouted through a PCV valve to the intake manifold.
2 The main components of the PCV system are the PCV valve, a fresh air filtered inlet and the vacuum hoses connecting these two components with the engine.
3 To maintain idle quality, the PCV valve restricts the flow when the intake manifold vacuum is high. During acceleration and high load conditions, vacuum is low, so the PCV passage opens more to facilitate flow.
4 Checking and replacement of the PCV valve and filter is covered in Chapter 1.

3 Evaporative emission control system

Refer to illustrations 3.2a, 3.2b, 3.2c and 3.2d
Warning: *Gasoline is extremely flammable, so take extra precautions when you work on any part of the fuel system. Don't smoke or allow open flames or bare light bulbs near the work area, and don't work in a garage where a natural gas-type appliance (such as a water heater or clothes dryer) with a pilot light is present. If you spill any fuel on your skin, rinse it off immediately with soap and water. When you perform any kind of work on the fuel system, wear safety glasses and have a Class B type fire extinguisher on hand.*
1 The evaporative emission control system is designed to prevent the escape of fuel vapors from the fuel system into the atmosphere.
2 The typical evaporative emission control system consists of a canister, a bowl vent valve, a purge control system, an overfill limiter, a thermo valve, a fuel check valve and a specially designed fuel filler cap **(see illustrations)**.

Canister
3 When the engine is inoperative, fuel vapors generated inside the fuel tank and the carburetor float chamber are absorbed and stored in the canister. When the engine is running, the fuel vapors absorbed in the canister are drawn into the intake manifold through the purge control.

Bowl vent valve (carbureted models)
4 The bowl vent valve controls the carburetor bowl vapors. When the engine is running, intake vacuum acts on a diaphragm to close the bowl vent valve so that the bowl is connected to an air vent. During engine operation, the diaphragm is kept open by a solenoid valve – even when intake manifold vacuum falls to a value equal to atmospheric pressure – as long as the ignition key is turned on. When the key is turned off, the solenoid valve closes, the bowl vent valve opens to connect the carburetor bowl to the canister, allowing fuel vapors to flow to the canister where they are stored until the next time the key is turned on.

Purge control system
5 The purge control system is closed during idle to prevent vaporized fuel from entering the intake manifold. At higher engine speeds, the purge control valve opens.

Thermo valve
6 The thermo valve, which monitors the engine coolant temperature at the intake manifold, controls the purge control sytem when the engine coolant temperature is lower than the pre-set temperature. This reduces CO and HC emissions under engine warmup conditions. The thermo valve opens the purge control system when the engine coolant temperature is above the pre-set temperature. The thermo valve also controls the choke breaker, EGR valve and choke opener.

Fuel filler cap
7 The fuel filler cap is equipped with a vacuum relief valve to prevent the escape of fuel vapor into the atmosphere.

Overfill limiter
8 The typical overfill limiter consists of two valves:
  a) The pressure valve, which opens when fuel tank internal pressure exceeds normal pressure.
  b) The vacuum valve, which opens when fuel tank internal pressure is lower than normal pressure.

Fuel check valve
9 The fuel check valve, prevents fuel leaks in the event of a vehicle rollover. The valve contains two balls. Under normal conditions, the fuel vapor passage in the valve is open, but, should the vehicle roll over, one of the balls closes the fuel passage, preventing fuel leakage.

4 Heated Air Intake (HAI) system (carbureted models)

Refer to illustration 4.2
1 Carbureted models are equipped with a Heated Air Intake (HAI) system to permit leaner carburetion, which reduces HC and CO emissions, improves warm-up characteristics and minimizes carburetor icing.
2 The air cleaner is equipped with an air control valve inside the snorkel to regulate the temperature of intake air, which is admitted to the carburetor through the fresh air duct, the heat cowl and air duct, or through both routes **(see illustration)**.

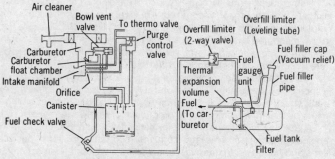

3.2a Details of the evaporative emission control system for carbureted models (typical)

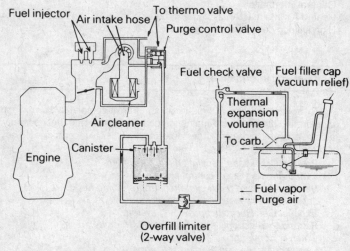

3.2b Details of the evaporative emission control system for TBI models (typical)

Chapter 6 Emissions control systems

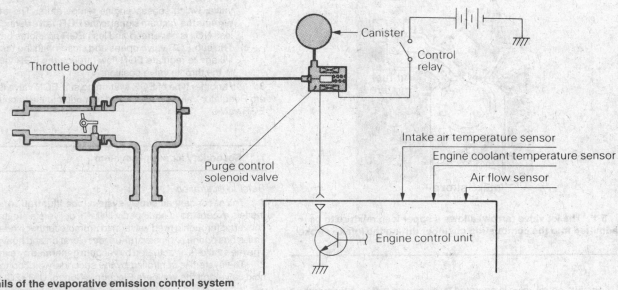

3.2c Details of the evaporative emission control system for 1.5L MPI models (typical)

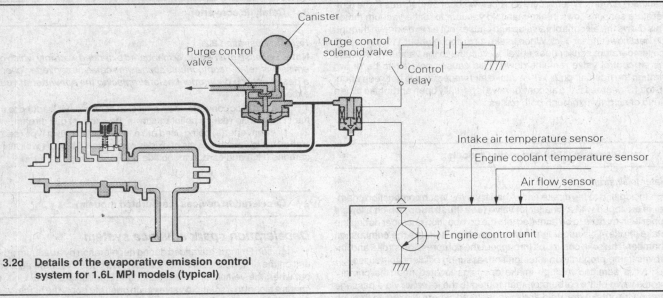

3.2d Details of the evaporative emission control system for 1.6L MPI models (typical)

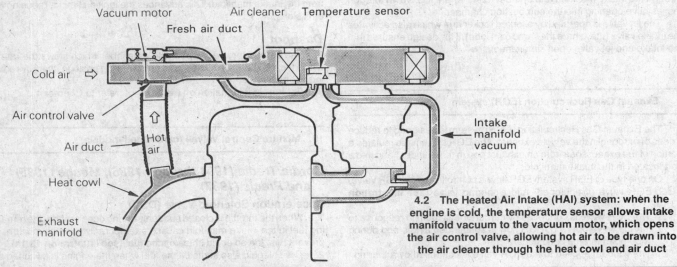

4.2 The Heated Air Intake (HAI) system: when the engine is cold, the temperature sensor allows intake manifold vacuum to the vacuum motor, which opens the air control valve, allowing hot air to be drawn into the air cleaner through the heat cowl and air duct

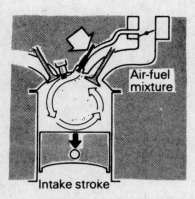

5.1 The jet valve (arrow) allows a super lean mixture to be admitted into the combustion chamber during the intake stroke

3 The air control valve is operated by a vacuum motor. A temperature sensor in the air cleaner controls when the vacuum motor receives intake manifold vacuum. When intake air is below about 86-degrees F, the temperature sensor allows intake manifold vacuum to to the vacuum motor. This draws the air control valve up and causes hot air to be drawn through the heat cowl and air duct. When intake air is above about 113-degrees F, the temperature sensor does not allow vacuum to the vacuum motor and the air control valve remains down. This causes cold air to be drawn through the fresh air duct. When intake air temperature is between about 86 to 113-degrees F, the air control valve is partially open and intake air is a blend of air drawn through both routes.

5 Jet air system (1988 and earlier models)

Refer to illustration 5.1

1 In addition to the intake and exhaust valves, each combustion chamber is equipped with a smaller jet valve **(see illustration)** which allows a super lean mixture to be admitted into the combustion chamber during the intake stroke. This super lean mixture swirls as it enters the combustion chamber. The swirl continues throughout the compression stroke and improves flame propagation after ignition, assuring efficient combustion.
2 Air is admitted through intake openings located near the primary throttle valve of the carburetor, then routed to the jet valve via a passage through the intake manifold and cylinder head, where it is drawn through the jet valve opening into the combustion chamber.
3 The jet valve is operated by a forked rocker arm which also activates the intake valve (they share the same cam lobe). This design ensures that the intake and jet valve open simultaneously.

6 Exhaust Gas Recirculation (EGR) system

1 The Exhaust Gas Recirculation (EGR) system is designed to reduce oxides of nitrogen in the vehicle exhaust. The EGR system recirculates a portion of the exhaust gas from an exhaust port in the cylinder head into a port located in the intake manifold.
2 On one type of EGR system EGR flow is controlled by an EGR valve, a sub-EGR valve (if equipped) and a thermo valve **(see illustration 13.1a)**.
  a) The EGR valve is controlled by carburetor vacuum in response to throttle valve opening; EGR flow is suspended at idle and during wide open throttle conditions.
  b) The vacuum applied to the EGR valve is controlled by a thermo valve, which senses engine temperature. The thermo valve interrupts the vacuum signal to the EGR valve during warm-up when less NOx is generated and less EGR promotes better driveability.
  c) The sub-EGR valve opens and closes with the throttle valve via a linkage to regulate EGR flow through the EGR valve in response to the throttle valve position.
3 On another type of EGR system no sub-EGR valve is used. A Vacuum Regulator Valve (VRV) is used to modulate the vacuum signal to the EGR valve.

7 Secondary air supply system

Refer to illustration 7.1

1 The secondary air supply system **(see illustration)**, installed on all models except those equipped with MPI, delivers air to the front catalytic converter through a reed valve(s) to promote further oxidation of exhaust emissions during engine warm-up, deceleration and heavy engine loads. The reed valve(s) is actuated by vacuum generated by exhaust pulsation.
2 The system is controlled by the secondary air control valve, a solenoid valve and the vehicle's Electronic Control Unit (ECU).

8 Catalytic converter

Refer to illustration 8.2

Note: *Because of a Federally mandated extended warranty which covers emissions-related components such as the catalytic converter, check with a dealer service department before replacing the converter at your own expense.*

1 The catalytic converter is an emission control device added to the exhaust system to reduce pollutants from the exhaust gas stream.
2 The converter(s) being used are a three-way catalyst type **(see illustration)**. It lowers the levels of oxides of nitrogen (NOx) as well as hydrocarbons (HC) and carbon monoxide (CO).

9 Deceleration devices (carbureted models)

Deceleration spark advance system

1 Ignition timing is advanced during deceleration by a vacuum advance unit on the distributor. This unit normally has ported vacuum applied to it, but when the vehicle is decelerating, the Electronic Control Unit (ECU) sends a signal to a solenoid valve which opens to apply the higher vacuum from the intake manifold. This advances the ignition timing, reducing HC emissions.

Dashpot

2 The carburetor is equipped with a dashpot which slows the rate at which the throttle valve closes to its normal idling position, thereby reducing HC emissions.
3 For further information on the dashpot, refer to Chapter 4.

10 Mixture Control Valves (carbureted models)

Cordia/Tredia (1984 through 1986), Mirage (1985) and Precis (1987)

Deceleration Solenoid Valve (DSV)

1 When the throttle is closed suddenly during deceleration, the remaining fuel in the intake manifold causes a temporarily overrich mixture. To prevent this, the solenoid shuts off the fuel **(see illustration 13.1a)**.
2 The solenoid also shuts off the fuel when the engine is switched off.

Chapter 6 Emissions control systems

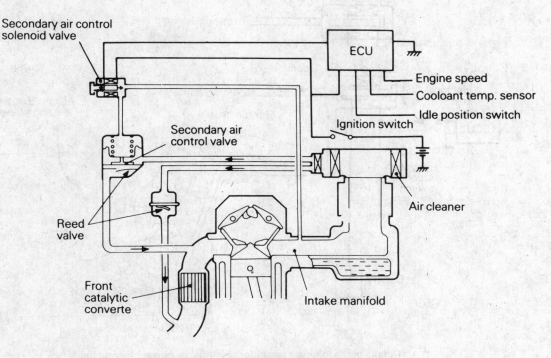

7.1 Details of the secondary air supply system

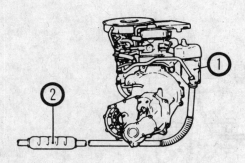

8.2 Some models use two catalytic converters, others only use one

1. Front converter (if equipped)
2. Rear converter

Enrichment Solenoid Valve (ESV)

3 This solenoid is used to supply extra fuel that is needed when the engine is cold or heavy acceleration is required **(see illustration 13.1a)**.

Jet Solenoid Valve (JSV)

4 This solenoid is the main air to fuel ratio controller **(see illustration 13.1a)**. The amount of time it is energized will be increased as more fuel is needed.

Cordia/Tredia (1987 and 1988), Mirage (1986 through 1988) and Precis (1988 and 1989)

Feedback Solenoid Valve (FBSV)

5 This solenoid works to maintain the proper air to fuel ratio. When it is energized the vehicle will run leaner **(see illustration 13.1b)**.

Slow Cut Solenoid Valve (SCSV)

6 This solenoid is used to reduce the rich mixture during deceleration **(see illustration 13.1b)**.

Mixture Control Valve (MCV) (1988 and 1989 Precis)

7 When the throttle is closed suddenly during deceleration, the remaining fuel in the intake manifold causes a temporarily overrich mixture. To prevent this, the solenoid temporarily supplies air from another passage to correct the air to fuel ratio and reduce emissions.

11 Idle-up system (carbureted models)

Electrical load or power steering load

Refer to illustration 11.1

1 This idle-up system consists of a dashpot assembly, a solenoid valve, electrical load sensing circuitry and an oil pump pressure switch in the power steering system **(see illustration)**.

2 When there is an electrical load or power steering load at idle that requires a greater engine speed, the solenoid valve is opened, allowing vacuum from the intake manifold to act on the dashpot. This vacuum acting on the dashpot opens the throttle valve slightly via the idle-up lever on the throttle shaft.

Air conditioning load

Refer to illustration 11.3

3 This idle-up system consists of a throttle opener assembly, a solenoid valve, an air conditioning switch and the ECU **(see illustration)**.

4 When the air conditioning compressor is engaged at idle, a greater engine speed is required, the solenoid valve is opened, allowing vacuum from the intake manifold to act on the throttle opener. This vacuum acting on the throttle opener causes the engine speed to increase.

Chapter 6 Emissions control systems

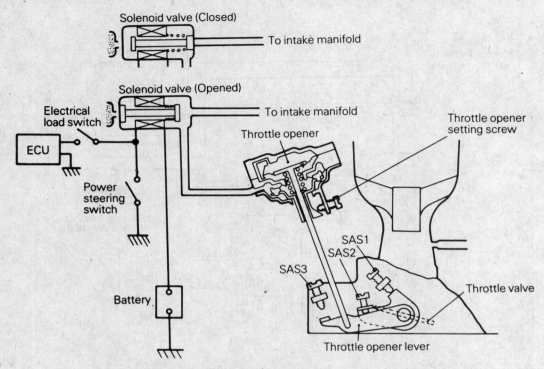

11.1 Details of the idle-up system for electrical or power steering load

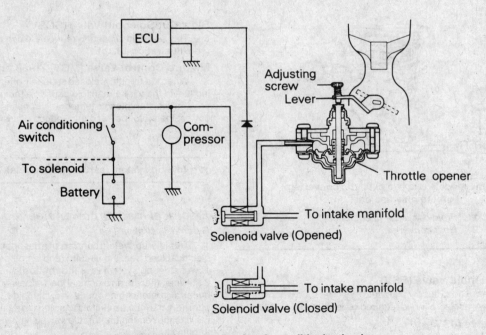

11.3 Details of idle-up system for air conditioning load

12 High altitude compensation system

Feedback Carbureted (FBC) models

1 To meet emission laws at all altitudes, all models using a feedback carburetor are equipped with a high altitude compensation system **(see illustrations 13.1a and 13.1b)**. This system consists of a high altitude compensator (HAC) valve, a vacuum switching valve (49-state vehicles), a check valve and a distributor equipped with a high altitude vacuum advance device.

2 At high altitude, the system maintains the air-fuel mixture at its sea level ratio by supplying additional air into the carburetor. The system also supplies additional vacuum to the high altitude vacuum advance device on the distributor to compensate for high altitude.

Chapter 6 Emissions control systems

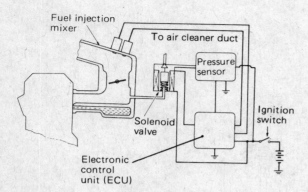

12.3 Details of the BARO/MAP pressure sensor

Fuel-injected models

Refer to illustration 12.3

3   Both the TBI and MPI systems use a BARO/MAP sensor in conjunction with a solenoid **(see illustration)**. The solenoid controls whether the sensor will read barometric pressure or manifold vacuum. The solenoid allows barometric pressure reading when the ignition switch is turned to the "ON" position and periodically after that. The manifold pressure reading mode is in effect any time the barometric pressure reading mode is not.
4   The input signal from the BARO/MAP sensor is sent to the ECU and is used to adjust the amount of time the fuel injectors are held open according to barometric pressure and manifold vacuum.

13   Computerized Engine Controls (CEC)

Refer to illustration 13.1a, 13.1b and 13.1c

1   The computerized engine control system consists of three types of components **(see illustrations)**:
   a) Information sensors (inputs information to the ECU)
   b) Electronic Control Unit (ECU) (makes decisions based on input)
   c) Output actuators (follows orders from the ECU)

Typical information sensors
Carbureted models
   Exhaust oxygen sensor
   Coolant temperature sensor
   Engine speed sensor
   Throttle position sensor
   Closed throttle (idle) switch
   Intake air temperature sensor (if equipped)
Fuel-injected models
   MPI
   Exhaust oxygen sensor
   Coolant temperature sensor
   Crank angle sensor
   Number one cylinder TDC sensor
   Throttle Position Sensor (TPS)
   Closed throttle (idle) switch
   Intake air temperature sensor
   BAP/MAP sensor
   Air Flow Sensor (AFS)
   Knock sensor
   TBI
   Throttle Position Sensor (TPS)
   BAP/MAP sensor
   Coolant Temperature Sensor (CTS)
   Intake air temperature sensor
   Closed throttle (idle) switch
   Exhaust oxygen sensor
   Air Flow Sensor (AFS)
   Vehicle Speed Sensor (VSS)
   Engine speed signal

Electronic Control Unit (ECU)

2   The ECU processes the constant information it receives from the information sensors, compares this information to the values stored in its memory. The ECU makes calculations necessary to maintain good driveability along with low emissions. The calculations are sent to the output actuators.

Output actuators

3   The output actuators adjust the operating conditions. When energized by signals from the ECU, actuators turn on or off.
Carbureted models (typical)
   Cordia and Tredia (1984 through 1986), Mirage (1985), Precis (1987)
      Enrichment Solenoid Valve (ESV)
      Deceleration Solenoid Valve (DSV)
      Jet Mixture Solenoid Valve (JSV)
      Idle-up (throttle opener) control solenoid valve
      Secondary air control solenoid valve
      Distributor advance control solenoid valve
   Cordia and Tredia (1987 and 1988), Mirage (1986 through 1988), Precis (1988 and 1989)
      Feedback Solenoid Valve (FBSV)
      Slow Cut Solenoid Valve (SCSV)
      Distributor advance control solenoid valve
      Distributor cold advance control solenoid valve (Precis only)
      Secondary air control solenoid valve
      Idle-up (throttle opener) control solenoid valve
      Mixture control valve (Precis only)
      Mixture heater relay (Precis only)
Fuel-injected models (typical)
   MPI
   Purge control solenoid valve
   Fuel pressure solenoid valve
   Idle Speed Control (ISC) servo
   Fuel pump control relay
   Air conditioner power relay
   Ignition timing control
   TBI
   Secondary air control solenoid valve
   EGR control solenoid valve (if equipped)
   Fuel pump relay
   Idle Speed Control (ISC) servo motor
   A/C power relay
   Fuel injectors
   Knock sensor (turbo models)

4   The CEC system (all sensors, actuators the ECU and all components related to the system) is covered by a Federally mandated extended warranty (see your dealer for details). Therefore, all system malfunctions can and should be referred to your dealer as long as the vehicle is still under warranty.
5   Because of the complexity of the CEC system and because of the special electronic instrument needed to check and troubleshoot the system in detail, most servicing is beyond the scope of the home mechanic. However, if you have a 1989 or later fuel-injected model, you can probably obtain "trouble codes" from the computer memory using an analog (needle-type) voltmeter. These codes can be very helpful in diagnosis. A list of codes, as well as procedures for obtaining them from the computer memory, are in the *Haynes Automotive Emissions Control* manual.
6   Where possible, the procedures for checking some of the components described above have been included in Chapter 4. However, because of the interrelationship and complexity of these devices, we do not recommend that you tackle CEC related problems unless you have the proper equipment and experience.

Chapter 6 Emissions control systems

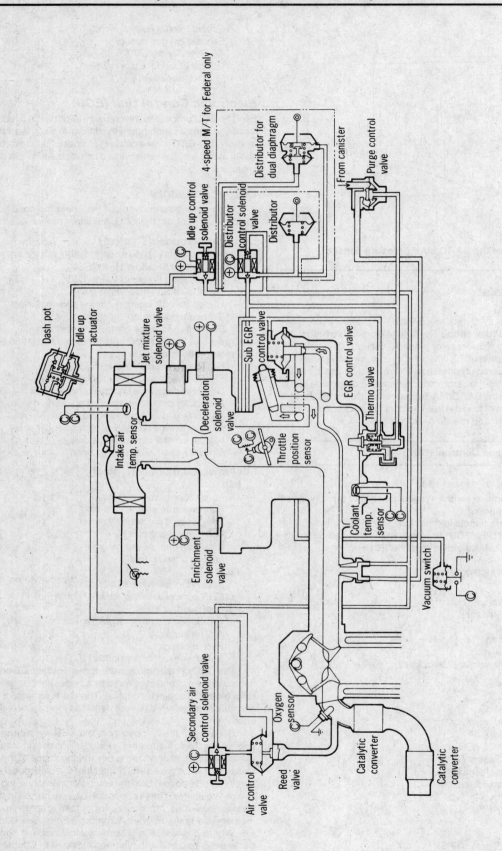

13.1a Typical feedback carburetor system – early models

Chapter 6 Emissions control systems

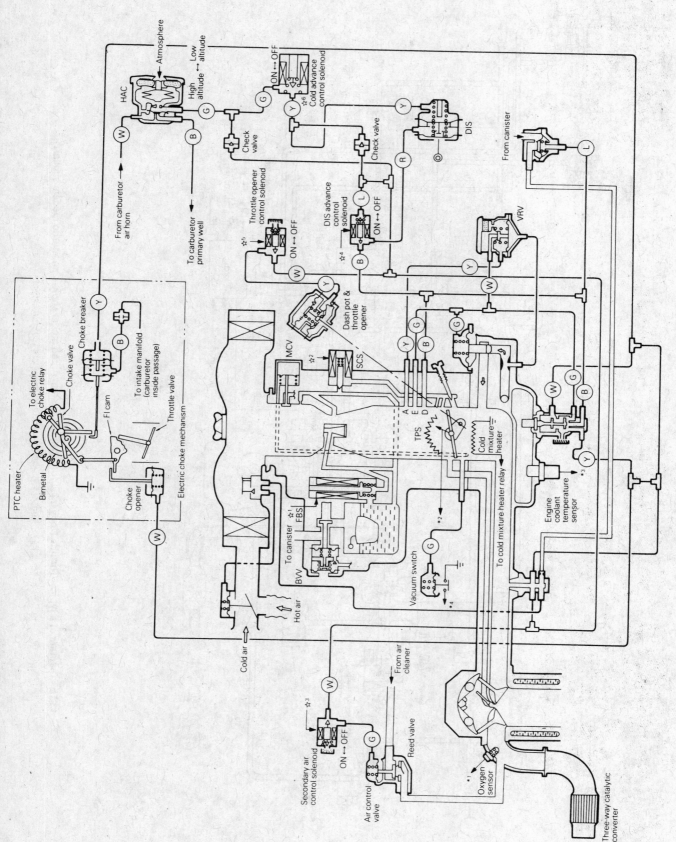

13.1b Typical feedback carburetor system – later models

Chapter 6 Emissions control systems

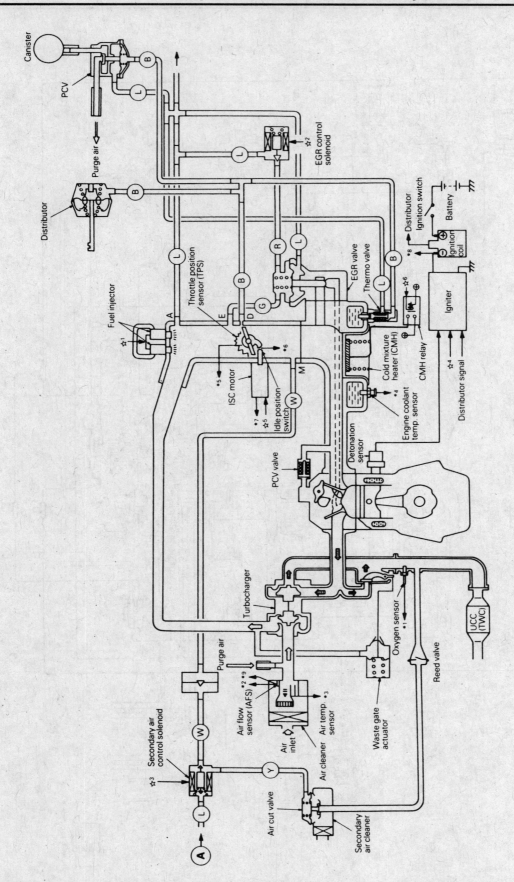

13.1c Diagram of a typical Throttle Body Injection (TBI) fuel system (turbo model shown, non-turbo models similar)

Chapter 7 Part A Manual transaxle

Contents

General information 1	Lubricant change .... See Chapter 1
Manual transaxle overhaul – general information 4	Lubricant level check .... See Chapter 1
Manual transaxle – removal and installation 3	Oil seal replacement .... See Chapter 7B
Manual transaxle shift assembly – removal, installation and adjustment 2	Transaxle mount – check and replacement .... See Chapter 7B

Specifications

Torque specifications
Ft-lbs

Transaxle-to-engine bolts
  1988 and earlier Cordia/Tredia and Precis, 1986 and earlier Mirage
    8 x 14 mm .... 7 to 9
    8 x 20 mm .... 11 to 16
    8 x 60 mm .... 22 to 25
    10 x 40 mm .... 31 to 40
    10 x 55 mm .... 16 to 23
    10 x 65 mm .... 31 to 40
  Galant, 1987 and later Mirage and 1989 and later Precis
    12 mm .... 32 to 39
    10 mm .... 22 to 25
    8 mm .... 7 to 9
Drain and filler plugs .... See Chapter 1
Shift rod set screw 24
Starter motor bolts .... 16 to 23

7A-2   Chapter 7 Part A   Manual transaxle

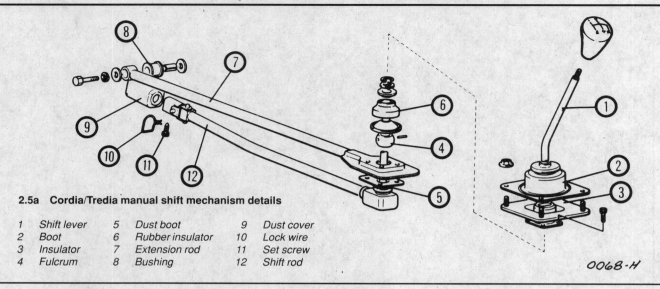

2.5a Cordia/Tredia manual shift mechanism details

1. Shift lever
2. Boot
3. Insulator
4. Fulcrum
5. Dust boot
6. Rubber insulator
7. Extension rod
8. Bushing
9. Dust cover
10. Lock wire
11. Set screw
12. Shift rod

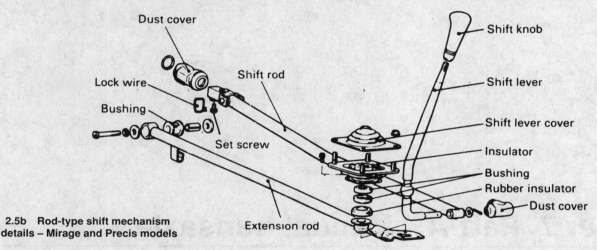

2.5b Rod-type shift mechanism details – Mirage and Precis models

1 General information

The vehicles covered by this manual are equipped with either a four- or five-speed manual transaxle or a three- or four-speed automatic transaxle. Information on the manual transaxle is included in this Part of Chapter 7. Service procedures for the automatic transaxle are contained in Chapter 7, Part B.

The manual transaxle is a compact, two piece, lightweight aluminum alloy housing containing both the transmission and differential assemblies.

Because of the complexity, unavailability of replacement parts and special tools required, internal repair of the manual transaxle by the home mechanic is not recommended. For readers who wish to tackle a transaxle rebuild, exploded views and a brief Manual transaxle overhaul – general information Section are provided. The bulk of information in this Chapter is devoted to removal and installation procedures.

2 Manual transaxle shift assembly – removal, installation and adjustment

Rod-type

Refer to illustrations 2.5a and 2.5b

1. Remove the shift knob (if required) and center console.
2. On Precis models so equipped, remove the rear heat floor duct.
3. On Mirage and Cordia/Tredia models, remove the shift assembly bracket nuts.
4. Raise the vehicle and support it securely on jackstands.
5. Remove the bolt and disconnect the extension rod from the transaxle **(see illustrations)**.
6. Remove the lock wire from the shift rod set screw. Loosen (Cordia/Tredia models) or remove (Mirage and Precis models) the set screw and detach the shift rod from the transaxle.
7. On some Cordia/Tredia models, remove the heat shield from the body for clearance when removing the shift assembly.
8. Remove the extension rod and shift lever assembly by lowering it from the vehicle.
9. Installation is the reverse of removal. No adjustment is possible on these models.

Cable-type

Refer to illustrations 2.11a, 2.11b, 2.15a, 2.15b and 2.17

Removal

10. Remove the shift knob and center console.
11. Make sure the shift lever and transaxle shift and select cable levers are in Neutral. Remove the cotter pins and clips and detach the shift cables from the shift lever assembly **(see illustrations)**.
12. Remove the bolts or nuts and detach the cable retainers.
13. In the engine compartment, remove the cotter pins and clips, then detach the cables from the transaxle.

Chapter 7 Part A Manual transaxle

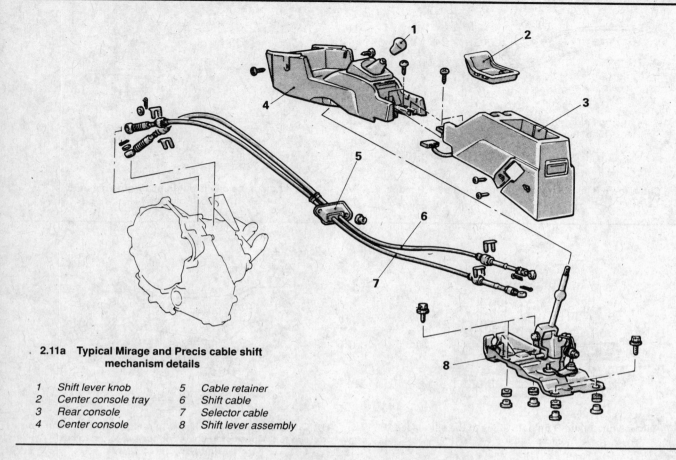

2.11a Typical Mirage and Precis cable shift mechanism details

1 Shift lever knob
2 Center console tray
3 Rear console
4 Center console
5 Cable retainer
6 Shift cable
7 Selector cable
8 Shift lever assembly

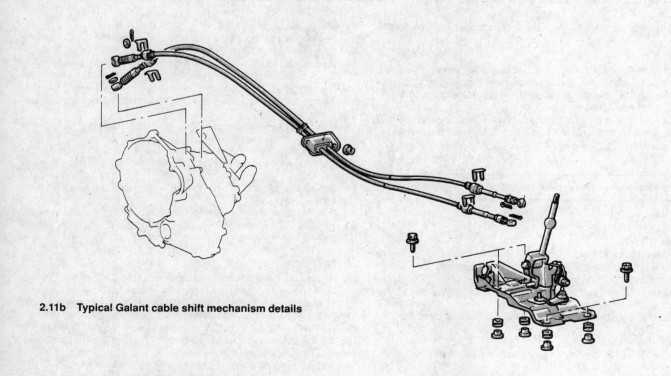

2.11b Typical Galant cable shift mechanism details

Chapter 7 Part A Manual transaxle

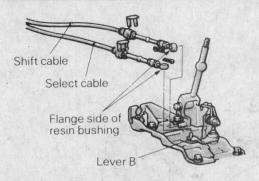

2.15a Cable-to-shift lever connection details – Lever B must be in Neutral

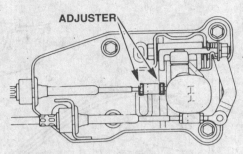

2.15b Adjust the cable length by turning the nuts (labelled "Adjuster")

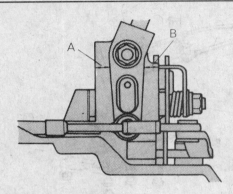

2.17 Dimensions A and B on both sides of the shift lever must be equal

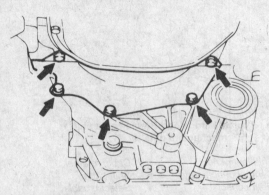

3.11 On most models you'll have to remove the bolts and detach the bellhousing cover for access to the lower transaxle-to-engine bolts

14 Remove the cable assembly by pulling it into the passenger compartment.

Installation

15 With the shift lever and transaxle shift and selector levers in Neutral, connect the cables to the shift lever assembly **(see illustration)**. If necessary, turn the adjusters located at the shift lever end of the cable until lever B is in Neutral **(see illustration)**.
16 Connect the cables to the transaxle and install the retainers.
17 After installation, dimensions A and B on both sides of the shift lever should be equal **(see illustration)**. If it isn't, adjust the cable as necessary.

3 Manual transaxle – removal and installation

Refer to illustration 3.11

Removal

1 Disconnect the cables from the battery. Remove the battery and tray.
2 Raise the vehicle and support it securely on jackstands.
3 Drain the transaxle lubricant (see Chapter 1).
4 Disconnect the shift (see Section 2) and clutch linkage from the transaxle.
5 Detach the speedometer cable and electrical connectors from the transaxle.
6 Remove the starter motor.
7 Remove the exhaust system components as necessary for clearance.
8 Support the engine. This can be done from above with an engine hoist, or by placing a jack (with a block of wood as an insulator) under the engine oil pan. The engine must remain supported at all times while the transaxle is out of the vehicle!
9 Remove any chassis or suspension components that will interfere with transaxle removal (see Chapter 10).
10 Disconnect the driveaxles from the transaxle (see Chapter 8).
11 Support the transaxle with a jack, then remove the bolts securing the transaxle to the engine. On some models you'll have to remove the bellhousing cover to gain access to the lower transaxle-to-engine bolts **(see illustration)**.
12 Remove the transaxle mount nuts and bolts. Unbolt and remove the transaxle mount bracket.
13 Make a final check that all wires and hoses have been disconnected from the transaxle, then carefully pull the transaxle and jack away from the engine.
14 Once the input shaft is clear, lower the transaxle and remove it from under the vehicle.
15 With the transaxle removed, the clutch components are now accessible and can be inspected. In most cases, new clutch components should be routinely installed when the transaxle is removed.

Installation

16 If removed, install the clutch components (see Chapter 8).
17 With the transaxle secured to the jack with a chain, raise it into position behind the engine, then carefully slide it forward, engaging the input shaft with the clutch plate hub splines. Do not use excessive force to install the transaxle – if the input shaft does not slide into place, readjust the angle of the transaxle so it is level and/or turn the input shaft so the splines engage properly with the clutch plate hub.
18 Install the transaxle-to-engine bolts. Tighten the bolts securely.
19 Install the transaxle mount bracket and the mount nuts or bolts.
20 Install the chassis and suspension components which were removed. Tighten all nuts and bolts securely.

Chapter 7 Part A Manual transaxle

21 Remove the jacks supporting the transaxle and engine.
22 Install the various items removed previously, referring to Chapter 8 for installation of the driveaxles and Chapter 4 for information regarding the exhaust system components.
23 Make a final check that all wires, hoses, linkages and the speedometer cable have been connected and that the transaxle has been filled with lubricant to the proper level (see Chapter 1).
24 Install the battery and tray and connect the cables (negative cable last). Road test the vehicle for proper operation and check for leaks.

4 Manual transaxle overhaul – general information

Refer to illustrations 4.4a and 4.4b

Overhauling a manual transaxle is a difficult job for the do-it-yourselfer. It involves the disassembly and reassembly of many small parts. Numerous clearances must be precisely measured and, if necessary, changed with select fit spacers and snap-rings. As a result, if transaxle problems arise, it can be removed and installed by a competent do-it-yourselfer, but overhaul should be left to a transmission repair shop. Rebuilt transaxles may be available – check with your dealer parts department and auto parts stores. At any rate, the time and money involved in an overhaul is almost sure to exceed the cost of a rebuilt unit.

Nevertheless, it's not impossible for an inexperienced mechanic to rebuild a transaxle if the special tools are available and the job is done in a deliberate step-by-step manner so nothing is overlooked.

The tools necessary for an overhaul include internal and external snap-ring pliers, a bearing puller, a slide hammer, a set of pin punches, a dial indicator and possibly a hydraulic press. In addition, a large, sturdy workbench and a vise or transaxle stand will be required.

During disassembly of the transaxle, make careful notes of how each piece comes off, where it fits in relation to other pieces and what holds it in place. Exploded views are included **(see illustrations)** to show where

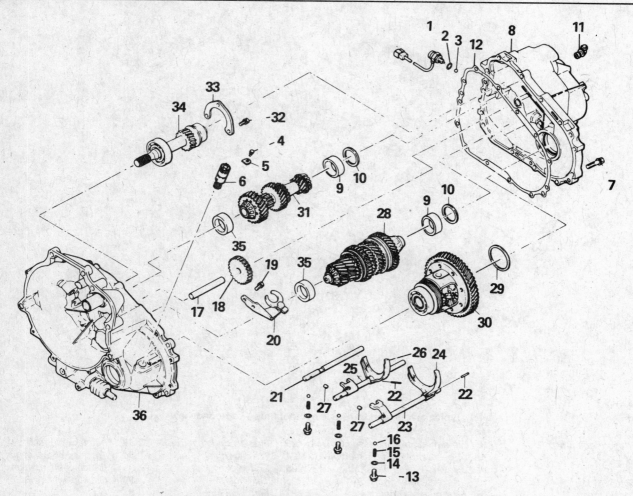

4.4a Typical four-speed transaxle components – exploded view

1	Back-up light switch	10	Spacers	19	Bolts
2	Gasket	11	Breather	20	Reverse shift lever assembly
3	Steel ball	12	Gasket	21	Reverse shift rail
4	Bolt	13	Poppet plugs	22	Spring pins
5	Locking plate	14	Gasket	23	First/second shift rail
6	Speedometer gear assembly	15	Poppet springs	24	First/second shift fork
7	Bolt	16	Poppet balls	25	3rd/4th shift rail
8	Transaxle case	17	Reverse idler gear shaft	26	3rd/4th shift fork
9	Outer bearing race	18	Reverse idler gear	27	Interlock plungers

28	Output shaft assembly
29	Spacer
30	Differential assembly
31	Intermediate gear assembly
32	Bolts
33	Bearing retainer
34	Input shaft assembly
35	Bearing outer race
36	Clutch housing

the parts go – but actually noting how they are installed when you remove the parts will make it much easier to get the transaxle back together.

Before taking the transaxle apart for repair, it will help if you have some idea what area of the transaxle is malfunctioning. Certain problems can be closely tied to specific areas in the transaxle, which can make component examination and replacement easier. Refer to the Troubleshooting section at the front of this manual for information regarding possible sources of trouble.

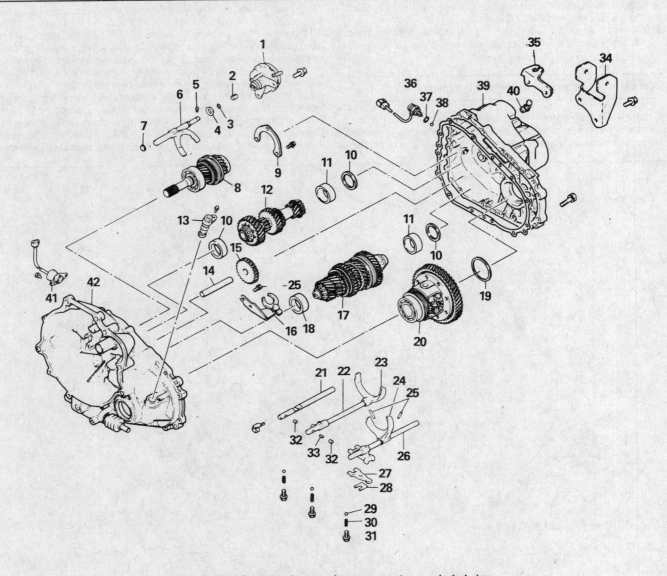

4.4b Typical five-speed transaxle components – exploded view

1 5th gear actuator	12 Intermediate gear assembly	23 First/second shift fork	33 Interlock plunger
2 Collar	13 Speedometer gear assembly	24 Third/fourth shift fork	34 Transaxle bracket
3 O-ring	14 Reverse idler gear shaft	25 Spring pin	35 Clutch cable bracket
4 Seat	15 Reverse idler gear	26 Third/fourth shift rail	36 Backup light switch
5 Locking plate	16 Reverse shift gear	27 Fifth gear lug	37 Gasket
6 Select rail fork assembly	17 Output shaft assembly	28 Selector spacer	38 Steel ball
7 Seat	18 Outer bearing race	29 Poppet ball	39 Transaxle case
8 Input shaft assembly	19 Spacer	30 Poppet spring	40 Breather
9 Bearing retainer	20 Differential assembly	31 Plug	41 Fifth gear actuator switch
10 Spacer	21 Reverse shift rail	32 Interlock plunger	42 Clutch housing
11 Bearing outer race	22 First/second shift rail		

Chapter 7 Part B Automatic transaxle

Contents

Automatic transaxle fluid and filter change .... See Chapter 1	Oil seal replacement 3
Automatic transaxle fluid level check .... See Chapter 1	Selector linkage – removal, installation and adjustment 5
Automatic transaxle – removal and installation 8	Throttle valve (TV) cable (three-speed models) – check
Diagnosis – general 2	and adjustment 6
General information 1	Transaxle mount – check and replacement 4
Neutral start switch – check, replacement and adjustment 7	

Specifications

Clearances

Selector lever assembly adjusting sleeve-to-lever end clearance
  Galant .... 0.598 to 0.625 in (15.2 to 15.9 mm)
  All other models
    1986 and earlier .... 0.677 to 0.705 in (17.2 to 17.9 mm)
    1987 on .... 0.598 to 0.625 in (15.2 to 15.9 mm)
Throttle valve (TV) cable clearance .... 0.04 ± 0.02 in (1 ± 0.5 mm)

Torque specifications

Ft-lbs

Transaxle-to-engine bolts
  6 x 12 mm .... 7 to 9
  8 x 12 mm .... 22 to 36
  8 x 20 mm .... 11 to 13
  8 x 60 mm .... 22 to 25
  10 x 40 mm .... 31 to 40
  10 x 55 mm .... 16 to 23
  10 x 65 mm .... 31 to 40
Torque converter-to-driveplate bolts .... 30 to 35
Drain and filler plugs .... See Chapter 1
Throttle valve lower cable bracket bolt 10
Neutral start switch bolts 8
Fluid pan bolts .... See Chapter 1
Starter motor bolts .... 20 to 25

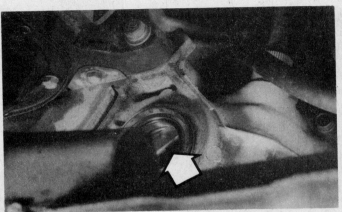

3.4 Insert the tip of a large screwdriver (arrow) behind the oil seal and very carefully pry it out

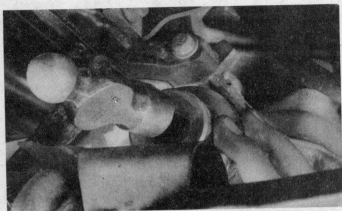

3.6 Apply a thin layer of grease to the outer edge of the new seal and carefully tap it into the bore with a large socket and hammer

1 General information

All vehicles covered in this manual come equipped with either a four or five speed manual transaxle or a three- or four-speed automatic transaxle. All information on the automatic transaxle is included in this Part of Chapter 7. Information for the manual transaxle can be found in Part A of this Chapter.

Due to the complexity of the automatic transaxle and the need for specialized equipment to perform most service operations, this Chapter contains only general diagnosis, routine maintenance, adjustment and removal and installation procedures.

If the transaxle requires major repair work, it should be left to a dealer service department or an automotive or transmission repair shop. You can, however, remove and install the transaxle yourself and save the expense, even if the repair work is done by a transmission shop.

2 Diagnosis – general

Note: *Automatic transaxle malfunctions may be caused by four general conditions: poor engine performance, improper adjustments, hydraulic malfunctions or mechanical malfunctions. Diagnosis of these problems should always begin with a check of the easily repaired items: fluid level and condition (see Chapter 1), selector linkage adjustment and throttle linkage adjustment. Next, perform a road test to determine if the problem has been corrected or if more diagnosis is necessary. If the problem persists after the preliminary tests and corrections are completed, additional diagnosis should be done by a dealer service department or transmission repair shop. Refer to the Troubleshooting section at the front of this manual for transaxle problem diagnosis.*

Preliminary checks

1 Drive the vehicle to warm the transaxle to normal operating temperature.
2 Check the fluid level as described in Chapter 1:
  a) If the fluid level is unusually low, add enough fluid to bring the level within the designated area of the dipstick, then check for external leaks.
  b) If the fluid level is abnormally high, drain off the excess, then check the drained fluid for contamination by coolant. The presence of engine coolant in the automatic transmission fluid indicates that a failure has occurred in the internal radiator walls that separate the coolant from the transmission fluid (see Chapter 3).
  c) If the fluid is foaming, drain it and refill the transaxle, then check for coolant in the fluid or a high fluid level.
3 Check the engine idle speed. **Note:** *If the engine is malfunctioning, do not proceed with the preliminary checks until it has been repaired and runs normally.*
4 Check the throttle valve cable (if equipped) for freedom of movement. Adjust it if necessary (see Section 6). **Note:** *The throttle valve cable may function properly when the engine is shut off and cold, but it may malfunction once the engine is hot. Check it cold and at normal engine operating temperature.*
5 Inspect the selector control cable (see Section 5). Make sure that it's properly adjusted and that the linkage operates smoothly.

Fluid leak diagnosis

6 Most fluid leaks are easy to locate visually. Repair usually consists of replacing a seal or gasket. If a leak is difficult to find, the following procedure may help.
7 Identify the fluid. Make sure it's transmission fluid and not engine oil or brake fluid (automatic transmission fluid is a deep red color).
8 Try to pinpoint the source of the leak. Drive the vehicle several miles, then park it over a large sheet of cardboard. After a minute or two, you should be able to locate the leak by determining the source of the fluid dripping onto the cardboard.
9 Make a careful visual inspection of the suspected component and the area immediately around it. Pay particular attention to gasket mating surfaces. A mirror is often helpful for finding leaks in areas that are hard to see.
10 If the leak still cannot be found, clean the suspected area thoroughly with a degreaser or solvent, then dry it.
11 Drive the vehicle for several miles at normal operating temperature and varying speeds. After driving the vehicle, visually inspect the suspected component again.
12 Once the leak has been located, the cause must be determined before it can be properly repaired. If a gasket is replaced but the sealing flange is bent, the new gasket will not stop the leak. The bent flange must be straightened.
13 Before attempting to repair a leak, check to make sure that the following conditions are corrected or they may cause another leak. **Note:** *Some of the following conditions cannot be fixed without highly specialized tools and expertise. Such problems must be referred to a transmission shop or a dealer service department.*

Gasket leaks

14 Check the pan periodically. Make sure the bolts are tight, no bolts are missing, the gasket is in good condition and the pan is flat (dents in the pan may indicate damage to the valve body inside).
15 If the pan gasket is leaking, the fluid level or the fluid pressure may be too high, the vent may be plugged, the pan bolts may be too tight, the pan sealing flange may be warped, the sealing surface of the transaxle housing may be damaged, the gasket may be damaged or the transaxle casting may be cracked or porous. If sealant instead of gasket material has been

Chapter 7 Part B Automatic transaxle

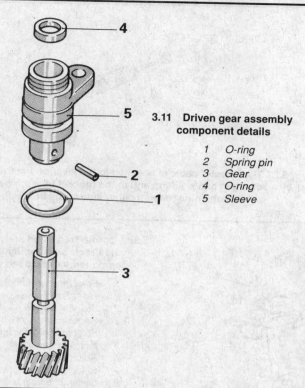

3.11 Driven gear assembly component details
1 O-ring
2 Spring pin
3 Gear
4 O-ring
5 Sleeve

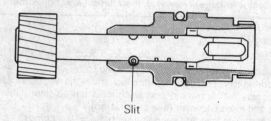

3.12 When assembling the driven gear to the sleeve, make sure the spring pin is installed with the slit facing away from the shaft

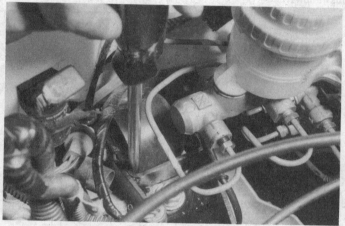

4.1 Pry on the transaxle mount with a large screwdriver to check for movement

used to form a seal between the pan and the transaxle housing, it may be the wrong sealant.

Seal leaks
16 If a transaxle seal is leaking, the fluid level or pressure may be too high, the vent may be plugged, the seal bore may be damaged, the seal itself may be damaged or improperly installed, the surface of the shaft protruding through the seal may be damaged or a loose bearing may be causing excessive shaft movement.
17 Make sure the dipstick tube seal is in good condition and the tube is properly seated. Periodically check the area around the speedometer gear or sensor for leakage. If transmission fluid is evident, check the O-ring for damage. Also inspect the side gear shaft oil seals for leakage.

Case leaks
18 If the case itself appears to be leaking, the casting is porous and will have to be repaired or replaced.
19 Make sure the oil cooler hose fittings are tight and in good condition.

Fluid comes out vent pipe or fill tube
20 If this condition occurs, the transaxle is overfilled, there is coolant in the fluid, the case is porous, the dipstick is incorrect, the vent is plugged or the drain back holes are plugged.

3 Oil seal replacement

Refer to illustrations 3.4, 3.6, 3.11 and 3.12

1 Oil leaks frequently occur due to wear of the driveaxle oil seals, and/or the speedometer driven gear O-rings. Replacement of these seals is relatively easy, since the repairs can usually be performed without removing the transaxle from the vehicle.
2 The driveaxle oil seals are located at the sides of the transaxle, where the driveaxles are attached. If leakage at the seal is suspected, raise the vehicle and support it securely on jackstands. If the seal is leaking, lubricant will be found on the sides of the transaxle.
3 Refer to Chapter 8 and remove the driveaxles.
4 Using a screwdriver or prybar, carefully pry the oil seal out of the transaxle bore (see illustration).

5 If the oil seal cannot be removed with a screwdriver or pry bar, a special oil seal removal tool (available at auto parts stores) will be required.
6 Using a large section of pipe or a large deep socket as a drift, install the new oil seal (see illustration). Drive it into the bore squarely and make sure it's completely seated.
7 Install the driveaxle(s). Be careful not to damage the lip of the new seal.
8 The speedometer cable and driven gear housing is located on the transaxle housing. Look for lubricant around the cable housing to determine if the O-ring is leaking.
9 Disconnect the speedometer cable from the transaxle.
10 Unscrew the retaining bolt and remove the speedometer gear assembly.
11 Drive out the spring pin and separate the driven gear from the sleeve (see illustration).
12 Install new O-rings on the driven gear and reassemble the housing using a new spring pin (see illustration).
13 Install the speedometer gear and cable assembly.

4 Transaxle mount – check and replacement

Refer to illustration 4.1

1 Insert a large screwdriver or pry bar between the mount and transaxle bracket and pry it back and forth (see illustration).
2 The transaxle bracket should not move away from the mount. If it does, replace the mount.
3 To replace a mount, support the transaxle with a jack, remove the nut and through bolt and the bracket-to-transaxle bolts, then detach the mount. It may be necessary to lower the transaxle slightly to provide enough clearance to remove the mount.
4 Installation is the reverse of removal.

5 Selector linkage – removal, installation and adjustment

Removal

Refer to illustrations 5.2, 5.5a, 5.5b, 5.5c, 5.5d, 5.5e, 5.5f and 5.5g

1. Disconnect the negative cable from the battery.
2. Working in the engine compartment, remove the air cleaner assembly (if necessary for clearance) and disconnect the selector cable at the bracket and transaxle lever **(see illustration)**.
3. Raise the vehicle and support it securely on jackstands.
4. Working under the vehicle, remove the selector cable assembly mounting bolt and nuts.
5. Remove the selector lever assembly mounting nuts or bolts **(see illustrations)**.

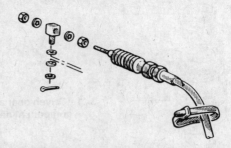

5.2 The selector cable is attached to the lever on the transaxle by a cotter pin and washers and to the bracket by a spring clip or two large nuts (one on either side of the bracket)

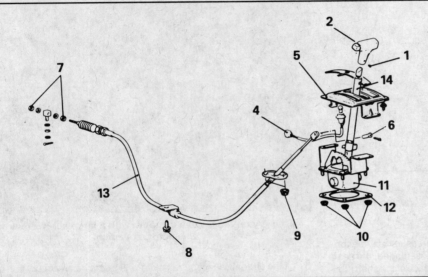

5.5a Cordia/Tredia and early Mirage and Precis selector linkage components – exploded view

1. Selector lever handle set screw
2. Selector handle
3. Console (not shown)
4. Selector indicator light bulb
5. Indicator panel
6. Bolt
7. Selector cable adjusting nuts
8. Cable mounting bolt
9. Cable mounting bracket nuts
10. Selector lever mounting nuts
11. Selector lever assembly mounting nuts
12. Gasket
13. Selector cable
14. Selector lever

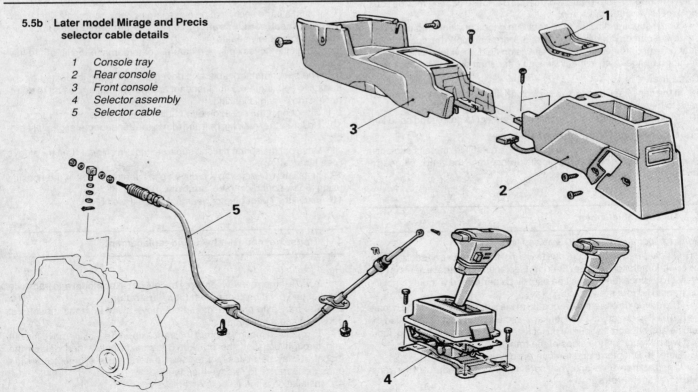

5.5b Later model Mirage and Precis selector cable details

1. Console tray
2. Rear console
3. Front console
4. Selector assembly
5. Selector cable

Chapter 7 Part B Automatic transaxle

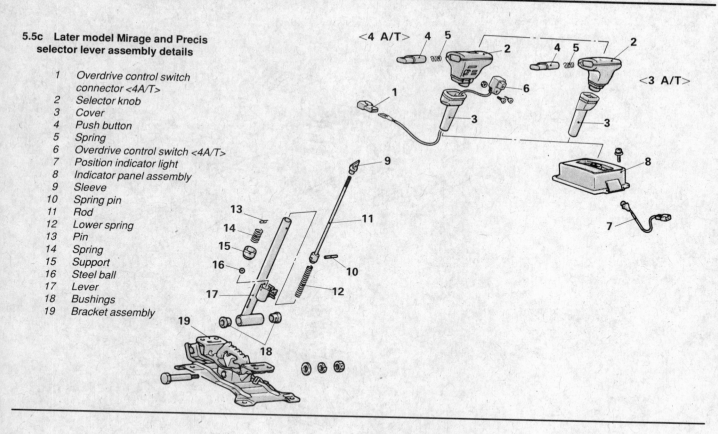

5.5c Later model Mirage and Precis selector lever assembly details

1. Overdrive control switch connector <4A/T>
2. Selector knob
3. Cover
4. Push button
5. Spring
6. Overdrive control switch <4A/T>
7. Position indicator light
8. Indicator panel assembly
9. Sleeve
10. Spring pin
11. Rod
12. Lower spring
13. Pin
14. Spring
15. Support
16. Steel ball
17. Lever
18. Bushings
19. Bracket assembly

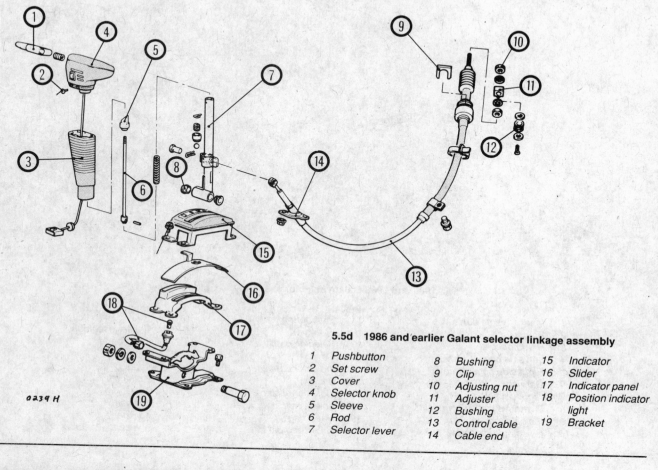

5.5d 1986 and earlier Galant selector linkage assembly

1. Pushbutton
2. Set screw
3. Cover
4. Selector knob
5. Sleeve
6. Rod
7. Selector lever
8. Bushing
9. Clip
10. Adjusting nut
11. Adjuster
12. Bushing
13. Control cable
14. Cable end
15. Indicator
16. Slider
17. Indicator panel
18. Position indicator light
19. Bracket

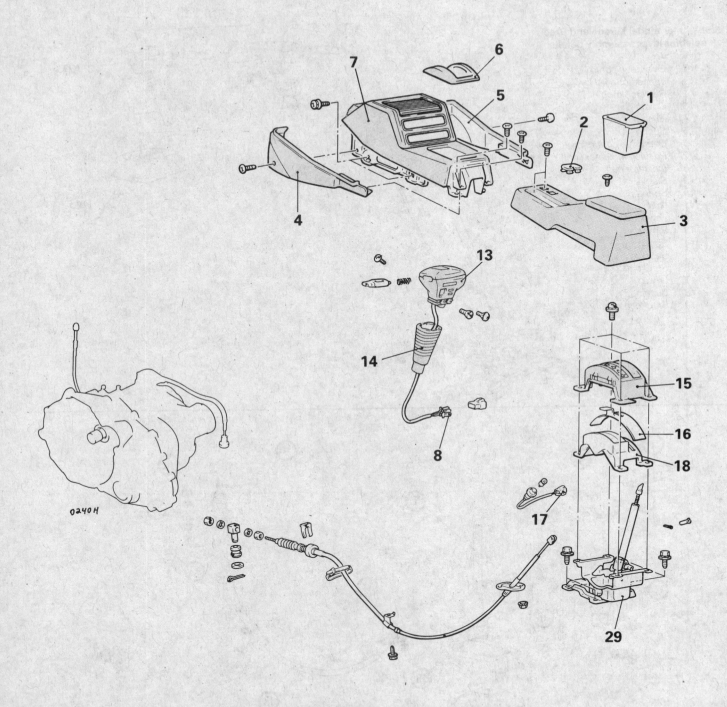

5.5e 1987 and 1988 Galant selector assembly – exploded view

1	Inner console box	11	Pushbutton	21	Air filter assembly
2	Plug	12	Spring	22	Cable band
3	Rear console	13	Selector knob	23	Clip
4	Left side console cover	14	Cover	24	Cotter pin
5	Right side console cover	15	Indicator panel	25	Bushing
6	Garnish	16	Slider	26	Cable
7	Front console	17	Light connector	27	Adjusting nut
8	Overdrive switch connector	18	Lower indicator panel	28	Adjuster
9	Indicator screw	19	Snap-ring	29	Selector lever assembly
10	Selector knob screw	20	Cable end pin		

Chapter 7 Part B Automatic transaxle

7B–7

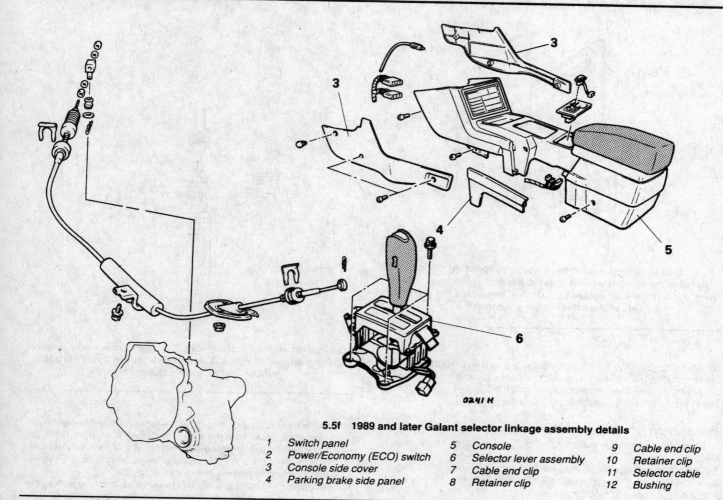

5.5f 1989 and later Galant selector linkage assembly details

1 Switch panel
2 Power/Economy (ECO) switch
3 Console side cover
4 Parking brake side panel
5 Console
6 Selector lever assembly
7 Cable end clip
8 Retainer clip
9 Cable end clip
10 Retainer clip
11 Selector cable
12 Bushing

5.5g 1989 and later Galant selector assembly – exploded view

1 Switch connector
2 Cover
3 Selector knob
4 Overdrive switch button
5 Overdrive control switch
6 Pin
7 Pushbutton
8 Spring
9 Indicator panel
10 Slider
11 Lower indicator panel
12 Socket assembly
13 Sleeve
14 Lever assembly
15 Bushing
16 Bracket assembly

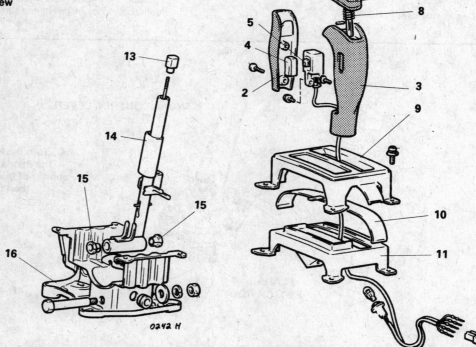

Chapter 7 Part B Automatic transaxle

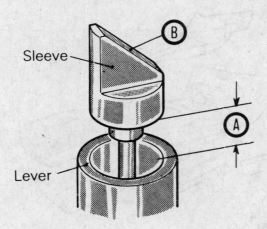

5.11 Turn the selector lever to achieve the specified clearance (listed at the front of this Chapter) to the end of the lever (A) – most models have an angled face (B) that must face the driver's side of the vehicle after adjustment

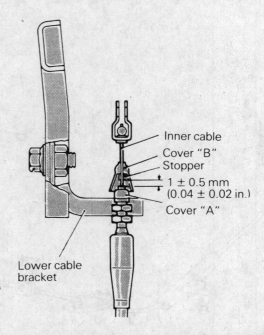

6.3 Pull up on the Throttle Valve (TV) inner cable cover B and check the stopper-to-cover A clearance – loosen the lower cable bracket nut to make adjustments

6 Working inside the vehicle, remove the selector lever handle and the center console. Disconnect the selector cable and remove the selector lever assembly **(see illustrations 5.5a through 5.5g)**.

7 Remove the selector cable from under the vehicle.

Installation

8 Install the selector lever assembly. Tighten the mounting nuts or bolts securely.

9 Install the selector cable and connect it to the transaxle and selector lever. On early models, make sure the toothed washer is correctly positioned on the transaxle mounting bracket.

Adjustment

Refer to illustration 5.11

10 Place the selector lever in Neutral.

11 Turn the lever adjusting sleeve to achieve the specified sleeve-to-lever end clearance. Make sure the angled surface of the sleeve faces the driver's side after adjustment **(see illustration)**. Apply a small dab of multi-purpose grease to the outer edge and the angled surface of the sleeve.

12 Install the console and the selector lever handle.

13 If necessary, eliminate any slack in the selector cable by turning the adjusting nuts on the transaxle selector lever **(see illustrations 5.5a through 5.5g)**.

14 Check the operation of the transaxle in each selector lever position (try to start the engine in each gear – the starter should operate in Park and Neutral only). Adjust the Neutral start switch if necessary (see Section 7).

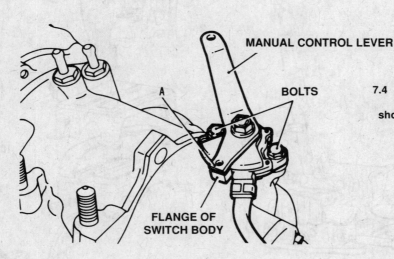

7.4 Turn the Neutral start switch until the flange is lined up with the shorter of the two control levers (A), then tighten the bolts

Adjustment

11 Rotate the Neutral start switch until the shorter of the two shift levers is aligned with the flange on the switch body, then tighten the bolts to the torque in this Chapter's Specifications **(see illustration 7.4)**.
12 Check the operation of the transaxle in each selector lever position, readjusting the switch if necessary.

8 Automatic transaxle – removal and installation

Refer to illustration 8.5

Removal

1 Disconnect the cables and remove the battery and tray. On fuel-injected models, remove the air cleaner assembly.
2 Raise the vehicle and support it securely on jackstands.
3 Drain the transaxle fluid (see Chapter 1).
4 Remove the torque converter cover.
5 Mark the torque converter and the driveplate with white paint so they can be installed in the same position **(see illustration)**.
6 Remove the three torque converter-to-driveplate bolts. Turn the crankshaft pulley bolt for access to each bolt.
7 Remove the starter motor (see Chapter 5).
8 Disconnect the driveaxles from the transaxle (see Chapter 8).
9 Disconnect the speedometer cable.
10 Disconnect the wire harness from the transaxle.
11 On models so equipped, disconnect the vacuum hose(s).
12 Remove any exhaust components which will interfere with transaxle removal (see Chapter 4).
13 Disconnect the TV cable (if equipped).
14 Disconnect the selector linkage.
15 Support the engine using a hoist from above or a jack and a block of wood under the oil pan to spread the load.
16 Support the transaxle with a jack – preferably a special jack made for this purpose. Safety chains will help steady the transaxle on the jack.
17 Remove any chassis or suspension components which will interfere with transaxle removal.
18 Remove the bolts securing the transaxle to the engine.
19 Remove the nuts and bolts and detach the transaxle mount.
20 Lower the transaxle slightly and disconnect and plug the transaxle cooler lines.
21 Move the transaxle back to disengage it from the engine block dowel pins and make sure the torque converter is detached from the driveplate. Secure the torque converter to the transaxle so it will not fall out during removal. Lower the transaxle from the vehicle.

Installation

22 Prior to installation, make sure that the torque converter hub is securely engaged in the pump.
23 With the transaxle secured to the jack, raise it into position. Be sure to keep it level so the torque converter does not slide out. Connect the fluid cooler lines.
24 Turn the torque converter to line up the holes with the holes in the driveplate. The white paint mark on the torque converter and the driveplate made in Step 5 must line up.
25 Move the transaxle forward carefully until the dowel pins and the torque converter are engaged.
26 Install the transaxle-to-engine bolts. Tighten them securely. Install the transaxle mount.
27 Install the torque converter-to-driveplate bolts. Tighten the bolts to the torque listed in this Chapter's Specifications.
28 Install the transaxle and any suspension and chassis components which were removed. Tighten the bolts and nuts to the specified torque.
29 Remove the jacks supporting the transaxle and the engine.
30 Install the starter motor (see Chapter 5).
31 Connect the vacuum hose(s) (if equipped).
32 Connect the selector and TV linkage.

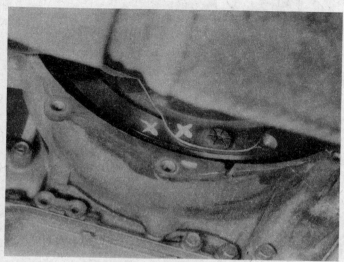

8.5 Mark the driveplate and the torque converter with white paint so they can be reinstalled in the same relationship

6 Throttle valve (TV) cable (three-speed models) – check and adjustment

Refer to illustration 6.3

1 The throttle valve (TV) cable adjustment is very important to proper transaxle operation. The cable positions a valve inside the transaxle which controls shift speed, shift quality and part throttle downshift sensitivity. If the cable is adjusted too short, early shifts and slippage between shifts may occur. If the cable is adjusted too long, shifts may be delayed and part throttle downshifts may be erratic.
2 Start and run the engine until it reaches normal operating temperature. Make sure the choke is off and the carburetor throttle lever is at the normal curb idle position. Turn off the engine.
3 Pull up on the cable cover B **(see illustration)** to expose the stopper and cover A, then loosen the lower cable bracket bolt.
4 Move the lower cable until the stopper is the distance from cover A listed in this Chapter's Specifications.
5 Tighten the cable bracket bolt to the specified torque and recheck the clearance.
6 Open the throttle lever to the wide open position and make sure the cable does not bind.

7 Neutral start switch – check, replacement and adjustment

Refer to illustration 7.4

Check

1 Try to start the engine in each gear – the starter should operate in Park and Neutral only.

Replacement

2 Remove the battery, and, on fuel-injected models, the air cleaner assembly.
3 Place the selector lever in Neutral.
4 Remove the nut and separate the manual control lever from the shaft on the transaxle **(see illustration)**.
5 Unplug the Neutral start switch electrical connector.
6 Remove the two Neutral start switch mounting bolts.
7 Lift the switch off the transaxle.
8 Place the new switch in position and install the mounting bolts loosely.
9 Install the manual control lever and nut. Tighten the nut securely.
10 Plug in the electrical connector.

33 Plug in the transaxle electrical connectors.
34 Install the torque converter cover.
35 Connect the driveaxles (see Chapter 8).
36 Connect the speedometer cable.
37 Adjust the selector linkage (see Section 5).
38 Install any exhaust system components that were removed or disconnected.
39 Lower the vehicle.
40 Fill the transaxle (see Chapter 1), run the vehicle and check for fluid leaks.

Chapter 8 Clutch and driveaxles

Contents

Clutch cable – removal, installation and adjustment	5
Clutch components – removal, inspection and installation	3
Clutch – description and check	2
Clutch hydraulic system – bleeding	8
Clutch master cylinder – removal, overhaul and installation	6
Clutch pedal freeplay check and adjustment	See Chapter 1
Clutch release bearing, fork and shaft – removal and installation	4
Clutch release cylinder – removal, overhaul and installation	7
Driveaxle boot check	See Chapter 1
Driveaxle boot replacement and Constant Velocity (CV) joint overhaul	11
Driveaxle oil seal replacement	See Chapter 7B
Driveaxles – general information and inspection	9
Driveaxles – removal and installation	10
Flywheel – removal and installation	See Chapter 2A
General information	1

Specifications

Clutch

Pedal freeplay	See Chapter 1
Clutch disc lining minimum thickness	0.012 in above rivet heads

Driveaxles

Inner CV joint assembled length (approx.)

Cordia and Tredia	3.45 in
Galant	3.45 in
Mirage	3.20 in
Precis	3.0 in

Torque specifications

	Ft-lbs
Pressure plate-to-flywheel bolts	11 to 15
Driveaxle hub nut	144 to 187
Wheel lug nuts	See Chapter 1

1 General information

The information in this Chapter deals with the components from the rear of the engine to the drive wheels, except for the transaxle, which is dealt with in the previous Chapter. For the purposes of this Chapter, these components are grouped into two categories; clutch and driveaxles. Separate Sections within this Chapter offer general descriptions and checking procedures for components in each of the two groups.

Since nearly all the procedures covered in this Chapter involve working under the vehicle, make sure it's securely supported on sturdy jackstands or on a hoist where the vehicle can be easily raised and lowered.

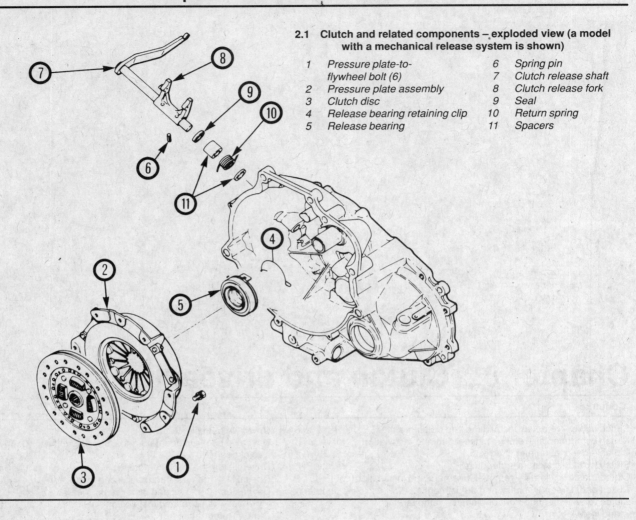

2.1 Clutch and related components – exploded view (a model with a mechanical release system is shown)

1. Pressure plate-to-flywheel bolt (6)
2. Pressure plate assembly
3. Clutch disc
4. Release bearing retaining clip
5. Release bearing
6. Spring pin
7. Clutch release shaft
8. Clutch release fork
9. Seal
10. Return spring
11. Spacers

2 Clutch – description and check

Refer to illustration 2.1

1 All vehicles with a manual transaxle use a single dry plate, diaphragm spring type clutch **(see illustration)**. The clutch disc has a splined hub which allows it to slide along the splines of the transaxle input shaft. The clutch and pressure plate are held in contact by spring pressure exerted by the diaphragm in the pressure plate.

2 The clutch release system is operated by hydraulic pressure on some models, while on others a mechanical system is used. The hydraulic release system consists of the clutch pedal, a master cylinder and fluid reservoir, the hydraulic line, a release (or slave) cylinder which actuates the clutch release lever and the clutch release (or throwout) bearing. The mechanical release system includes the clutch pedal with adjuster mechanism, a clutch cable which actuates the clutch release lever and the release bearing.

3 When pressure is applied to the clutch pedal to release the clutch, hydraulic or mechanical pressure is exerted against the outer end of the clutch release lever. As the lever pivots the shaft fingers push against the release bearing. The bearing pushes against the fingers of the diaphragm spring of the pressure plate assembly, which in turn releases the clutch plate.

4 Terminology can be a problem when discussing the clutch components because common names are in some cases different from those used by the manufacturer. For example, the driven plate is also called the clutch plate or disc, the clutch release bearing is sometimes called a throwout bearing, the release cylinder is sometimes called the operating or slave cylinder.

5 Other than to replace components with obvious damage, some preliminary checks should be performed to diagnose clutch problems.

  a) The first check should be of the fluid level in the clutch master cylinder (hydraulic release systems only). If the fluid level is low, add fluid as necessary and inspect the hydraulic system for leaks. If the master cylinder reservoir has run dry, bleed the system as described in Section 8 and retest the clutch operation.

  b) To check "clutch spin down time," run the engine at normal idle speed with the transaxle in Neutral (clutch pedal up – engaged). Disengage the clutch (pedal down), wait several seconds and shift the transaxle into Reverse. No grinding noise should be heard. A grinding noise would most likely indicate a problem in the pressure plate or the clutch disc.

  c) To check for complete clutch release, run the engine (with the parking brake applied to prevent movement) and hold the clutch pedal approximately 1/2-inch from the floor. Shift the transaxle between 1st gear and Reverse several times. If the shift is hard or the transaxle grinds, component failure is indicated. On vehicles with a hydraulic release system, check the release cylinder pushrod travel. With the clutch pedal depressed completely, the release cylinder pushrod should extend substantially. If it doesn't, check the fluid level in the clutch master cylinder and bleed the system (see Section 8).

  d) Visually inspect the pivot bushing at the top of the clutch pedal to make sure there is no binding or excessive play.

  e) On vehicles with mechanical release systems, a clutch pedal that is difficult to operate is most likely caused by a faulty clutch cable. Check the cable where it enters the housing for frayed wires, rust

Chapter 8 Clutch and driveaxles

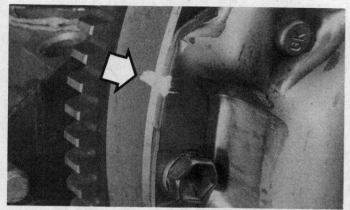

3.5 Make an index mark across the pressure plate and flywheel (just in case you're going to reuse the same pressure plate)

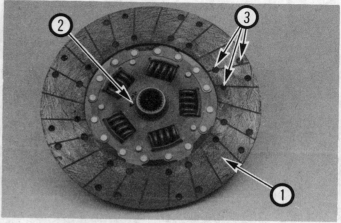

3.9 The clutch disc

1 Lining – this will wear down in use
2 Marks – "Flywheel Side" or something similar
3 Rivets – secure the lining and will damage the pressure plate if allowed to contact it

and other signs of corrosion. If it looks good, lubricate the cable with penetrating oil. If pedal operation improves, the cable is worn out and should be replaced.
  f) Crawl under the vehicle and make sure the clutch release lever is solidly mounted on the ball stud.

3 Clutch components – removal, inspection and installation

Warning: *Dust produced by clutch wear and deposited on clutch components may contain asbestos, which is hazardous to your health. DO NOT blow it out with compressed air and DO NOT inhale it. DO NOT use gasoline or petroleum-based solvents to remove the dust. Brake system cleaner should be used to flush the dust into a drain pan. After the clutch components are wiped clean with a rag, dispose of the contaminated rags and cleaner in a covered, marked container.*

Removal
Refer to illustration 3.5

1 Access to the clutch components is normally accomplished by removing the transaxle, leaving the engine in the vehicle. If, of course, the engine is being removed for major overhaul, then check the clutch for wear and replace worn components as necessary. However, the relatively low cost of the clutch components compared to the time and trouble spent gaining access to them warrants their replacement anytime the engine or transaxle is removed, unless they are new or in near perfect condition. The following procedures are based on the assumption the engine will stay in place.
2 Referring to Chapter 7 Part A, remove the transaxle from the vehicle. Support the engine while the transaxle is out. Preferably, an engine hoist should be used to support it from above. However, if a jack is used underneath the engine, make sure a piece of wood is positioned between the jack and oil pan to spread the load. **Caution:** *The pickup for the oil pump is very close to the bottom of the oil pan. If the pan is bent or distorted in any way, engine oil starvation could occur.*
3 The clutch fork and release bearing can remain attached to the transaxle housing for the time being.
4 To support the clutch disc during removal, install a clutch alignment tool through the clutch disc hub.
5 Carefully inspect the flywheel and pressure plate for indexing marks. The marks are usually an X, an O or a white letter. If they cannot be found, scribe marks yourself so the pressure plate and the flywheel will be in the same alignment during installation (**see illustration**).
6 Turning each bolt only 1/4-turn at a time, loosen the pressure plate-to-flywheel bolts. Work in a criss-cross pattern until all spring pressure is relieved. Then hold the pressure plate securely and completely remove the bolts, followed by the pressure plate and clutch disc.

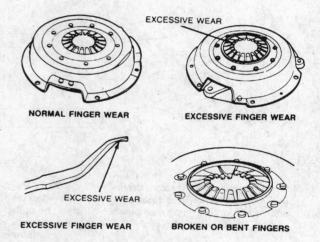

3.11a Replace the pressure plate if the fingers are worn excessively, broken or bent

Inspection
Refer to illustrations 3.9, 3.11a and 3.11b

7 Ordinarily, when a problem occurs in the clutch, it can be attributed to wear of the clutch driven plate assembly (clutch disc). However, all components should be inspected at this time.
8 Inspect the flywheel for cracks, heat checking, grooves and other obvious defects. If the imperfections are slight, a machine shop can machine the surface flat and smooth, which is highly recommended regardless of the surface appearance. Refer to Chapter 2, Part A for the flywheel removal and installation procedure.
9 Inspect the lining on the clutch disc. There should be at least 0.012 inch of lining above the rivet heads. Check for loose rivets, distortion, cracks, broken springs and other obvious damage (**see illustration**). As mentioned above, ordinarily the clutch disc is routinely replaced, so if in doubt about the condition, replace it with a new one.
10 The release bearing should also be replaced along with the clutch disc (see Section 4).
11 Check the machined surfaces and the diaphragm spring fingers of the pressure plate (**see illustrations**). If the surface is grooved or otherwise

3.11b Also examine the pressure plate friction surface for score marks, cracks and evidence of overheating (blue discolored areas)

3.13 Center the clutch disc in the pressure plate with a clutch alignment tool

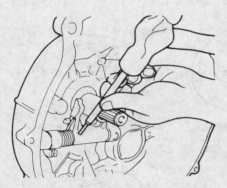

4.5a If there is enough room, drive out the spring pins with a small pin punch and hammer – use new pins for reassembly

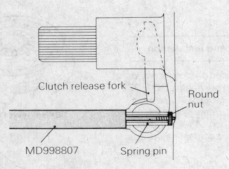

4.5b If there is not enough room to drive the pins through the shaft, they must be pulled out with a special puller – this cutaway view shows a properly set up Mitsubishi puller (MD998807)

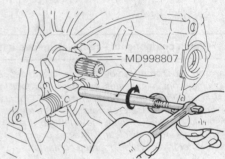

4.5c After setting up puller MD998807, turn the head clockwise to remove the spring pin

---

damaged, replace the pressure plate. Also check for obvious damage, distortion, cracking, etc. Light glazing can be removed with emery cloth. If a new pressure plate is required, new and factory-rebuilt units are available.

Installation

Refer to illustration 3.13

12 Before installation, clean the flywheel and pressure plate machined surfaces with lacquer thinner or acetone. It's important that no oil or grease is on these surfaces or the lining of the clutch disc. Handle the parts only with clean hands.

13 Position the clutch disc and pressure plate against the flywheel with the clutch held in place with an alignment tool **(see illustration)**. Make sure it's installed properly (most replacement clutch plates will be marked "flywheel side" or something similar – if not marked, install the clutch disc with the damper springs toward the transaxle).

14 Tighten the pressure plate-to-flywheel bolts only finger tight, working around the pressure plate.

15 Center the clutch disc by ensuring the alignment tool extends through the splined hub and into the pocket in the crankshaft. Wiggle the tool up, down or side-to-side as needed to center the disc. Tighten the pressure plate-to-flywheel bolts a little at a time, working in a criss-cross pattern to prevent distorting the cover. After all of the bolts are snug, tighten them to the torque listed in this Chapter's Specifications. Remove the alignment tool.

16 Using high temperature grease, lubricate the inner groove of the release bearing (refer to Section 4). Also place grease on the release lever contact areas and the transaxle input shaft bearing retainer.

17 Install the clutch release bearing as described in Section 4.

18 Install the transaxle and all components removed previously. Tighten all fasteners to the proper torque specifications.

4 Clutch release bearing, fork and shaft – removal and installation

Warning: *Dust produced by clutch wear and deposited on clutch components may contain asbestos, which is hazardous to your health. DO NOT blow it out with compressed air and DO NOT inhale it. DO NOT use gasoline or petroleum-based solvents to remove the dust. Brake system cleaner should be used to flush the dust into a drain pan. After the clutch components are wiped clean with a rag, dispose of the contaminated rags*

Chapter 8 Clutch and driveaxles

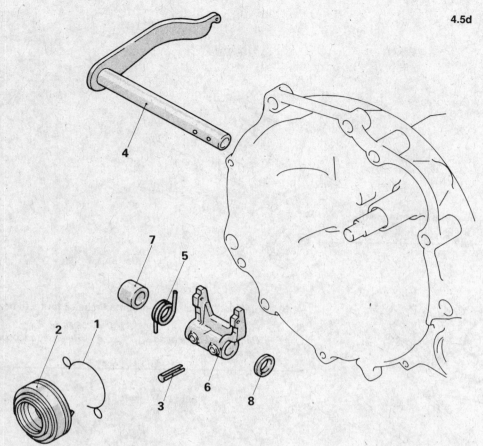

4.5d Details of the shaft-mounted release fork and bearing

1. Return clip
2. Clutch release bearing
3. Spring pin
4. Release fork shaft
5. Return spring
6. Release fork
7. Spacer
8. Spacer

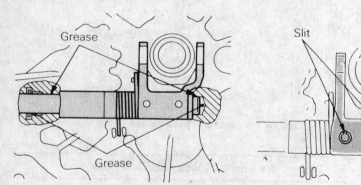

4.6a Apply high-temperature grease to the areas shown

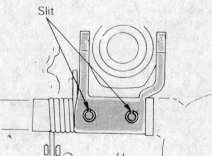

4.6b Drive the retaining pins in so the slits are facing up

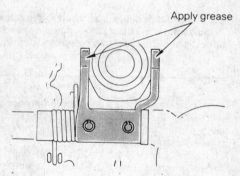

4.7a Apply high-temperature grease to the areas shown

and cleaner in a labeled, covered container.
1 Refer to Chapter 7, Part A, and remove the transaxle.
2 Using a pair of pliers, pull the ends of the bearing retaining clip out of the release fork **(see illustration 2.1)**.
3 Remove the bearing from the clutch housing. Be careful not to lose the retaining spring.
4 Hold the center of the bearing and turn the outer portion while applying pressure. If it doesn't turn smoothly or if it's noisy, it must be replaced. It's a good idea to replace it anyway, when you consider the time and effort spent on removing the transaxle. However, if you elect to reinstall the old bearing, wipe it off with a clean rag. Don't immerse the bearing in solvent – it's sealed for life and would be ruined by the solvent.

Shaft-mounted fork

Refer to illustrations 4.5a, 4.5b 4.5c, 4.5d, 4.6a, 4.6b, 4.7a and 4.7b

5 Check the release fork ends for excessive wear. If the fork must be replaced and room permits, drive the spring pins out of the fork and shaft with a small pin punch and a hammer **(see illustration)**. If their is not enough room to use a hammer and punch, use a pulling type tool such as the one shown **(see illustrations)**. Discard the spring pins, slide the shaft out of the clutch housing and remove the fork, spring and spacers **(see illustration)**.
6 Lubricate the release shaft bore in the clutch housing with high-temperature grease **(see illustration)**, slide the shaft part-way into the hous-

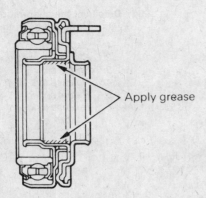

4.7b Pack the inner groove of the release bearing with high-temperature grease

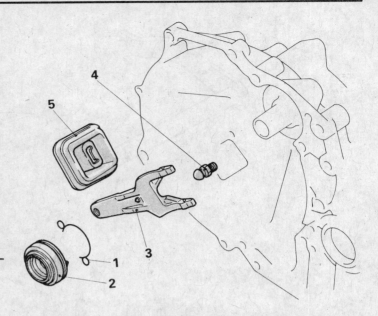

4.10 Details of the ball pivot-mounted release fork and bearing

1. Return clip
2. Clutch release bearing
3. Release fork
4. Ball pivot
5. Release fork boot

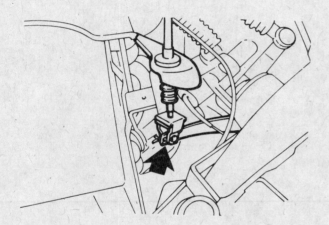

5.3 The clutch cable is secured to the release shaft lever by a clevis pin and cotter pin – once they're removed, the cable housing and grommet can be pulled (or pryed, if necessary) out of the bracket

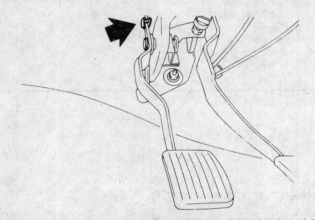

5.4 Unhook the cable end from the top of the clutch pedal

ing and place the spacers, spring and fork in position. Continue to slide the shaft through the components and align the spring pin holes. Insert new spring pins and drive them into position so the slits in the spring pins face up **(see illustration)**.

7 Lightly lubricate the release fork ends with a small amount of high-temperature grease **(see illustration)**. Pack the groove in the inner diameter of the release bearing with the same grease **(see illustration)** and position it against the release fork.

8 Install the release bearing retaining clip. Make sure the clip is secured in the bearing groove and the ends are inserted into the release fork properly.

9 Install the transaxle.

Ball pivot-mounted fork

Refer to illustration 4.10

10 Check the release fork ends for excessive wear. Replace the fork, if necessary. Pull the fork toward the rubber boot and detach it from the ball pivot **(see illustration)**.

11 Before installing the fork, use grease to lightly lubricate all metal-to-metal contact surfaces.

12 Installation is the reverse of removal. Be sure the clip ends on the release fork engage behind the ball pivot.

5 Clutch cable – removal, installation and adjustment

Refer to illustrations 5.3 and 5.4

Removal

1 Disconnect the negative battery cable (ground cable) from the battery.

2 Unscrew the adjuster wheel completely to produce slack in the cable.

3 Remove the cotter pin from the clutch release lever, pull out the clevis pin and disconnect the cable housing from the bracket **(see illustration)**.

4 Working under the dash, disconnect the cable end from the top of the clutch pedal **(see illustration)**.

5 Pull the cable through the firewall from the engine side. If the rubber grommet sticks in the firewall, pry it out with a screwdriver.

Installation and adjustment

6 When installing the cable, lightly lubricate the ends with multi-purpose grease. Install the cable by reversing the removal procedure, then adjust the clutch pedal freeplay by following the procedure in Chapter 1.

Chapter 8 Clutch and driveaxles

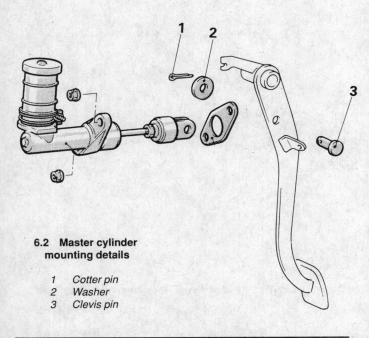

6.2 Master cylinder mounting details

1. Cotter pin
2. Washer
3. Clevis pin

6 Clutch master cylinder – removal, overhaul and installation

Note: *Before beginning this procedure, contact local parts stores and dealer service departments concerning the purchase of a rebuild kit or a new master cylinder. Availability and cost of the necessary parts may dictate whether the cylinder is rebuilt or replaced with a new one. If it's decided to rebuild the cylinder, inspect the bore as described in Step 12 before purchasing parts.*

Removal

Refer to illustration 6.2

1 Disconnect the negative cable from the battery.
2 Under the dashboard, disconnect the pushrod from the top of the clutch pedal **(see illustration)**.
3 Disconnect the hydraulic line at the clutch master cylinder. If available, use a flare-nut wrench on the fitting, which will prevent the fitting from being rounded off. Have rags handy as some fluid will be lost as the line is removed. **Caution:** *Don't allow brake fluid to come into contact with paint as it will damage the finish.*
4 Remove the two nuts which secure the master cylinder to the engine firewall. Remove the master cylinder, again being careful not to spill any of the fluid.

Overhaul

Refer to illustration 6.6

5 Remove the reservoir cap and drain all fluid from the master cylinder. Remove the clamp and disconnect the reservoir from the master cylinder body.
6 Pull back the rubber boot on the pushrod **(see illustration)** and remove the snap-ring.
7 Remove the pushrod from the cylinder.
8 Tap the master cylinder on a block of wood to eject the piston assembly from inside the bore. **Note:** *If the rebuild kit contains a complete piston assembly, ignore Steps 9, 11 and 14.*
9 Separate the spring from the piston.
10 Remove the spring support, seal and shim from the pushrod.
11 Carefully remove the seal from the piston.
12 Inspect the bore of the master cylinder for deep scratches, score marks and ridges. The surface must be smooth to the touch. If the bore isn't perfectly smooth, the master cylinder must be replaced with a new or factory rebuilt unit.
13 If the cylinder will be rebuilt, use the new parts contained in the rebuild kit and follow any specific instructions which may have accompanied the rebuild kit. Wash all parts to be re-used with brake cleaner, denatured alcohol or clean brake fluid. DO NOT use petroleum-based solvents.
14 Attach the piston seal to the piston. The seal lips must face away from the pushrod end of the piston.
15 Assemble the spring on the piston.
16 Lubricate the bore of the cylinder and the seals with plenty of fresh brake fluid (DOT 3).
17 Carefully guide the piston assembly into the bore, being careful not to damage the seals. Make sure the spring end is installed first, with the pushrod end of the piston closest to the opening.

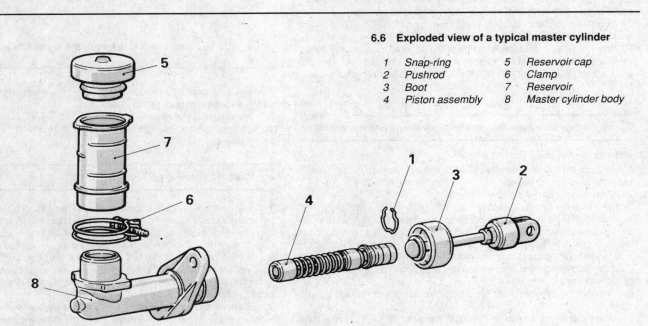

6.6 Exploded view of a typical master cylinder

1	Snap-ring	5	Reservoir cap
2	Pushrod	6	Clamp
3	Boot	7	Reservoir
4	Piston assembly	8	Master cylinder body

Chapter 8 Clutch and driveaxles

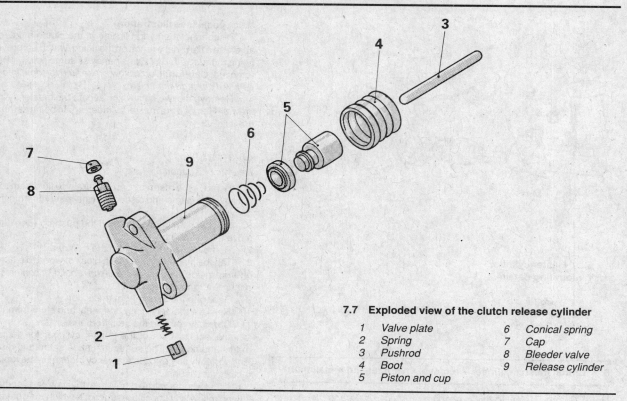

7.7 Exploded view of the clutch release cylinder

1	Valve plate	6	Conical spring
2	Spring	7	Cap
3	Pushrod	8	Bleeder valve
4	Boot	9	Release cylinder
5	Piston and cup		

18 Position the pushrod and retaining washer in the bore, compress the spring and install a new snap-ring.
19 Apply a liberal amount of rubber grease or equivalent to the inside of the dust cover and attach it to the master cylinder.

Installation

20 Attach the master cylinder to the firewall and install the mounting nuts finger tight.
21 Connect the hydraulic line to the master cylinder, moving the cylinder slightly as necessary to thread the fitting properly into the bore. Don't cross-thread the fitting as it's installed.
22 Tighten the mounting nuts to the torque listed in this Chapter's Specifications, then tighten the hydraulic line fitting.
23 Working inside the vehicle, connect the pushrod to the clutch pedal.
24 Fill the clutch master cylinder reservoir with brake fluid conforming to DOT 3 specifications and bleed the clutch system as described in Section 8.

7 Clutch release cylinder – removal, overhaul and installation

Refer to illustrations 7.7 and 7.8

1 Detach the hydraulic line from the release cylinder and plug the line to keep the fluid from draining.
2 Remove the clevis pin (if equipped).
3 Remove the release cylinder mounting bolts and the cylinder.
4 To remove the internal parts of the cylinder, wrap a rag around the cylinder piston end and apply a light amount of air pressure to the fluid inlet orifice of the cylinder. **Caution:** *The piston may be expelled with some force. Don't point the piston toward yourself, anyone else or anything that could be damaged if the piston becomes a projectile.*
5 Clean the cylinder walls.
6 Inspect the cylinder walls for pitting or scoring and replace if damaged.
7 If the cylinder is able to be rebuilt, completely strip the cylinder **(see illustration)**.
8 Using the new parts in the rebuild kit, assemble the components using

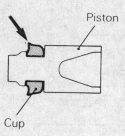

7.8 Be sure the cup is facing the proper direction

plenty of fresh brake fluid for lubrication. Note the installed direction of the spring and the cup **(see illustration)**.
9 Installation is the reverse of removal.
10 Be sure to bleed the clutch as outlined in Section 8.

8 Clutch hydraulic system – bleeding

1 The hydraulic system should be bled to remove all air whenever any part of the system has been removed or if the fluid level has been allowed to fall so low that air has been drawn into the master cylinder. The procedure is very similar to bleeding a brake system.
2 Fill the master cylinder with new brake fluid conforming to DOT 3 specifications. **Caution:** *Do not re-use any of the fluid coming from the system during the bleeding operation or use fluid which has been inside an open container for an extended period of time.*
3 Remove the dust cap which fits over the bleeder valve and push a length of plastic hose over the valve. Place the other end of the hose into a clear container with about two inches of brake fluid. The hose end must be in the fluid at the bottom of the container.

10.1 With the vehicle on the ground, loosen the driveaxle hub nut

10.4a If the driveaxle won't slide out of the hub easily, a puller can be used to push it out – DO NOT hammer on the end of the driveaxle to force it out!

4   Have an assistant depress the clutch pedal and hold it. Open the bleeder valve on the release cylinder, allowing fluid to flow through the hose. Close the bleeder valve when your assistant signals that the clutch pedal is at the bottom of its travel. Once closed, have your assistant release the pedal.
5   Continue this process until all air is evacuated from the system, indicated by a solid stream of fluid being ejected from the bleeder valve each time with no air bubbles in the hose or container. Keep a close watch on the fluid level inside the clutch master cylinder reservoir – if the level drops too low, air will be sucked back into the system and the process will have to be started all over again.
6   Install the dust cap and lower the vehicle. Check carefully for proper operation before placing the vehicle in normal service.

9   Driveaxles – general information and inspection

Power is transmitted from the transaxle to the wheels through a pair of driveaxles. The inner end of each driveaxle is splined to the differential. The outer ends of the driveaxles are splined to the axle hubs and locked in place by an axle nut.

The inner ends of the driveaxles are equipped with sliding Constant Velocity (CV) joints, which are capable of both angular and axial motion. Each inner joint assembly consists of an inner race, ball bearing and cage assembly and an outer race (housing) in which the inner bearing assembly is free to slide in and out as the driveaxle moves up and down with the wheel. The inner joints are rebuildable.

Each outer joint, which consists of ball bearings running between an inner race and an outer cage, is capable of angular but not axial movement. The outer joints are neither rebuildable nor removable. Should one of them fail, a new driveaxle/outboard joint assembly must be installed.

The boots should be periodically inspected for damage, leaking lubricant and cuts. Damaged CV joint boots must be replaced immediately or the joints can be damaged. Boot replacement involves removal of the driveaxle. **Note:** *Some auto parts stores carry "split" type replacement boots, which can be installed without removing the driveaxle from the vehicle. This is a convenient alternative; however, it's recommended that the driveaxle be removed and the CV joint disassembled and cleaned to ensure that the joint is free from contaminants such as moisture and dirt, which will accelerate CV joint wear.* The most common symptom of worn or damaged CV joints, besides lubricant leaks, is a clicking noise in turns, a clunk when accelerating from a coasting condition or vibration at highway speeds.

To check for wear in the CV joints and driveaxle shafts, grasp each axle (one at a time) and rotate it in both directions while holding the CV joint housings, inspecting for movement, indicating worn splines or sloppy CV joints. Also check the driveaxle shafts for cracks, dents, twisting and bending.

10   Driveaxles – removal and installation

Refer to illustrations 10.1, 10.4a, 10.4b, 10.5 and 10.7

Removal

1   Set the parking brake, remove the wheel cover and loosen the driveaxle hub nut **(see illustration)**. Loosen the wheel lug nuts, raise the front of the vehicle and support it securely on jackstands. Remove the front wheel.
2   Remove the driveaxle hub nut. To prevent the hub from turning, brace a screwdriver across two of the wheel studs, then remove the nut.
3   Remove the two bolts and separate the lower arm from the balljoint or separate the balljoint from the steering knuckle (refer to Chapter 10 if necessary).
4   Push the driveaxle from the hub with a puller (if it's stuck), then pull out on the steering knuckle and remove the outer end from the hub splines **(see illustrations)**.

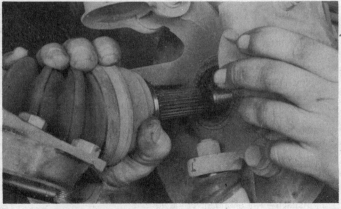

10.4b Pull out on the steering knuckle and remove the driveaxle from the hub splines

Chapter 8 Clutch and driveaxles

10.5 Be careful when prying the inner end of the axle out of the transaxle (the seal is very close to the edge of the joint and could be damaged if the prybar is inserted too far)

10.7 Always replace the circlip on the inner CV joint stub shaft before reinstalling the driveaxle

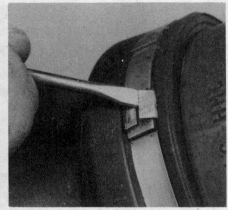

11.3 Pry the boot clamp retaining tabs up with a small screwdriver and slide the clamps off the boot

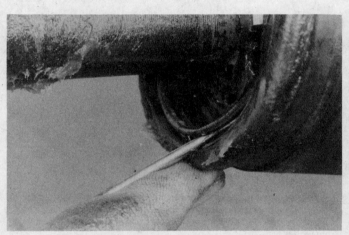

11.4 Pry the wire ring ball retainer out of the outer race, ...

11.5a ... then slide the outer race off the inner bearing assembly

5   Carefully pry the inner end of the axle from the transaxle, using a large prybar positioned between the transaxle housing and the CV joint housing **(see illustration)**.
6   Support the CV joints and carefully remove the driveaxle from the vehicle.

Installation

7   Pry the old circlip from the inner end of the driveaxle and install a new one **(see illustration)**. Lubricate the differential seal with multipurpose grease, raise the driveaxle into position while supporting the CV joints and insert the splined end of the inner CV joint into the the differential side gear. Seat the shaft into the side gear by pushing firmly on the driveaxle.
8   Apply a light coat of multi-purpose grease to the outer CV joint splines, pull out on the strut/steering knuckle assembly and install the stub axle in the hub.
9   Connect the balljoint to the lower arm or the steering knuckle (see Chapter 10).
10 Install the hub nut using a new large washer underneath it. Lock the disc so that it cannot turn and tighten the hub nut securely.
11 Grasp the inner CV joint housing (not the driveaxle) and pull out to make sure that the axle has seated securely in the transaxle.
12 Install the wheel and lower the vehicle.
13 Tighten the hub nut to the torque listed in this Chapter's Specifications and install the wheel cover.

11 Driveaxle boot replacement and Constant Velocity (CV) joint overhaul

Inner CV joint and boot

Double Offset Joint (DOJ)
Refer to illustrations 11.3, 11.4, 11.5a, 11.5b, 11.6, 11.7, 11.9, 11.10a, 11.10b, 11.11a, 11.11b, 11.13, 11.14, 11.17, 11.19, 11.20, 11.21a and 11.21b

Disassembly

1   Remove the driveaxle from the vehicle (see Section 10).
2   Mount the driveaxle in a vise. The jaws of the vise should be lined with wood or rags to prevent damage to the axleshaft.
3   Pry the boot clamp retaining tabs up with a small screwdriver and slide the clamps off the boot **(see illustration)**.
4   Slide the boot back on the axleshaft and pry the wire ring ball retainer from the outer race **(see illustration)**.
5   Slide the outer race off the inner bearing assembly **(see illustrations)**.
6   Remove the snap-ring from the groove in the axleshaft with a pair of snap-ring pliers **(see illustration)**.

Chapter 8 Clutch and driveaxles

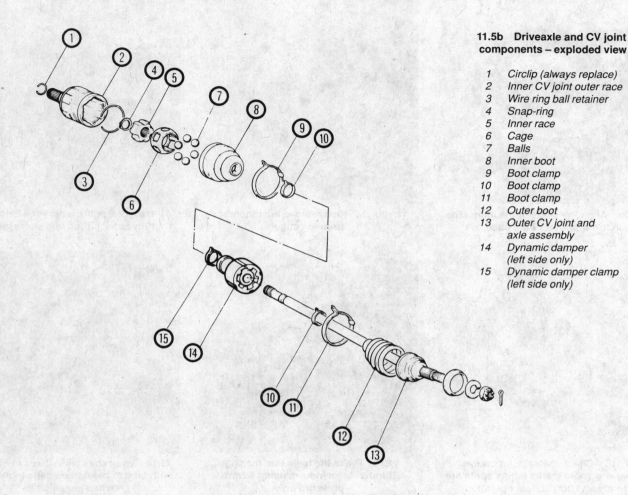

11.5b Driveaxle and CV joint components – exploded view

1. Circlip (always replace)
2. Inner CV joint outer race
3. Wire ring ball retainer
4. Snap-ring
5. Inner race
6. Cage
7. Balls
8. Inner boot
9. Boot clamp
10. Boot clamp
11. Boot clamp
12. Outer boot
13. Outer CV joint and axle assembly
14. Dynamic damper (left side only)
15. Dynamic damper clamp (left side only)

11.6 Using snap-ring pliers, remove the snap-ring from the end of the axle

11.7 Place match marks on the inner race and cage to identify which side must face the end of the shaft during reassembly

11.9 Pry the balls out of the cage with a screwdriver – be careful not to nick or scratch them

7 Mark the inner race and cage to be sure they are reassembled with the correct sides facing out (see illustration).
8 Slide the inner bearing assembly off the axleshaft.
9 Using a screwdriver or piece of wood, pry the balls from the cage (see illustration). Be careful not to scratch the inner race, the balls or the cage.
Note: *If the inner bearing assembly does not disassemble easily, do not disassemble it. Clean and regrease it as an assembly.*
10 Align the inner race lands with the cage windows and pull the race out

11.10a Align the lands of the inner race with the windows of the cage, . . .

11.10b . . . then remove the inner race from the cage

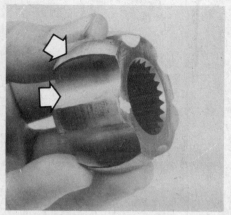

11.11a Check the inner race lands and grooves for pitting and score marks

11.11b Check the cage for cracks, pitting and score marks (shiny spots are normal and don't affect operation)

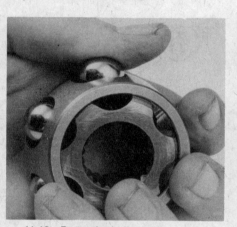

11.13 Press the balls into the cage through the windows using thumb pressure only

11.14 Wrap the splined area of the axle with tape to prevent damage to the boot when installing it

11.17 Pack the inner race and cage assembly full of CV joint grease (also note that the larger diameter side, or "bulge", is facing the axleshaft end)

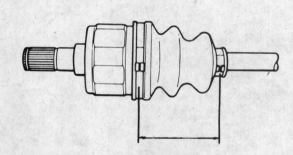

11.19 Adjust the length of the CV joint (distance between the boot bands) before tightening the clamps (see the Specifications in this Chapter)

of the cage **(see illustrations)**.

Inspection

11 Clean the components with solvent to remove all traces of grease. Inspect the cage and races for pitting, score marks, cracks and other signs of wear and damage. Shiny, polished spots are normal and will not adversely affect CV joint performance **(see illustrations)**.

Reassembly

12 Insert the inner race into the cage. Verify that the matchmarks are on the same side. However, it's not necessary for them to be in direct alignment with each other.

13 Press the balls into the cage windows with your thumbs **(see illustration)**.

14 Wrap the axleshaft splines with tape to avoid damaging the boot **(see illustration)**. Slide the small boot clamp and boot onto the axleshaft, then remove the tape.

15 Install the inner race and cage assembly on the axleshaft with the

Chapter 8 Clutch and driveaxles

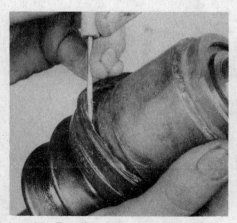

11.20 Equalize the pressure inside the boot by inserting a small, dull screwdriver between the boot and the outer race

11.21a To install the new clamps, bend the tang down and . . .

11.21b . . . fold the tabs over to hold it in place

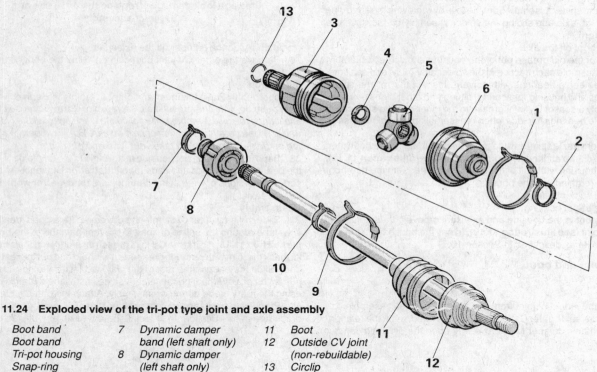

11.24 Exploded view of the tri-pot type joint and axle assembly

1 Boot band	7 Dynamic damper	11 Boot
2 Boot band	band (left shaft only)	12 Outside CV joint
3 Tri-pot housing	8 Dynamic damper	(non-rebuildable)
4 Snap-ring	(left shaft only)	13 Circlip
5 Spider assembly	9 Boot band	
6 Boot	10 Boot band	

larger diameter side or "bulge" of the cage (and the previously applied marks) facing the axleshaft end.
16 Install the snap-ring in the groove. Make sure it's completely seated by pushing on the inner race and cage assembly.
17 Using the CV joint grease included with the a new boot kit, pack the inner race and cage assembly with grease, by hand, until grease is worked completely into the assembly **(see illustration)**. Fill the outer race and boot with the remainder of the grease supplied.
18 Slide the outer race down onto the inner race and install the wire ring retainer.
19 Wipe any excess grease from the axle boot groove on the outer race. Seat the small diameter of the boot in the recessed area on the axleshaft. Push the other end of the boot onto the outer race and move the race in or

out to adjust the joint to the proper length **(see illustration)**.
20 With the axle set to the proper length, equalize the pressure in the boot by inserting a dull screwdriver between the boot and the outer race **(see illustration)**. Don't damage the boot with the tool.
21 Install the boot clamps **(see illustrations)**.
22 Install a new circlip on the inner CV joint stub axle **(see illustration 10.7)**.
23 Install the driveaxle as described in Section 10.

Tri-pot Joint (TJ) type
Refer to illustrations 11.24 and 11.26

24 Cut off the boot bands and slide the boot towards the center of the driveaxle **(see illustration)**. Mark the tri-pot housing and driveaxle so they

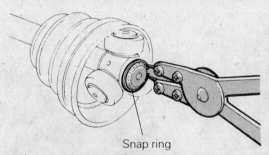

11.26 Remove the snap-ring and slide the spider assembly off the axle

11.38 After the old grease has been rinsed away and the solvent has been blown out with compressed air, rotate the outer joint housing through its full range of motion and inspect the bearing surfaces for wear and damage – if any of the balls, the race or the cage look damaged, replace the driveaxle and outer joint assembly

can be reinstalled in the same relative positions, then slide the housing off the axle.

25 Use tape or a cloth wrapped around the spider bearing assembly to retain the bearings during removal and installation.

26 Remove the spider assembly from the axle by removing the snap-ring on the end of the axle and sliding the spider assembly off the driveaxle **(see illustration)**.

27 Slide the boot off the axle.

28 Clean all of the old grease out of the housing and spider assembly. Carefully disassemble each section of the spider assembly, one at a time, and clean the needle bearings with solvent. Inspect the rollers, spider cross, bearings and housing for scoring, pitting and other signs of abnormal wear. Apply a coat of CV joint grease to the inner bearing surfaces to hold the needle bearings in place when reassembling the spider assembly.

29 Wrap the driveaxle splines with tape to avoid damaging the boot, then slide the small clamp and the boot onto the axle **(see illustration 11.14)**.

30 Pack the housing with half of the grease furnished with the new boot and place the remainder in the boot.

31 Install the spider bearing.

32 Install the tri-pot housing.

33 Seat the boot in the housing and axle seal grooves, then adjust the length of the joint **(see illustration 11.19)**. Install the boot bands, then install the driveaxle as described in Section 10.

Outer CV joint and boot

Refer to illustration 11.38

Disassembly

34 Remove the inner CV joint from the axleshaft and disassemble it.

35 If the left driveaxle outer CV joint is being serviced, mark the relationship of the dynamic damper to the axle. Pry open the dynamic damper clamp and slide the damper off the axleshaft.

36 Remove the outer CV joint boot clamps. Slide the boot off the axleshaft.

Inspection

37 Thoroughly wash the inner and outer CV joints in clean solvent and blow them dry with compressed air, if available. **Note:** *Because the outer joint cannot be disassembled, it is difficult to wash away all the old grease and to rid the bearing of solvent once it's clean. But it is imperative that the job be done thoroughly, so take your time and do it right.*

38 Bend the outer CV joint housing at an angle to the driveaxle to expose the bearings, inner race and cage **(see illustration)**. Inspect the bearing surfaces for signs of wear. If the bearings are damaged or worn, replace the driveaxle.

Reassembly

39 Slide the new outer boot onto the driveaxle. It's a good idea to wrap vinyl tape around the spline of shaft to prevent damage to the boot **(see illustration 11.14)**. Using the CV joint grease included with the a new boot kit, pack the joint with as much grease as it will hold and put the rest into the boot. Slide the boot on the rest of the way and install the new clamps.

40 If you are overhauling the left driveaxle, install the dynamic damper using the mark made during disassembly. Attach the damper clamp.

41 Proceed to clean and reassemble the inner CV joint by following the procedure in this Section, then install the driveaxle as outlined in Section 10.

Chapter 9 Brakes

Contents

Anti-lock Brake System (ABS) – general information	2
Brake check	See Chapter 1
Brake disc – inspection, removal and installation	5
Brake fluid level check	See Chapter 1
Brake light switch – removal, installation and adjustment	15
Brake hoses and lines – inspection and replacement	10
Brake system bleeding	11
Disc brake caliper – removal, overhaul and installation	4
Disc brake pads (front and rear) – replacement	3
Drum brake shoes – replacement	6
General information	1
Master cylinder – removal, overhaul and installation	8
Parking brake – adjustment	13
Parking brake cables – replacement	14
Power brake booster – check, removal and installation	12
Proportioning valve – removal and installation	9
Wheel cylinder – removal, overhaul and installation	7

Specifications

General

Brake fluid type	See Chapter 1
Power brake booster pushrod-to-master cylinder piston clearance	
Precis (1989 and earlier)	0.016 to 0.031 in (0.40 to 0.80 mm)
Precis (1990 on)	0 at 500 mm Hg vacuum
Galant (1988 and earlier)	0.059 to 0.075 in (1.5 to 1.9 mm)
Galant (1989 on)	
7 and 8 inch brake booster	0.020 to 0.028 in (0.5 to 0.7 mm)
9 inch brake booster	0.031 to 0.039 in (0.8 to 1.0 mm)
Cordia\Tredia (1983 and 1984 models)	0.004 to 0.020 in (0.1 to 0.5 mm)
Cordia\Tredia (1985 on)	0.016 to 0.031 in (0.4 to 0.8 mm)
Mirage (1988 and earlier)	0.016 to 0.031 in (0.4 to 0.80 mm)
Mirage (1989 on)	0.020 to 0.028 in (0.5 to 0.7 mm)
Parking brake lever travel	5 to 7 clicks

Disc brakes

Minimum brake pad thickness	See Chapter 1
Disc minimum thickness	Refer to marks cast into the disc
Disc runout	0.006 in (0.15 mm) maximum

Drum brakes

Minimum brake shoe lining thickness	See Chapter 1
Drum maximum diameter	Refer to marks cast into the drum
Drum out-of-round (maximum)	0.006 in (0.015 mm)

Torque specifications

Ft-lbs (unless otherwise indicated)

Caliper mounting bolts or pins	
8 mm diameter pin or bolt	23
10 mm diameter pin	35
14 mm diameter pin	62
Caliper inner-to-outer bolts (Sumitomo style calipers)	58 to 69
Caliper torque plate bolts	
Front	
Sumitomo style calipers	43 to 58
Tokico style calipers	47 to 54
All others	58 to 72
Rear	36 to 43
Inlet fitting bolt	
1990 and later Mirage	11 to 13
All others	18 to 22
Wheel cylinder mounting bolts	72 to 108 in-lbs
Brake backing plate-to-axle bolts	36 to 43
Master cylinder-to-brake booster nuts	72 to 108 in-lbs
Brake booster-to-firewall nuts	
1990 and later Precis	72 to 108 in-lbs
All others	96 to 144 in-lbs
Brake pedal support-to-firewall nuts	
1990 and later Mirage	12 to 15
All others	6 to 9
Wheel lug nuts	See Chapter 1

1 General information

The vehicles covered by this manual are equipped with hydraulically operated front and rear brake systems. The front brakes are disc type and the rear brakes are either disc or drum type. Both the front and rear brakes are self adjusting. The front disc brakes automatically compensate for pad wear, while the rear drum brakes incorporate an adjustment mechanism which is activated when the parking brake is applied and released and the brake pedal is pumped, alternately. Some models can be adjusted through the backing plate. Some later models are equipped with an optional Anti-lock brake system (ABS).

Hydraulic system

The hydraulic system consists of two separate diagonally-split circuits. The master cylinder has separate reservoirs for the two circuits and in the event of a leak or failure in one hydraulic circuit, the other circuit will remain operative. A visual warning of low fluid level is given by a warning light activated by a float switch in the master cylinder reservoir.

Proportioning valve

A proportioning valve, located in the engine compartment below the master cylinder, regulates outlet pressure to the rear brakes after a predetermined rear input pressure has been reached, preventing early rear wheel lock-up under heavy brake loads. The valve is also designed to assure full pressure to one brake system should the other system fail.

Power brake booster

The power brake booster, utilizing engine manifold vacuum and atmospheric pressure to provide assistance to the hydraulically operated brakes, is mounted on the firewall in the engine compartment.

Parking brake

The parking brake operates the rear brakes only, through a cable. It's activated by a lever mounted in the center console.

Service

After completing any operation involving disassembly of any part of the brake system, always test drive the vehicle to check for proper braking performance before resuming normal driving. When testing the brakes, perform the tests on a clean, dry flat surface. Conditions other than these can lead to inaccurate test results.

Test the brakes at various speeds with both light and heavy pedal pressure. The vehicle should stop evenly without pulling to one side or the other. Avoid locking the brakes because this slides the tires and diminishes braking efficiency and control of the vehicle.

Tires, vehicle load and front end alignment are factors which also affect braking performance.

2 Anti-lock Brake System (ABS) – general information

The optional Anti-lock Brake system is designed to maintain vehicle steerability, directional stability and optimum deceleration under severe braking conditions and on most road surfaces. It does so by monitoring the rotational speed of each wheel and controlling the brake line pressure to each wheel during braking. This prevents the wheel from locking-up and provides maximum vehicle controllability.

Components
Hydraulic assembly

The hydraulic assembly consists of the master cylinder, an electric hydraulic pump and accumulator assembly, a system of relays and a fluid reservoir.
a) The electric pump provides hydraulic pressure to charge the accumulator, which supplies pressure to the braking system. The pump and accumulator are mounted to the hydraulic assembly.
b) The solenoid valve body assembly modulates brake line pressure during ABS operation.

Wheel sensors

These sensors are located at each wheel and generate small electrical pulsations when the toothed sensor rings are turning, sending a signal to the electronic controller indicating wheel rotational speed.

The front wheel sensors are mounted to the front spindles in close relationship to the toothed sensor rings.

The rear wheel sensors bolt to the rear spindle adapters.

Electronic controller

The electronic controller is mounted on a bracket in the luggage compartment and is the "brain" for the ABS system. The function of the control module – consisting primarily of microprocessors and the related circuits needed for their operation – is to accept and process information received from the wheel speed sensors to control the hydraulic line pressure, avoiding wheel lock-up. The controller also constantly monitors the system, even under normal driving conditions, to find faults within the system.

If a problem develops within the system, the ABS warning light will glow on the dashboard.

Diagnosis and repair

If a dashboard warning light comes on and stays on while the vehicle is in operation, the ABS system requires attention.

Although special electronic ABS diagnostic testing methods are used to properly diagnose the system, the home mechanic can perform a few preliminary checks before taking the vehicle to a dealer who is equipped with the necessary equipment.
a) Check the brake fluid level in the reservoir.
b) Open the trunk lid and remove the left side panel. Check that the controller electrical connector is securely connected.
c) Check the electrical connectors at the hydraulic assembly.
d) Check the fuses.
e) Follow the wiring harness to each wheel and check that all connections are secure and that the wiring is not damaged.

If the above preliminary checks do not rectify the problem, the vehicle should be diagnosed by a dealer service department. Due to the rather complex nature of this system, all actual repair work must be done by a dealer.

3 Disc brake pads (front and rear) – replacement

Refer to illustrations 3.5a through 3.5f

Warning: *Disc brake pads must be replaced on both front wheels at the same time – never replace the pads on only one wheel. Also, the dust created by the brake system may contain asbestos, which is harmful to your health. Never blow it out with compressed air and don't inhale any of it. An approved filtering mask should be worn when working on the brakes. Do not, under any circumstances, use petroleum-based solvents to clean brake parts. Use brake cleaner or denatured alcohol only!*

Note: *When servicing the disc brakes, use only high quality, nationally recognized brand name pads.*

1 Remove the cap from the brake fluid reservoir (see Chapter 1) and siphon off about two-thirds of the fluid from the reservoir. Failing to do this could result in the reservoir overflowing when the caliper pistons are pressed into their bores.

2 Apply the parking brake. Loosen the wheel lug nuts, raise the vehicle and support it securely on jackstands. Place blocks behind the wheels that are still on the ground.

3 Remove the wheels. Work on one brake assembly at a time, using the assembled brake for reference, if necessary.

4 Inspect the brake disc carefully as outlined in Section 5. If machining is necessary, follow the information in that Section to remove the disc, at which time the pads can be removed from the calipers as well.

5 The pad replacement procedure varies, depending on whether your vehicle has a one-piece or a two-piece caliper (caliper body and caliper support) **(see illustrations)**.

Sumitomo style caliper (Precis models)

Refer to illustrations 3.6a through 3.6k

Chapter 9 Brakes

6 To replace the pads on this style caliper, follow illustrations 3.6a through 3.6k, being sure to stay in order and read the caption with each illustration. Then proceed to Step 13.

All other caliper designs

Refer to illustrations 3.7, 3.8, 3.9 and 3.10

7 Use a C-clamp to compress the piston in the caliper **(see illustration)**.
8 Remove the caliper lock pin or lower mounting bolt **(see illustration)**.
9 Swing the caliper up far enough to allow removal of the brake pads and secure it in the raised position with a piece of wire **(see illustration)**.
10 Pull the brake pads out of the torque plate **(see illustration)**. Remove the spring clips from each end of the torque plate (if equipped), noting their installed positions. Inspect each clip for distortion, replacing it if necessary.
11 Depress the caliper piston into the bore by pushing on it with a wooden tool, such as a hammer handle. This will provide room for the new pads.
12 Apply a coat of anti-squeal compound to the backing plates of the brake pads, following label instructions.
13 Place the spring clips on the torque plate (if equipped). Clip the anti-squeal shim on the outer pad **(see illustrations 3.5a through 3.5e)** and insert the pads into the carrier.
14 Swing the caliper body down over the pads and install the lower mounting bolt or lock pin, tightening it to the torque listed in this Chapter's Specifications.

All models

15 After replacing the pads, firmly depress the brake pedal a few times to bring the pads into contact with the disc. The pedal should be at normal height above the floorpan and firm.
16 Check the fluid level in the master cylinder and add fluid, if necessary (see Chapter 1). Check for fluid leakage and make sure the brakes operate normally before driving in traffic.

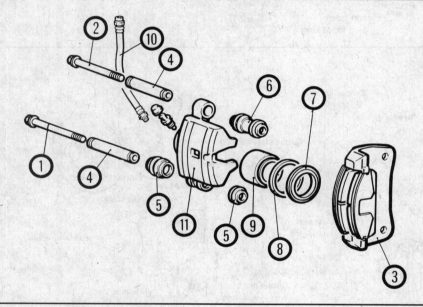

3.5a Front caliper assembly – PFS15 style (Mirage models)

1 Bolt
2 Bolt
3 Caliper torque plate (with pads, retainers and shims)
4 Sleeve
5 Sleeve boot
6 Bushing
7 Dust boot
8 Piston seal
9 Piston
10 Brake hose
11 Caliper body

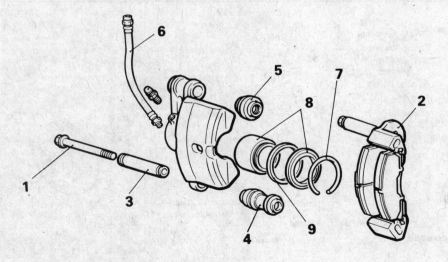

3.5b Front caliper assembly – AD54 style shown, AD60 similar (Cordia/Tredia/Galant non-turbo models)

1 Lock pin
2 Torque plate (with pads, shims, retainers)
3 Sleeve
4 Lock pin boot
5 Guide pin boot
6 Brake hose
7 Boot ring
8 Piston with piston boot
9 Piston seal

Chapter 9 Brakes

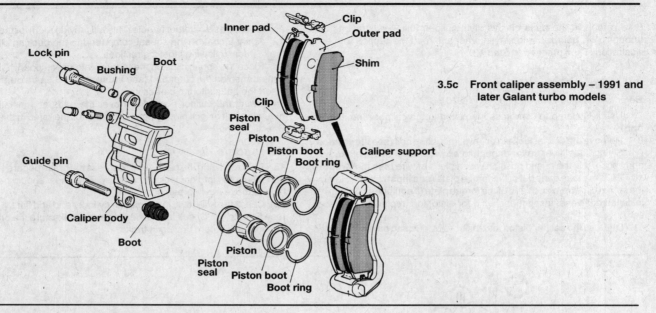

3.5c Front caliper assembly – 1991 and later Galant turbo models

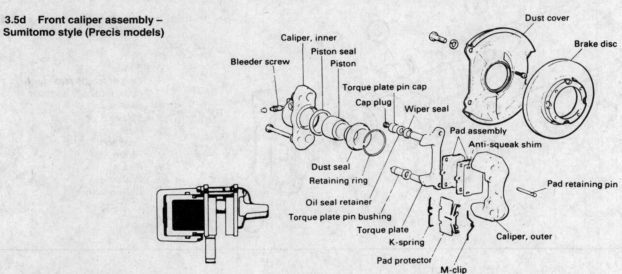

3.5d Front caliper assembly – Sumitomo style (Precis models)

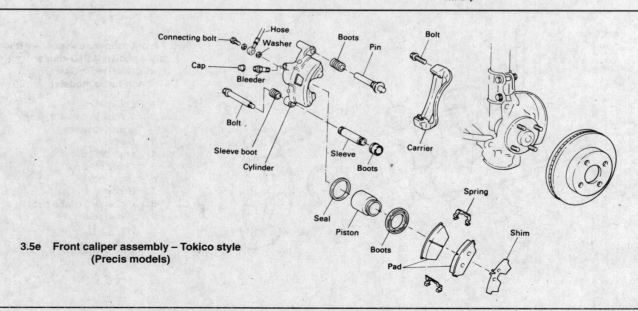

3.5e Front caliper assembly – Tokico style (Precis models)

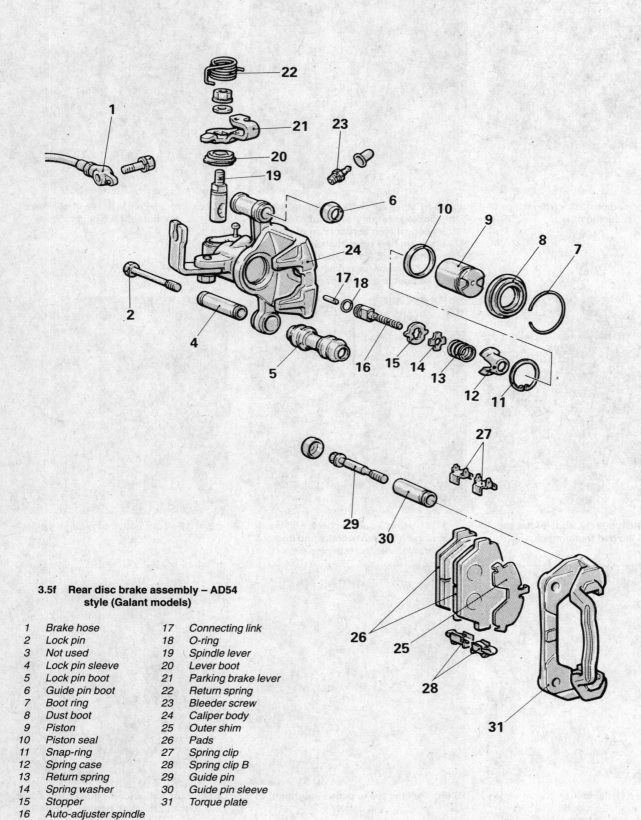

3.5f Rear disc brake assembly – AD54 style (Galant models)

1 Brake hose
2 Lock pin
3 Not used
4 Lock pin sleeve
5 Lock pin boot
6 Guide pin boot
7 Boot ring
8 Dust boot
9 Piston
10 Piston seal
11 Snap-ring
12 Spring case
13 Return spring
14 Spring washer
15 Stopper
16 Auto-adjuster spindle
17 Connecting link
18 O-ring
19 Spindle lever
20 Lever boot
21 Parking brake lever
22 Return spring
23 Bleeder screw
24 Caliper body
25 Outer shim
26 Pads
27 Spring clip
28 Spring clip B
29 Guide pin
30 Guide pin sleeve
31 Torque plate

3.6a Pry the pad protector from the pad retaining pins

3.6b Before removing anything, wash the brake assembly with brake cleaner (spraying with an aerosol can of cleaner is shown here) and allow it to dry – remember – NEVER blow off the brake dust with compressed air – asbestos dust is a health hazard!

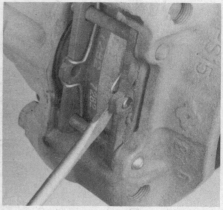

3.6c Pry the center of the M-clip out of the outer brake pad, . . .

3.6d . . . then pull the ends of the clip out of the pad retaining pins

3.6e Using a pair of pliers, pull the K-spring off the inner brake pad and out from under the retaining pins

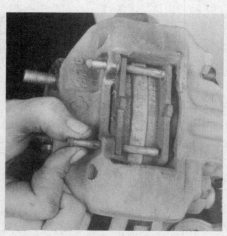

3.6f Pull out both pad retaining pins

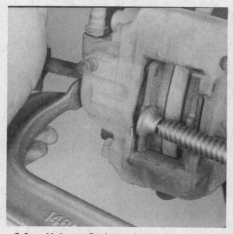

3.6g Using a C-clamp (or a pair of large pliers), push the inner brake pad towards the inside of the caliper, which will bottom the caliper piston in its bore, making room for the new pads

3.6h Pull the brake pads straight out of the caliper

3.6i If the anti-squeal shim hasn't stuck to the brake pad, remove it from the caliper

3.6j Before installing the new brake pads, it's a good idea to apply an anti-squeal compound to the pad backing plates, following label instructions

3.6k Install the new pads and clips, referring, if necessary, to the assembled brake on the other side of the car – when installing the M-clip, insert one end into the hole in one of the retaining pins, then rotate the other retaining pin so its hole is properly angled to insert the other end of the clip – push the center of the clip into the hole in the brake pad

3.7 Use a C-clamp to compress the piston and "free up" the brake pads

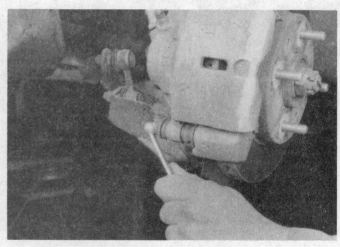

3.8 Remove the lower mounting bolt (or lock pin) from the caliper

3.9 Lift the caliper body from the caliper support and support it with a piece of wire

3.10 Remove the pads from the torque plate

Chapter 9 Brakes

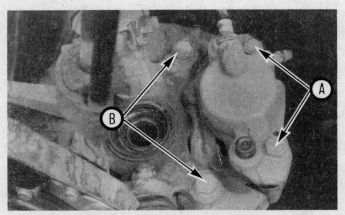

4.5 If a Sumitomo caliper is to be disassembled for overhaul, loosen the bolts that hold the inner and outer portions of the caliper together (A) – they are very tight and it's easier to break them loose before the caliper is removed – remove the caliper mounting bolts (B) to detach the caliper from the vehicle.

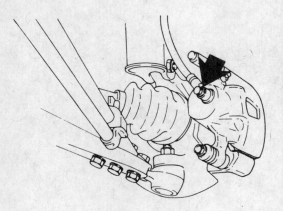

4.6 Remove the inlet fitting bolt (arrow) on models so equipped

4.9a With the Sumitomo caliper held in a vise, remove the bolts that secure the inner and outer halves of the caliper frame together, then separate them

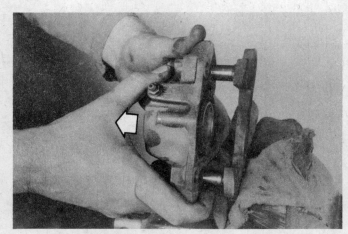

4.9b The inner half of the Sumitomo caliper can now be slid off of the torque plate

4 Disc brake caliper – removal, overhaul and installation

Warning: *Dust created by the brake system may contain asbestos, which is harmful to your health. Never blow it out with compressed air and don't inhale any of it. An approved filtering mask should be worn when working on the brakes. Do not, under any circumstances, use petroleum-based solvents to clean brake parts. Use brake cleaner or denatured alcohol only!*

Note: *If an overhaul is indicated (usually because of fluid leakage) explore all options before beginning the job. New and factory rebuilt calipers are available from auto parts stores on an exchange basis, which makes this job quite easy. If it's decided to rebuild the calipers, make sure a rebuild kit is available before proceeding. Always rebuild the calipers in pairs – never rebuild just one of them.*

Note: *See illustrations 3.5a through 3.5e for caliper identification*

Removal

1 Remove the cap from the brake fluid reservoir, siphon off about two-thirds of the fluid into a container and discard the fluid. Brake fluid will damage paint, so be careful not to spill any. **Note:** *The remainder of the removal procedure depends on whether your vehicle has One-piece or Two-piece calipers. To check which type it has, see if the calipers have pad protectors like those shown in illustration 3.6a. If they do, your vehicle has One-piece calipers. If not, your vehicle has Two-piece calipers.*

2 Apply the parking brake. Loosen the wheel lug nuts, raise the front of the vehicle and support it securely on jackstands. Remove the front wheels.

Sumitomo style caliper

Refer to illustration 4.5

3 Remove the brake pads from the caliper following the procedure in Section 3.
4 Loosen the hose fitting at the frame bracket and remove the hose securing clip (see Section 10). The brake hose can now be unscrewed from the caliper. Plug the end of the hose to prevent excessive fluid loss.
5 If the caliper is going to be overhauled, loosen the two bolts that secure the caliper frame together **(see illustration)**. Remove the two mounting bolts and lift the caliper off the vehicle.

All other calipers

Refer to illustration 4.6

6 Disconnect the brake line from the caliper by removing the inlet fitting bolt **(see illustration)**. Plug the end of the line to prevent excessive brake fluid loss.
7 Remove the caliper mounting bolts, or on some models, the lock pin **(see illustration 3.8)**. On models secured by a lock pin, swing the caliper up until it clears the disc and slide it off of the upper pin. The brake pads can stay in the torque plate.

4.10a On Sumitomo calipers, place the caliper piston-side down and direct compressed air into the brake hose inlet port to eject the piston

4.10b When ejecting the piston on other calipers, place a block of wood between the piston and caliper to prevent damage to the piston

4.11 Pry the wire retaining ring off the dust boot, then remove the boot

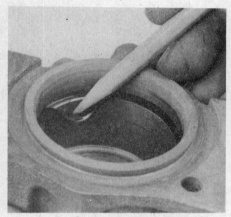

4.12 To avoid damaging the caliper bore or seal groove, remove the piston seal with a plastic or wooden tool – a pencil works well

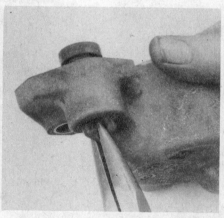

4.13 To remove the torque plate pin bushing, grab it with a pair of needle-nose pliers, twist it and push it through the caliper frame

4.18a On Sumitomo calipers, install the piston squarely in the caliper bore, then push it straight in until the dust boot groove is flush with the caliper face

Overhaul

Refer to illustrations 4.9a, 4.9b, 4.10a, 4.10b, 4.11, 4.12, 4.13, 4.18a, 4.18b, 4.18c, 4.19a, 4.19b, 4.20 and 4.21

8 Clean the exterior of the caliper with brake cleaner or denatured alcohol. **Warning:** *Never use gasoline, kerosene or petroleum-based cleaning solvents.*

9 If you are overhauling a Sumitomo caliper, remove the two bolts that secure the halves of the caliper frame together, then separate the halves **(see illustration)**. The torque plate can now be removed **(see illustration)**.

10 Two different techniques are used to eject the piston from the caliper, depending on caliper design:
   a) On a Sumitomo caliper, place the caliper piston-side down on the workbench, then use compressed air to remove the piston from the caliper **(see illustration)**.
   b) On all other calipers, place a block of wood or several shop rags in the caliper as a cushion **(see illustration)**, then apply compressed air to the inlet port to push the piston out.

Whichever method is used, apply only enough air pressure to ease the piston out of the bore. If the piston is blown out with much force, it could be damaged. **Warning:** *Never place your fingers in front of the piston in an attempt to catch or protect it when applying compressed air, as serious injury could occur.* **Note:** *On 1991 and later Galant Turbo models, only one of the two pistons will be ejected. Replace this piston loosely, hold it in place, then re-apply air pressure – the other piston will be ejected.*

11 Remove the dust boot. Pry the wire retaining ring off of the dust boot (on models so equipped), then remove the boot **(see illustration)**.

12 Using a wood or plastic tool, remove the piston seal from the groove in the caliper bore **(see illustration)**. Metal tools may damage the caliper bore.

13 Remove the bleeder screw. On Sumitomo calipers remove the torque plate pin bushing from the caliper housing **(see illustration)**. On other style calipers remove the pin boot, sleeve boots and sleeve, if equipped. Discard all rubber parts.

14 Clean the remaining parts with brake system cleaner or denatured alcohol then blow them dry with compressed air (make sure the air is filtered and unlubricated).

15 Carefully examine the piston for nicks and burrs. If surface defects are present, the parts must be replaced.

16 Check the caliper bore for nicks, burrs and corrosion. You can remove light corrosion and stains with crocus cloth. Discard the mounting bolts if they're corroded or damaged.

17 Lubricate the piston bore with clean brake fluid. Lubricate the new piston seal with the silicone grease supplied in the repair kit (or clean brake fluid if no silicone grease is supplied), then position the seal in the caliper bore groove.

18 On calipers that use a retaining ring to secure the dust boot, lubricate the piston with clean brake fluid. Insert the piston squarely into the caliper

4.18b Stretch the dust boot over the top of the piston, ...

4.18c ... then install the retaining ring

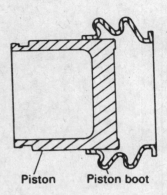

4.19a On models without a boot retaining ring, place the piston boot onto the bottom of the piston, ...

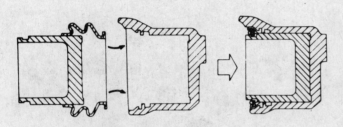

4.19b ... then tuck the fluted portion of the boot into the upper groove of the caliper bore and bottom the piston in the bore

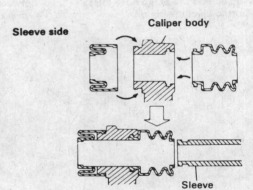

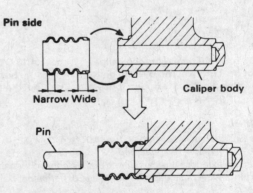

4.21 On other calipers, apply silicone grease to the sliding surfaces of the sleeve and pin, to the bores in the caliper body and to the lips of the rubber boots – on the sleeve side of the caliper, install the rubber boots first, then the sleeve

4.20 On Sumitomo calipers, lubricate the torque plate pins with silicone grease, then assemble the plate and the inner portion of the caliper

bore, then push down with your fingers to bottom it in the bore (see illustration). Smear the piston groove with the silicone grease supplied in the repair kit, then stretch the dust boot over the top of the piston and install the retaining ring (see illustrations).

19 On other calipers, lubricate the piston and piston boot lips with the silicone grease provided in the repair kit, then slide the piston boot onto the bottom (closed end) of the piston (see illustration). Lower the piston and boot assembly into the caliper bore and work the fluted portion of the dust boot into its groove in the bore (see illustration). Once the boot is seated around its entire circumference, the piston can be pushed into the bore.

20 On Sumitomo calipers, apply a coat of silicone grease to the torque plate pins and install the plate by sliding the pins through the inner half of the caliper body (see illustration). Place the caliper halves together, install the bolts and tighten them to the torque listed in this Chapter's Specifications.

21 On other calipers, lubricate the sleeve, pin bore and boots with silicone grease, then install them in their proper locations in the caliper body (see illustration).

Installation

22 On Sumitomo calipers, slide the caliper over the disc, aligning the threaded holes in the torque plate with the holes in the steering knuckle.

5.2 Suspend the caliper with a piece of wire whenever you have to move it – don't let it hang by the brake hose!

5.4a Check the disc runout with a dial indicator – if the reading exceeds the maximum allowable runout limit, the disc will have to be machined or replaced

5.4b Using a swirling motion, remove the glaze from the disc with emery cloth or sandpaper

Install the mounting bolts, tightening them to the torque listed in this Chapter's Specifications. The brake hose and pads can now be installed.

23 On other calipers, ensure the brake pads are in place, then slide the caliper over the upper mounting pin (on models so equipped). Pivot the caliper down over the pads and disc and install the mounting bolts (or lock bolt), tightening them to the torque listed in this Chapter's Specifications. Connect the brake hose, using new copper washers on each side of the inlet fitting bolt. Tighten the inlet fitting bolt to the torque listed in this Chapter's Specifications .

24 Bleed the brakes following the procedure in Section 11.

25 Install the wheels and lower the vehicle. Tighten the lug nuts to the torque listed in the Chapter 1 Specifications.

26 Firmly depress the brake pedal a few times to bring the pads into contact with the disc.

27 Inspect the caliper for fluid leaks and check brake operation before driving the vehicle in traffic.

5   Brake disc – inspection, removal and installation

Refer to illustrations 5.2, 5.4a, 5.4b, 5.5a and 5.5b

1 Loosen the wheel lug nuts, raise the vehicle and support it securely on jackstands. Remove the wheel.

2 Remove the brake caliper, pads and torque plate as outlined in Section 4. It's not necessary to disconnect the brake hose. After removing the caliper, suspend it out of the way with a piece of wire **(see illustration)**. *Note: On some models the brake disc can be slid off the hub after the caliper and torque plate have been removed. If the vehicle you are working on is of this design, reinstall the lug nuts and tighten them securely before performing the following runout check.*

Inspection

3 Visually inspect the disc surface for scoring and other damage. Light scratches and shallow grooves are normal after use and may not always be detrimental to brake operation, but deep scoring – over 0.015-inch (0.38 mm) – requires disc removal and refinishing by an automotive machine shop. Be sure to check both sides of the disc. If pulsating has been noticed during application of the brakes, suspect disc runout.

4 To check disc runout, place a dial indicator at a point about 1/2-inch from the outer edge of the disc **(see illustration)**. Set the indicator to zero and turn the disc. The difference between the highest and lowest indicator reading should not exceed the specified allowable runout maximum. If it does, the disc should be refinished by an automotive machine shop. *Note: Professionals recommend resurfacing of brake discs regardless of the dial indicator reading (to produce a smooth, flat surface that will eliminate

5.5a The minimum allowable disc thickness is stamped on the inner diameter of the disc

5.5b Check the thickness of the disc with a micrometer at several points

brake pedal pulsations and other undesirable symptoms related to questionable discs). At the very least, if you elect not to have the discs resurfaced, deglaze them with medium-grit emery cloth (use a swirling motion to ensure a non-directional finish)* **(see illustration)**.

5 It is absolutely critical that the disc not be machined to a thickness under the specified minimum allowable thickness. This thickness is stamped onto the inner diameter of the disc **(see illustration)**. The disc thickness can be checked with a micrometer **(see illustration)**.

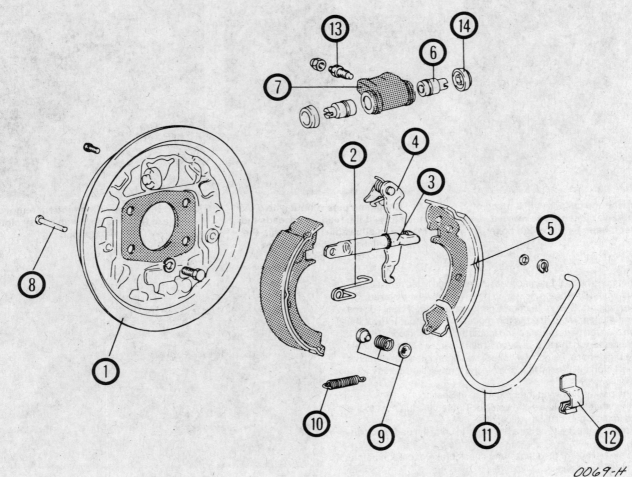

6.4a Rear drum brake assembly (1988 and earlier Mirage and Precis models)

1	Backing plate	6	Piston
2	Spring	7	Wheel cylinder body
3	Adjuster	8	Shoe hold down spring pin
4	Parking brake lever	9	Shoe hold down spring
5	Brake shoe	10	Shoe retainer spring

11	Shoe-to-shoe spring
12	Clip spring
13	Bleeder screw
14	Wheel cylinder boot

6.4b Before removing any internal drum brake components, wash them off with brake cleaner and allow them to dry – position a drain pan under the brake to catch the residue – DO NOT USE COMPRESSED AIR TO BLOW THE BRAKE DUST FROM THE PARTS!

6.4c Pry the clip spring off the anchor point at the bottom of the brake shoes

Chapter 9 Brakes

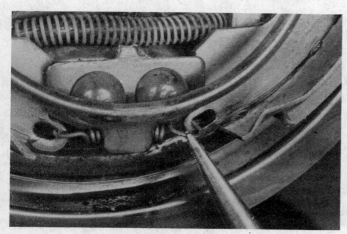

6.4d Using a pair of needle-nose pliers, unhook the shoe retainer spring from the bottom of the shoes

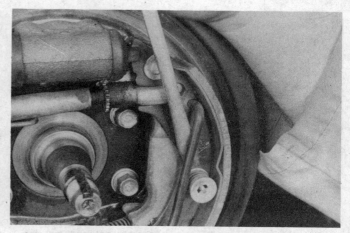

6.4e Pry the shoe-to-shoe spring out from the holes in the brake shoes

Removal and installation

6   As stated above, on some models the brake disc can be slid off the hub after the caliper and torque plate have been removed. If the vehicle you are working on is of this design, remove the lug nuts that were installed to perform the above checks and slide the disc off the hub.
7   Following the procedure outlined in Chapter 10, remove the steering knuckle and hub assembly from the vehicle.
8   Due to the need for special tools to remove and install the hub and disc on the steering knuckle and to properly adjust the hub bearing preload, the steering knuckle and hub assembly must be taken to a dealer service department or other shop equipped with the necessary tools.
9   Install the steering knuckle and hub assembly, referring to the procedure described in Chapter 10.
10   Install the caliper, torque plate and brake pad assembly over the disc and position it on the steering knuckle (refer to Section 4 for the caliper installation procedure, if necessary). Tighten the caliper bolts or lock pin to the torque listed in this Chapter's Specifications.
11   Install the wheel, then lower the vehicle to the ground. Depress the brake pedal a few times to bring the brake pads into contact with the disc. Bleeding of the system will not be necessary unless the brake hose was disconnected from the caliper. Check the operation of the brakes carefully before placing the vehicle into normal service.

6.4f Remove the shoe hold-down springs from each brake shoe – this is done by pushing the retainer cap down, turning it 90-degrees to align the slot with the pin and then releasing it

6 Drum brake shoes – replacement

Warning: *Drum brake shoes must be replaced on both wheels at the same time – never replace the shoes on only one wheel. Also, the dust created by the brake system may contain asbestos, which is harmful to your health. Never blow it out with compressed air and don't inhale any of it. An approved filtering mask should be worn when working on the brakes. Do not, under any circumstances, use petroleum-based solvents to clean brake parts. Use brake cleaner or denatured alcohol only!*
Caution: *Whenever the brake shoes are replaced, the shoe-to-shoe, automatic adjuster (if equipped) and hold-down springs should also be replaced. Due to the continuous heating/cooling cycle that the springs are subjected to, they lose their tension over a period of time and may allow the shoes to drag on the drum and wear at a much faster rate than normal. When replacing the rear brake shoes, use only high quality, nationally recognized brand-name parts.*

1   Loosen the wheel lug nuts, raise the rear of the vehicle and support it securely on jackstands. Block the front wheels to keep the vehicle from rolling.
2   Release the parking brake.
3   Remove the wheel and brake drum (see Chapter 1 for the brake drum removal procedure). **Note:** *All four rear brake shoes must be replaced at the same time, but to avoid mixing up parts, work on only one brake assembly at a time.*

6.4g Remove the shoes and adjuster from the backing plate as an assembly

Precis and Mirage (1988 and earlier only)
Refer to illustrations 6.4a through 6.4w

4   Refer to the accompanying illustrations (6.4a through 6.4w) for the inspection and replacement of the brake shoes. Be sure to stay in order and read the caption under each illustration. Then proceed to Step 19.

Chapter 9 Brakes

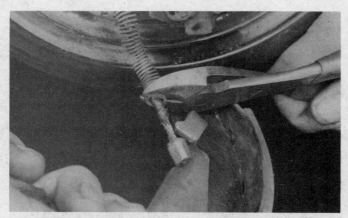

6.4h Disconnect the parking brake cable from the parking brake lever – this is most easily accomplished by pulling the spring back on the cable with a pair of diagonal cutting pliers and gripping the cable lightly with the pliers, then disengaging the cable from the lever – be careful not to nick the cable

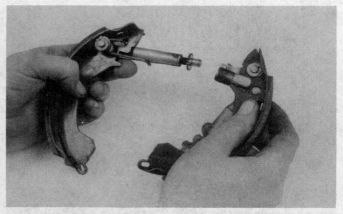

6.4i Separate the trailing shoe (the one with the parking brake lever attached) from the leading shoe

6.4j Pry the C-clip off of the parking brake lever pivot pin (there's a clip on each side, but only remove the one on the side opposite the lever)

6.4k Pry the spring off the adjuster actuator, . . .

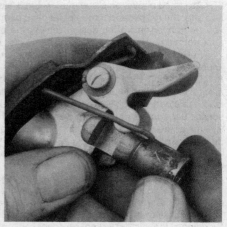

6.4l . . . then remove the end of the adjuster strut

6.4m Separate the parking brake lever from the trailing shoe – transfer the lever to the new trailing shoe and secure it in place with the C-clip, crimping it closed with pliers – attach the end of the adjuster to the shoe and pry the spring back over the actuator (reverse steps k and l)

6.4n Pry the spring on the adjuster screw up and over the screw

6.4o Pry off the C-clip from the back side of the leading shoe, then remove the adjuster screw

Chapter 9 Brakes

6.4p Clean the adjuster screw and apply multi-purpose grease to the threads and end – attach the adjuster to the new leading shoe and secure it by crimping the C-clip shut around the pivot pin – pry the spring over the adjuster so it's positioned on the underside

6.4q Apply a LIGHT coat of high-temperature grease to the brake shoe contact areas on the backing plate – don't overdo it, though, as grease on the friction surface of the shoes will ruin them

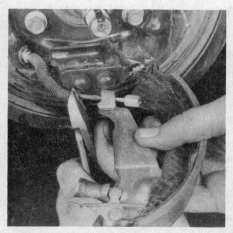

6.4r Attach the parking brake cable to the parking brake lever, once again holding the spring back on the cable with a pair of diagonal cutting pliers

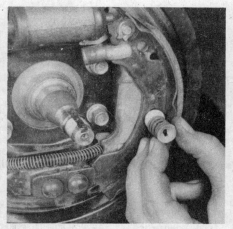

6.4s Position the trailing shoe on the backing plate and install the hold-down spring – make sure the shoe engages properly with the wheel cylinder

6.4t Mount the leading shoe to the backing plate, connecting the adjuster screw with its other end (on the trailing shoe) – install the hold-down spring and ensure the shoe is properly engaged with the wheel cylinder

6.4u Insert one end of the shoe-to-shoe spring into its hole, then push the other end into its corresponding hole – make sure the ends are pushed in completely

6.4v Connect the shoe retainer spring across the bottom of each shoe – it must be installed behind the anchor, as shown

6.4w Hook the clip spring under the shoe anchor, then pull it up over the top of the anchor far enough for the tang on the anchor (arrow) to snap into the slot in the clip

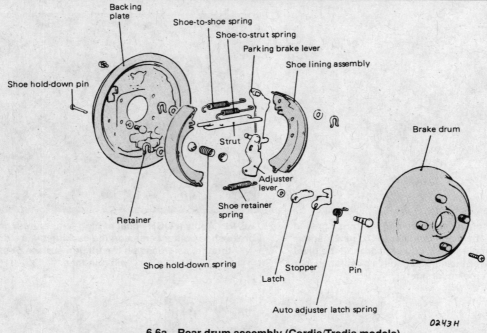

6.6a Rear drum assembly (Cordia/Tredia models)

Precis and Mirage (1989 on), Cordia/Tredia, and Galant

Refer to illustrations 6.6a, 6.6b, 6.6c and 6.17

5 Refer to illustration 6.4b and clean the brake assembly.
6 Using a pair of pliers, grip the shoe-to-strut spring (Cordia/Tredia models) or the shoe-to-lever spring (all other models), unhook it from the adjuster lever, then remove the spring and lever **(see illustrations)**.
7 Spread the shoes apart and remove the adjuster or strut.
8 Grip the shoe-to-shoe spring with a pair of pliers and unhook it.
9 Remove the hold-down springs from each shoe, using the technique shown in illustration 6.4f.
10 Remove the leading shoe from the brake backing plate, then unhook the parking brake cable from the parking brake lever.
11 Clean the adjuster screw and lubricate its threads with multi-purpose grease **(see illustration 6.4p)**.
12 Lubricate the brake shoe contact areas on the backing plate with multi-purpose grease **(see illustration 6.4q)**.
13 Transfer the parking brake lever from the old trailing shoe to the new one. It will be necessary to pry the C-clip from the lever pivot pin and crimp it into position once the lever is attached to the new shoe **(see illustration 6.4j)**.
14 Connect the parking brake lever to the parking brake cable and position the trailing shoe against the backing plate. Install the hold-down spring.
15 Place the leading shoe on the backing plate and install the hold-down spring.
16 Connect the shoe-to-shoe spring between the bottom of the two shoes.
17 Spread the tops of the shoes apart and install the adjuster or strut. Make sure the slotted ends of the adjuster or strut slide over the slots in the shoes **(see illustration)**.
18 Install the adjuster lever and automatic adjuster spring.

All models

Refer to illustration 6.19

19 Before reinstalling the drum it should be checked for cracks, score marks, deep scratches and hard spots, which will appear as small discolored areas. If the hard spots cannot be removed with fine emery cloth or if any of the other conditions listed above exist, the drum must be taken to an automotive machine shop to have it turned. **Note:** *Professionals recommend resurfacing the drums whenever a brake job is done. Resurfacing will eliminate the possibility of out-of-round drums. If the drums are worn so much that they can't be resurfaced without exceeding the maximum allowable diameter (cast into the drum)* **(see illustration)**, *then new ones will be required. At the very least, if you elect not to have the drums resurfaced, remove the glazing from the surface with medium-grit emery cloth using a swirling motion.*
20 Install the brake drum and adjust the rear wheel bearings (see Chapter 1).
21 Install the wheel and lug nuts, then lower the vehicle. Tighten the lug nuts to the torque listed in the Chapter 1 Specifications.
22 Alternately cycle the parking brake lever and pump the brake pedal several times to adjust the brake shoes to the drum.
23 Check brake operation before driving the vehicle in traffic.

7 Wheel cylinder – removal, overhaul and installation

Note: *If an overhaul is indicated (usually because of fluid leakage or sticky operation) explore all options before beginning the job. New wheel cylinders are available, which makes this job quite easy. If it's decided to rebuild the wheel cylinder, make sure that a rebuild kit is available before proceeding. Never overhaul only one wheel cylinder – always rebuild both of them at the same time.*

Removal

Refer to illustration 7.4

1 Raise the rear of the vehicle and support it securely on jackstands. Block the front wheels to keep the vehicle from rolling.
2 Remove the brake shoe assembly (see Section 6).
3 Remove all dirt and foreign material from around the wheel cylinder.
4 Unscrew the brake line fitting from the inner side of the brake backing plate **(see illustration)**. Use a flare nut wrench, if available, to prevent rounding off the corners of the fitting. Don't pull the brake line away from the wheel cylinder.
5 Remove the wheel cylinder mounting bolts from the inner side of the brake backing plate.
6 Detach the wheel cylinder from the brake backing plate and place it on a clean workbench. Immediately plug the brake line to prevent fluid loss and contamination. **Note:** *If the brake shoe linings are contaminated with brake fluid, install new brake shoes.*

Chapter 9 Brakes

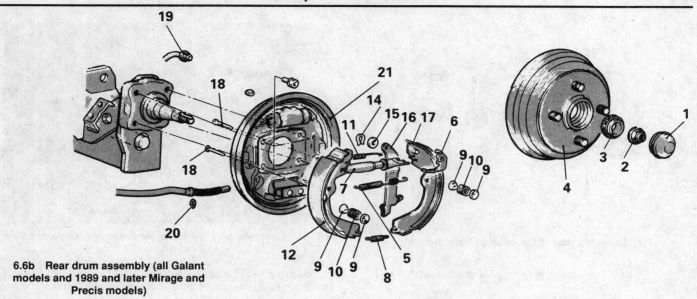

6.6b Rear drum assembly (all Galant models and 1989 and later Mirage and Precis models)

1 Hub cap
2 Wheel bearing nut
3 Outer wheel bearing
4 Brake drum
5 Shoe-to-lever spring
6 Adjuster lever
7 Auto adjuster assembly
8 Retainer spring
9 Hold-down cups
10 Hold-down springs
11 Shoe-to-shoe spring
12 Brake shoe
13 Not used
14 C-clip
15 Wave washer
16 Parking brake lever
17 Brake shoe
18 Hold-down pins
19 Brake line
20 Snap-ring
21 Backing plate

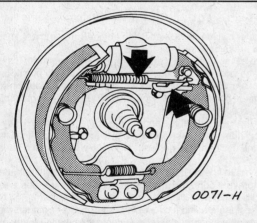

6.6c Unhook the adjuster spring, then remove the spring and lever (arrows)

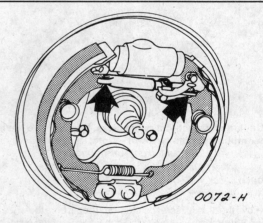

6.17 Make sure the slotted ends of the adjuster (or strut) slide over the slots in the shoes

6.19 The maximum allowable diameter is cast into the drum

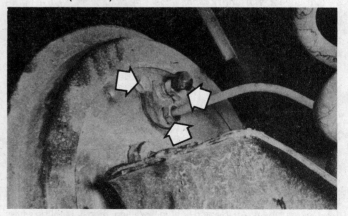

7.4 Unscrew the brake line fitting, then remove the two wheel cylinder mounting bolts (arrows)

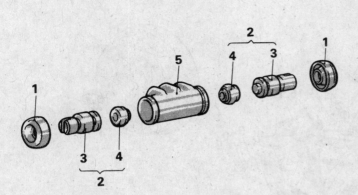

7.7 Exploded view of the wheel cylinder (typical)

1 Boot
2 Piston assembly
3 Piston
4 Piston cup
5 Cylinder body

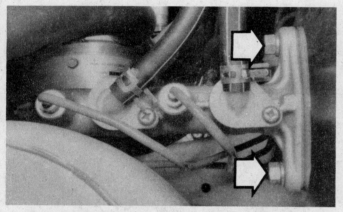

8.6 Unplug any electrical connectors, then remove the master cylinder mounting nuts (arrows) – Note: *The reservoir is remotely mounted on some models*

Overhaul

Refer to illustration 7.7

7 Remove the bleeder screw, pistons and boots from the wheel cylinder body **(see illustration)**. Remove the cup seals from the pistons.
8 Clean the wheel cylinder with brake fluid, denatured alcohol or brake system cleaner. **Warning:** *Do not, under any circumstances, use petroleum based solvents to clean brake parts!*
9 Use compressed air to remove excess fluid from the wheel cylinder and to blow out the passages.
10 Check the cylinder bore for corrosion and score marks. Crocus cloth can be used to remove light corrosion and stains, but the cylinder must be replaced with a new one if the defects cannot be removed easily, or if the bore is scored.
11 Lubricate the new cup seals with brake fluid. Apply corrosion prevention compound (included in the repair kit) to the inside of the cylinder.
12 Assemble the wheel cylinder components. Make sure the cup lips face toward the center of the cylinder, or they won't seal.

Installation

13 Apply a thin coat of silicone sealant to the backing plate where the wheel cylinder seats, place the cylinder in position and install the mounting bolts (but don't tighten them yet).
14 Connect the brake line and tighten the fitting, then tighten the wheel cylinder bolts to the torque listed in this Chapter's Specifications. Install the brake shoe assembly.

8.4 Unscrew the brake line fitting tube nuts (arrows) with a flare nut wrench

15 Bleed the brakes (see Section 11).
16 Check brake operation before driving the vehicle in traffic.

8 Master cylinder – removal, overhaul and installation

Note: *If the vehicle is equipped with an Anti-Lock Brake System (ABS), have the master cylinder rebuilt at a dealer service department. Also, before deciding to overhaul the master cylinder, check on the availability and cost of a new or factory rebuilt unit and also the availability of a rebuild kit.*

Removal

Refer to illustrations 8.4 and 8.6

1 The master cylinder is located in the engine compartment, mounted to the power brake booster.
2 Remove as much fluid as you can from the reservoir with a syringe.
3 Place rags under the fluid fittings and prepare caps or plastic bags to cover the ends of the lines once they are disconnected. **Caution:** *Brake fluid will damage paint. Cover all body parts and be careful not to spill fluid during this procedure.*
4 Loosen the tube nuts at the ends of the brake lines where they enter the master cylinder **(see illustration)**. To prevent rounding off the flats on these nuts, the use of a flare nut wrench, which wraps around the nut, is preferred.
5 Pull the brake lines slightly away from the master cylinder and plug the ends to prevent contamination. If the master cylinder has a remote fluid reservoir, disconnect the hose from the master cylinder and plug it.
6 Disconnect the electrical connector at the master cylinder, then remove the nuts attaching the master cylinder to the power booster **(see illustration)**. Pull the master cylinder off the studs and out of the engine compartment. Again, be careful not to spill the fluid as this is done.

Overhaul

Refer to illustrations 8.8a, 8.8b, 8.9, 8.10, 8.11a, 8.11b and 8.11c

7 Before attempting the overhaul of the master cylinder, obtain the proper rebuild kit, which will contain the necessary replacement parts and also any instructions which may be specific to your model.
8 Inspect the reservoir grommets for indications of leakage near the base of the reservoir. Remove the reservoir **(see illustrations)**.
9 Place the cylinder in a vise and use a punch or Phillips screwdriver to depress the pistons until they bottom against the other end of the master cylinder **(see illustration)**. Hold the pistons in this position and remove the stop screw on the side of the master cylinder.
10 Carefully remove the snap-ring at the end of the master cylinder **(see illustration)**.
11 The internal components can now be removed from the cylinder bore **(see illustrations)**. Make a note of the proper order of the components so they can be returned to their original locations. **Note:** *The two springs are of different tension, so pay particular attention to their order. Also, do not disassemble the piston components – they are serviced as an assembly.*

8.8a Remove the reservoir mounting screw (arrow), . . .

8.8b . . . then pull the reservoir out of the grommets

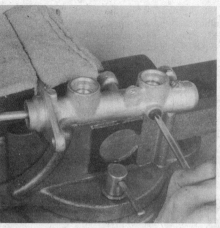

8.9 Push the pistons all the way in and remove the stop screw

8.10 Push the pistons in and use snap-ring pliers to remove the snap-ring

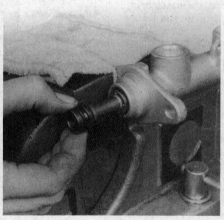

8.11a Remove the primary piston assembly from the master cylinder bore

8.11b To remove the secondary piston assembly, tap the master cylinder firmly against a block of wood

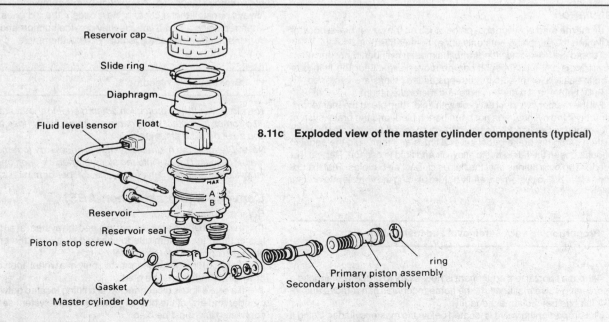

8.11c Exploded view of the master cylinder components (typical)

12   Carefully inspect the bore of the master cylinder. Any deep scoring or other damage will mean a new master cylinder is required.
13   Replace all parts included in the rebuild kit, following any instructions in the kit. Clean all reused parts with clean brake fluid or denatured alcohol. Do not use any petroleum-based cleaners. During assembly, lubricate all parts liberally with clean brake fluid.
14   Push the assembled components into the bore, bottoming them against the end of the master cylinder, then install the stop screw. Be sure to install a new gasket on the stop screw.
15   Install the new snap-ring, making sure it is seated properly in the groove.
16   Before installing the master cylinder it should be bench bled. Because it will be necessary to apply pressure to the master cylinder piston and, at the same time, control flow from the brake line outlets, it is recommended that the master cylinder be mounted in a vise, with the jaws of the vise clamping on the mounting flange. The master cylinder is aluminum, so be careful not to damage it.
17   Insert threaded plugs into the brake line outlet holes and snug them down so that there will be no air leakage past them, but not so tight that they cannot be easily loosened.
18   Fill the reservoir with brake fluid of the recommended type (see Chapter 1).
19   Remove one plug and push the piston assembly into the master cylinder bore to expel the air from the master cylinder. A large Phillips screwdriver can be used to push on the piston assembly.
20   To prevent air from being drawn back into the master cylinder the plug must be replaced and snugged down before releasing the pressure on the piston assembly.
21   Repeat the procedure until only brake fluid is expelled from the brake line outlet hole. When only brake fluid is expelled, repeat the procedure with the other outlet hole and plug. Be sure to keep the master cylinder reservoir filled with brake fluid to prevent the introduction of air into the system.
22   Since high pressure is not involved in the bench bleeding procedure, an alternative to the removal and replacement of the plugs with each stroke of the piston assembly is available. Before pushing in on the piston assembly, remove the plug as described in Step 19. Before releasing the piston, however, instead of replacing the plug, simply put your finger tightly over the hole to keep air from being drawn back into the master cylinder. Wait several seconds for brake fluid to be drawn from the reservoir into the piston bore, then depress the piston again, removing your finger as brake fluid is expelled. Be sure to put your finger back over the hole each time before releasing the piston, and when the bleeding procedure is complete for that outlet, replace the plug and snug it before going on to the other port.

Installation

23   Install the master cylinder over the studs on the power brake booster and tighten the attaching nuts only finger tight at this time.
24   Thread the brake line fittings into the master cylinder. Since the master cylinder is still a bit loose, it can be moved slightly for the fittings to thread in easily. Do not strip the threads as the fittings are tightened.
25   Fully tighten the mounting nuts and the brake fittings.
26   Fill the master cylinder reservoir with fluid, then bleed the master cylinder (only if the cylinder has not been bench bled) and the brake system as described in Section 11. To bleed the cylinder on the vehicle, have an assistant pump the brake pedal several times and then hold the pedal to the floor. Loosen the fitting nut to allow air and fluid to escape. Repeat this procedure on both fittings until the fluid is clear of air bubbles. Test the operation of the brake system carefully before placing the vehicle into normal service.

9   Proportioning valve – removal and installation

Removal

1   Although special test equipment is necessary to properly diagnose a proportioning valve malfunction, the home mechanic can remove and install it if it has been diagnosed faulty.
2   The proportioning valve is located below the master cylinder. Using a flare nut wrench to avoid rounding off the tube nuts, unscrew the tube nuts at the ends of the brake lines where they enter the valve. Gently pull the lines away from the valve and plug the ends of the lines or wrap plastic bags tightly around them to prevent excessive leakage and brake system contamination.
3   Remove the mounting nut from the center of the valve and maneuver the valve out from the brake lines.

Installation

4   Position the valve on the bracket and insert the brake lines into the holes. Screw all of the tube nuts into the valve by hand, being careful not to cross thread them. Install the mounting nut and tighten it securely.
5   Tighten the tube nuts securely, again using the flare nut wrench.
6   Bleed the entire brake system as described in Section 11. If the ends of the lines were not capped securely and all of the fluid in the master cylinder reservoir has drained out, the master cylinder will have to be bled first. The master cylinder bleeding procedure (on vehicle) can be found in Section 8.

10   Brake hoses and lines – inspection and replacement

Refer to illustrations 10.4a and 10.4b

1   About every six months the flexible hoses which connect the steel brake lines with the rear brakes and front calipers should be inspected for cracks, chafing of the outer cover, leaks, blisters, and other damage (see Chapter 1).
2   Replacement steel and flexible brake lines are commonly available from dealer parts departments and auto parts stores. Do not, under any circumstances, use anything other than genuine steel lines or approved flexible brake hoses as replacement items.
3   When installing the brake line, leave at least 3/4-inch (19 mm) clearance between the line and any moving or vibrating parts.
4   When disconnecting a hose and line, first remove the spring clip **(see illustration)**. Then, using an open end wrench to hold the hose and a flare nut wrench to hold the hose, unscrew the fitting **(see illustration)**. Use the wrenches in the same manner when reconnecting the hose and line, then install the clip. **Note:** *Make sure the tube passes through the center of its grommet.*
5   When disconnecting two hoses, use open end wrenches on the hose fittings. When connecting two hoses, make sure they are not bent, twisted or strained.
6   Steel brake lines are usually retained along their span with clips. Always remove these clips completely before removing a fixed brake line. Always reinstall these clips, or new ones if the old ones are damaged, when replacing a brake line, as they provide support and keep the lines from vibrating, which can eventually break them.

11   Brake system bleeding

Warning: *Wear eye protection when bleeding the brake system. If the fluid comes in contact with your eyes, immediately rinse them with water and seek medical attention.*

Note: *Bleeding the hydraulic system is necessary to remove any air that manages to find its way into the system when it's been opened during removal and installation of a hose, line, caliper or master cylinder.*

Conventional brakes (non-ABS)

Refer to illustration 11.8

1   It will probably be necessary to bleed the system at all four brakes if air has entered the system due to low fluid level, or if the brake lines have been disconnected at the master cylinder.
2   If a brake line was disconnected only at a wheel, then only that caliper or wheel cylinder must be bled.
3   If a brake line is disconnected at a fitting located between the master cylinder and any of the brakes, that part of the system served by the disconnected line must be bled.

Chapter 9 Brakes

9–21

10.4a Remove the spring clip from the brake hose fitting

10.4b A backup wrench must be used to keep the hose from turning, otherwise the steel brake line will twist – if available, use a flare nut wrench on the line fitting to prevent rounding off the corners

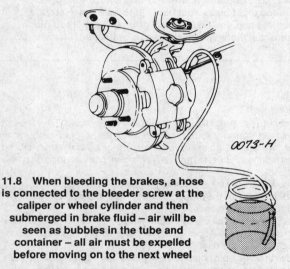

11.8 When bleeding the brakes, a hose is connected to the bleeder screw at the caliper or wheel cylinder and then submerged in brake fluid – air will be seen as bubbles in the tube and container – all air must be expelled before moving on to the next wheel

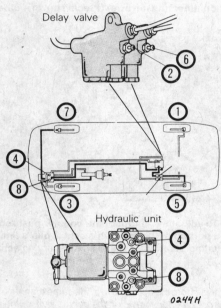

11.15 Be sure to follow the bleeding pattern in the exact order (bleeder screw no. two is for the left rear brake – no. six is for the right rear brake)

4   Remove any residual vacuum from the brake power booster by applying the brake several times with the engine off.
5   Remove the master cylinder reservoir cover and fill the reservoir with brake fluid. Reinstall the cover. **Note:** *Check the fluid level often during the bleeding operation and add fluid as necessary to prevent the fluid level from falling low enough to allow air bubbles into the master cylinder.*
6   Have an assistant on hand, as well as a supply of new brake fluid, a clear container partially filled with clean brake fluid, a length of 3/16-inch plastic, rubber or vinyl tubing to fit over the bleeder screw and a wrench to open and close the bleeder screw.
7   Beginning at the left rear wheel, loosen the bleeder screw slightly, then tighten it to a point where it is snug but can still be loosened quickly and easily.
8   Place one end of the tubing over the bleeder screw and submerge the other end in brake fluid in the container **(see illustration)**.
9   Have the assistant pump the brakes slowly a few times to get pressure in the system, then hold the pedal firmly depressed.
10 While the pedal is held depressed, open the bleeder screw just enough to allow a flow of fluid to occur. Watch for air bubbles to exit the submerged end of the tube. When the fluid flow slows after a couple of seconds, tighten the screw and have your assistant release the pedal.
11 Repeat Steps 9 and 10 until no more air is seen leaving the tube, then tighten the bleeder screw and proceed to the right front wheel, the right rear wheel and the left front wheel, in that order, and perform the same procedure. Be sure to check the fluid in the master cylinder reservoir frequently.
12 Never use old brake fluid. It contains moisture which will deteriorate the brake system components.
13 Refill the master cylinder with fluid at the end of the operation.
14 Check the operation of the brakes. The pedal should feel solid when depressed, with no sponginess. If necessary, repeat the entire process. **Warning:** *Do not operate the vehicle if you are in doubt about the effectiveness of the brake system.*

Anti-lock brake system (ABS)

Refer to illustration 11.15
Warning: *In order to effectively bleed the brake system on models equipped with ABS, the engine must be running. Be sure the vehicle is supported properly before going underneath to bleed the brake system. Raise the front and rear end of the vehicle and support it securely with jackstands. Be sure the work area is level.*
15 Start the engine. Follow Steps 8 and 10 and bleed the air from the calipers and the delay valve(s). Be sure to follow the exact order of bleeding (alternating between the wheel cylinders and the delay valves) **(see illustration)**. **Note:** *When bleeding the rear brakes, the brake pedal will have a heavier feeling than when bleeding the front brakes. This is the result of air constriction of fluid pressure within the delay valve and is not a malfunction.*

12.7 Remove the cotter pin and clevis pin, then unscrew the booster mounting nuts (arrows) (Precis shown – others similar)

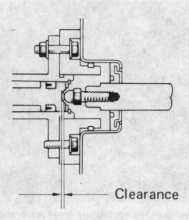

12.14a The booster pushrod-to-master cylinder piston clearance must be as specified – if there is interference between the two, the brakes may drag; if there is too much clearance, there will be excessive brake pedal travel

12.14b To adjust the length of the booster pushrod, hold the serrated portion of the rod with a pair of pliers and turn the adjusting screw in or out, as necessary, to achieve the desired setting

16 Bleed the air from the delay valve in the same manner as the calipers. **Note:** *There are bleeder screws on the delay valve for the left rear brake and the right rear brake* **(see illustration)**. *Don't mix them up!*
17 Tighten the delay valve bleeder screw securely.
18 Attach a box wrench to the bleeder screw on the hydraulic unit, located under the hood near the firewall and bleed the air from the hydraulic unit **(see illustration 11.15)**.
19 Refill the master cylinder with fluid at the end of the operation.

12 Power brake booster – check, removal and installation

Operating check

1 Depress the brake pedal several times with the engine off and make sure that there is no change in the pedal reserve distance.
2 Depress the pedal and start the engine. If the pedal goes down slightly, operation is normal.

Air tightness check

3 Start the engine and turn it off after one or two minutes. Depress the brake pedal several times slowly. If the pedal goes down farther the first time but gradually rises after the second or third depression, the booster is air tight.
4 Depress the brake pedal while the engine is running, then stop the engine with the pedal depressed. If there is no change in the pedal reserve travel after holding the pedal for 30 seconds, the booster is air tight.

Removal

Refer to illustration 12.7

5 Power brake booster units should not be disassembled. They require special tools not normally found in most automotive repair stations or shops. They are fairly complex and because of their critical relationship to brake performance it is best to replace a defective booster unit with a new or rebuilt one.
6 To remove the booster, first remove the brake master cylinder as described in Section 8. Pull the proportioning valve and bracket forward, being careful not to kink the lines.
7 Locate the pushrod clevis connecting the booster to the brake pedal **(see illustration)**. This is accessible from the interior in front of the passenger's seat, under the dash.
8 Remove the clevis pin cotter pin with pliers and pull out the clevis pin.
9 Holding the clevis with pliers, disconnect the clevis locknut with a wrench. The clevis is now loose.
10 Disconnect the hose leading from the engine to the booster. Be careful not to damage the hose when removing it from the booster fitting.
11 Remove the four nuts and washers holding the brake booster to the firewall. You may need a light to see them, as they are under the dash **(see illustration 12.7)**.
12 Slide the booster straight out from the firewall until the studs clear the holes, and pull the booster, brackets and gaskets from the engine compartment area.

Installation

Refer to illustrations 12.14a and 12.14b

13 Installation procedures are basically the reverse of those for removal. Tighten the clevis locknut securely and the booster mounting nuts to the torque listed in this Chapter's Specifications.
14 If the power booster unit is being replaced, the clearance between the master cylinder piston and the pushrod in the vacuum booster must be measured. Using a depth micrometer or vernier calipers, measure the distance from the seat (recessed area) in the master cylinder to the master cylinder mounting flange. Next, measure the distance from the end of the vacuum booster pushrod to the mounting face of the booster (including gasket) where the master cylinder mounting flange seats. Subtract the two measurements to get the clearance **(see illustration)**. If the clearance is more or less than specified, turn the adjusting screw on the end of the power booster pushrod until the clearance is within the specified limit **(see illustration)**.
15 After the final installation of the master cylinder and brake lines, the system must be bled (see Section 11).

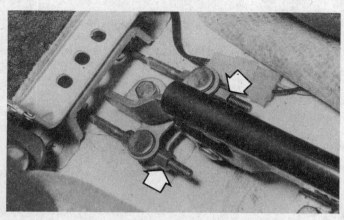

13.3a On some models, the parking brake cable adjusting nuts are located under the center console – when adjusting the cables, turn each adjusting nut (arrows) an equal amount to obtain the correct lever travel (Precis shown)

13.3b On other models, the parking brake cables are adjusted with a single nut. Turn the adjusting nut (arrow) until the correct lever travel is attained (Galant shown)

13.4a On some models, to adjust the parking brake light switch, loosen the mounting screw (arrow) and, with the handle in the released position, move the switch as necessary to turn the light off – tighten the mounting screw (Precis shown)

13.4b On other models, to adjust the parking brake light switch, bend the tab that sits directly on the switch (arrow) to attain the correct adjustment (Galant)

14.6 Pry off the parking brake E-clip, then push the cable through the backing plate

13 Parking brake – adjustment

Refer to illustrations 13.3a, 13.3b, 13.4a and 13.4b

1 Pull the parking brake lever up as far as possible, counting the number of clicks as you go. The travel should be five to seven clicks. If you are able to raise the lever higher, the parking brake might not hold the vehicle on an incline. If the travel is less than specified, the automatic adjusters on the rear brakes could be rendered useless. If the parking brake lever travel is not five to seven clicks, adjust the parking brake, as described below.
2 Remove the center console.
3 Turn the cable adjusting nut(s) on the equalizer until the specified lever travel is obtained **(see illustration)**. Tighten the nut(s) evenly to avoid over-adjusting one cable, leaving the other slack.
4 If necessary, adjust the parking brake indicator light switch so the Brake light on the instrument panel is off when the parking brake lever is all the way down. To adjust the switch, loosen the switch mounting screw (directly in front of the lever quadrant teeth) and move the switch up or down, as necessary or bend the tab **(see illustrations)**. After the switch has been adjusted, tighten the screw securely.
5 Raise the rear of the vehicle and turn the wheels to make sure the brakes don't drag when the vehicle is driven.

14 Parking brake cables – replacement

Refer to illustrations 14.6, 14.7 and 14.8
Note: *This procedure applies to both the right and left cables.*

1 Remove the center console (see Chapter 11). On some models it may be necessary to remove the floor console and the rear seats.
2 Completely unscrew the cable adjusting nut from the cable to be removed.
3 Loosen the rear wheel lug nuts on the side of the vehicle on which the cable is to be removed. Raise the rear of the vehicle and support it securely on jackstands. Place blocks in front of the front wheels. Remove the wheel.
4 Remove the brake drum and shoes, following the procedure outlined in Section 6.
5 Unhook the cable end from the parking brake lever on the brake shoe.
6 Remove the E-clip that secures the cable to the brake backing plate **(see illustration)** and pull the cable through the plate.

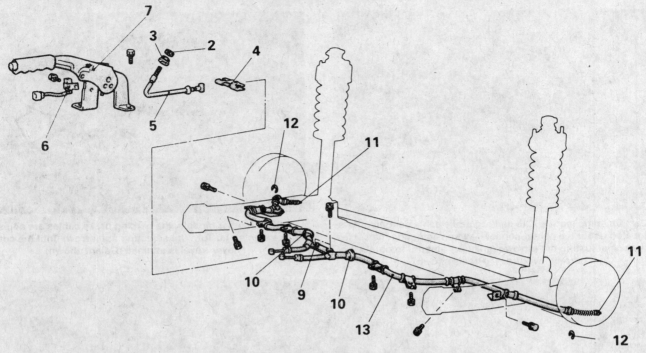

14.7 Routing details of the parking brake cable (typical)

1 Not used	5 Front cable	9 Cable clamp
2 Lock nut	6 Parking brake light switch	10 Grommet
3 Adjusting nut	7 Parking brake lever	11 Cable end
4 Equalizer	8 Not used	12 E-clip
		13 Rear cable

7 Unhook the cable from the clips on the suspension trailing arm **(see illustration)**.

8 Remove the rear seat cushion. Lift up the carpet, unscrew the cable clamp bolt and free the cable from the clamp **(see illustration)**. Pull the cable and grommet through the floorpan into the interior.

9 Installation is the reverse of the removal procedure. After the cable is installed, adjust the parking brake (see Section 13).

15 Brake light switch – removal, installation and adjustment

1 Remove the lower dash panel trim under the steering column (Chapter 11).

2 Locate the brake light switch, which is mounted at the top of the left side brake pedal support. Follow the switch wiring to the electrical connector and unplug it.

3 Unscrew the locknut on the switch and unscrew the switch from its bracket.

4 Installation is the reverse of removal. To adjust the switch, loosen the locknut and turn the switch in or out to allow the switch plunger to be depressed when the brake pedal is at rest. Tighten the locknut when the correct adjustment is obtained.

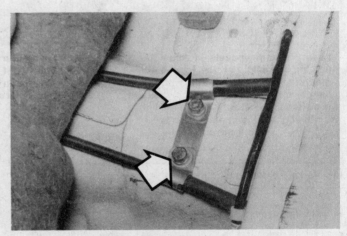

14.8 The rear seat cushion must be removed and the carpet peeled back to gain access to the cable clamp bolts (arrows) – you only need to remove the bolt that holds the cable being replaced (Precis shown)

Chapter 10
Suspension and steering systems

Contents

Balljoints – check and replacement	7
Control arm – removal, inspection and installation	6
Front axle hub and bearing – replacement	9
Front end alignment – general information	24
Front stabilizer bar – removal and installation	2
Front strut and coil spring assembly – removal, inspection and installation	4
General information	1
Intermediate shaft – removal and installation	17
Lateral rod – removal and installation	11
Power steering fluid level check	See Chapter 1
Power steering pump – removal and installation	21
Power steering system – bleeding	22
Rear axle assembly – removal, overhaul and installation	14
Rear coil spring – removal and installation	12
Rear shock absorber – removal and installation	13
Rear stabilizer bar (Precis only) – removal and installation	10
Rear wheel bearing check, repack and adjustment	See Chapter 1
Steering and suspension check	See Chapter 1
Steering gear boots – replacement	20
Steering gear – removal and installation	18
Steering knuckle and hub assembly – removal and installation	8
Steering system – general information	15
Steering wheel – removal and installation	16
Strut bar – removal and installation	3
Strut cartridge – replacement	5
Tie-rod ends – removal and installation	19
Tire and tire pressure checks	See Chapter 1
Tire rotation	See Chapter 1
Wheels and tires – general information	23

Specifications

Torque specifications

Front suspension

	Ft-lbs
Front strut assembly upper mounting nuts	
Precis	11 to 14
Galant	29 to 36
Mirage	25 to 33
Cordia/Tredia	22 to 29
Strut piston rod-to-upper insulator nut	
Precis	29 to 36
Galant	43 to 51
Mirage	44 to 51
Cordia/Tredia	43 to 50
Balljoint-to-control arm bolts/nuts	69 to 87
Balljoint-to-steering knuckle nut	43 to 52
Strut bar-to-control arm bolts/nuts	69 to 87
Strut bar-to-crossmember nut	54 to 61
Steering knuckle-to-strut assembly bolts/nuts	
Precis	
Through 1992	54 to 65
1993	65 to 76
Galant	65 to 76
Mirage	80 to 94
Cordia/Tredia	54 to 65
Support bracket-to-crossmember	51 to 58
Control arm clamp-to-crossmember nut	25 to 34
Control arm clamp-to-crossmember bolt	
Mark 7 on bolt head	58 to 72
Mark 10 on bolt head	72 to 87

Torque specifications (continued)

Ft-lbs

Front suspension (continued)
Control arm pivot bolt nut	72 to 87
Stabilizer bar bracket-to-crossmember	22 to 30
Stabilizer link	25 to 33

Rear suspension
Fixture-to-floorpan bolts	
1991 and 1992 Precis	65 to 80
1993 Precis	94 to 108
All others	36 to 51
Fixture-to-suspension arm nut	
1990 and 1991 Precis	65 to 79
All others	36 to 51
Torsion axle arm mounting bolt	72 to 90
Lateral rod mounting bolt (axle beam side)	
1990 and later Mirage	58 to 72
All others	72 to 87
Lateral rod mounting bolt (body side)	58 to 72
Shock absorber piston rod nut	14 to 22
Backing plate bolts	36 to 43

Steering
Tie-rod end-to-steering knuckle nut	17 to 25
Steering wheel nut	25 to 32
Steering gear-to-crossmember bolts	
1989 and earlier	35 to 45
1990 and later	43 to 58
Intermediate shaft pinch bolts	11 to 14
Wheel lug nuts	See Chapter 1

1 General information

Refer to illustrations 1.1a, 1.1b, 1.2a and 1.2b

The front suspension is a MacPherson strut design **(see illustration)**. The steering knuckle is located by a control arm. A stabilizer bar, mounted to the front crossmember and connecting the control arms, minimizes body lean.

The rear suspension consists of either the trailing arm suspension or the torsion axle suspension **(see illustrations)**. Some models are equipped with a stabilizer bar.

The rack-and-pinion steering gear is located behind the engine/transaxle assembly on the firewall and actuates the steering arms, which are integral with the steering knuckles. The steering column is designed to collapse in the event of an accident.

The Active-Electronic Control Suspension (ECS) system, installed as an option on some later models, changes the shock absorber valving to firm-up the suspension to suit road conditions and driving style. The system incorporates a switch, mounted on the center console, which allows the driver to select three different positions – SPORT, AUTO and SOFT.

The ECS system receives information from various sensors located throughout the vehicle. These sensors send inputs to the control module, which in turn decides whether or not to adjust the shock absorbers. The adjustable shock absorber system is controlled by the Electronic Control Unit (ECU) and the switch selection.

Due to the rather complex nature of these systems, all troubleshooting and repairs, with the exception of searching for loose connectors, wires and blown fuses, should be left to a qualified dealer service department technician.

Frequently, when working on the suspension or steering system components, you may come across fasteners which seem impossible to loosen. These fasteners on the underside of the vehicle are continually subjected to water, road grime, mud, etc., and can become rusted or "frozen," making them extremely difficult to remove. In order to unscrew these stubborn fasteners without damaging them (or other components), be sure to use lots of penetrating oil and allow it to soak in for a while. Using a wire brush to clean exposed threads will also ease removal of the nut or bolt and prevent damage to the threads. Sometimes a sharp blow with a hammer and punch is effective in breaking the bond between a nut and bolt threads, but care must be taken to prevent the punch from slipping off the fastener and ruining the threads. Heating the stuck fastener and surrounding area with a torch sometimes helps too, but isn't recommended because of the obvious dangers associated with fire. Long breaker bars and extension, or "cheater," pipes will increase leverage, but never use an extension pipe on a ratchet – the ratcheting mechanism could be damaged. Sometimes, turning the nut or bolt in the tightening (clockwise) direction first will help to break it loose. Fasteners that require drastic measures to unscrew should always be replaced with new ones.

Since most of the procedures that are dealt with in this Chapter involve jacking up the vehicle and working underneath it, a good pair of jackstands will be needed. A hydraulic floor jack is the preferred type of jack to lift the vehicle, and it can also be used to support certain components during various operations. **Warning:** *Never, under any circumstances, rely on a jack to support the vehicle while working on it. Whenever any of the suspension or steering fasteners are loosened or removed they must be inspected and, if necessary, replaced with new ones of the same part number or of original equipment quality and design. Torque specifications must be followed for proper reassembly and component retention. Never attempt to heat or straighten any suspension or steering components. Instead, replace bent or damaged parts with new ones.*

2 Front stabilizer bar – removal and installation

Refer to illustrations 2.4, 2.6a, 2.6b and 2.7

Removal

1   Raise the front of the vehicle and support it securely on jackstands. Apply the parking brake and position blocks behind the rear wheels.

2   Remove the front exhaust pipe (see Chapter 4).

3   On Mirage models, separate the tie-rod ends from the steering knuckle to provide extra clearance (see Section 19).

Chapter 10 Suspension and steering systems

1.1a Underside view of the front suspension and steering components (Galant shown, Mirage, Tredia and Cordia similar)

1. Balljoint
2. Stabilizer bar
3. Support bracket
4. Steering gear
5. Driveaxle
6. Control arm
7. Strut and coil spring assembly
8. Driveaxle boot

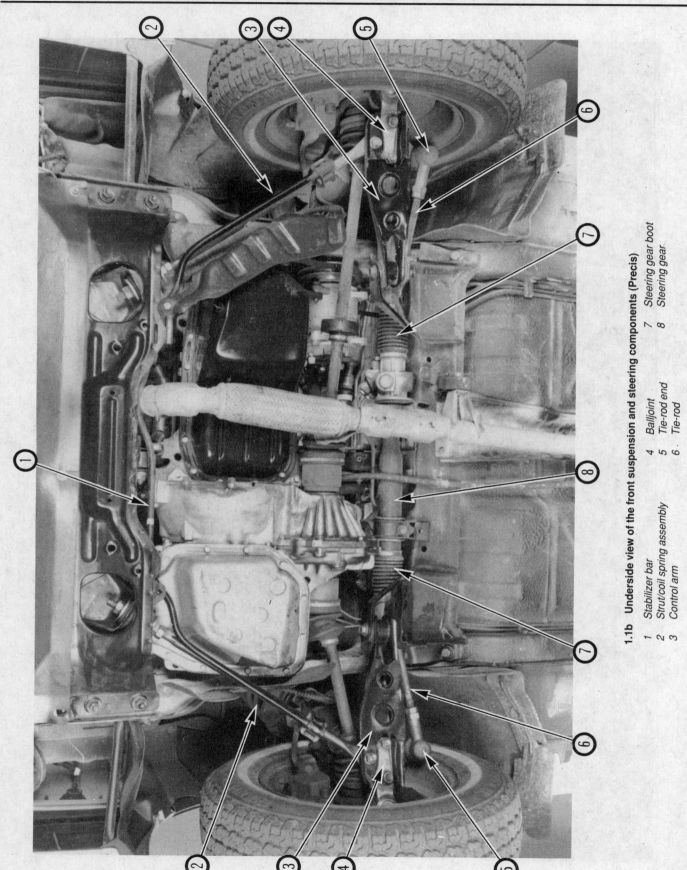

1.1b Underside view of the front suspension and steering components (Precis)

1. Stabilizer bar
2. Strut/coil spring assembly
3. Control arm
4. Balljoint
5. Tie-rod end
6. Tie-rod
7. Steering gear boot
8. Steering gear

Chapter 10 Suspension and steering systems

1.2a Underside view of the rear suspension components (Galant shown, Mirage, Tredia and Cordia similar)

1 Shock absorber and coil spring assembly
2 Torsion axle
3 Lateral rod
4 Torsion axle arm

1.2b Underside view of the rear suspension components (Precis)

1. Suspension trailing arm
2. Stabilizer bar
3. Coil spring

Chapter 10 Suspension and steering systems   10-7

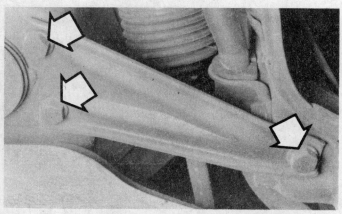

2.4 Remove the three bolts (arrows) that hold the support bracket

4 Remove the bolts that retain the support bracket (if equipped) and remove the bracket **(see illustration)**.

5 Remove the center member. On some models the center member has a dynamic damper attached. Remove the damper along with the center member.
6 Remove the stabilizer bar-to-control arm bolts and clamps, noting how the clamp halves are positioned **(see illustrations)**. **Note:** *On some models, the stabilizer bar is mounted to a stabilizer link that is attached to the control arm. If the link is equipped with a self-locking nut, use an Allen wrench and a box-end wrench to unlock the nut.*
7 Remove the stabilizer bar bracket bolts and detach the bar from the vehicle **(see illustration)**.
8 Pull the brackets off the stabilizer bar and inspect the bushings for cracks, hardening and other signs of deterioration. If the bushings are damaged, replace them.

Installation

9 Position the stabilizer bar bushings on the bar with the slits facing the front of the vehicle.
10 Insert the lower ends of the brackets into their slots, push the brackets over the bushings and raise the bar up to the frame. Install the bracket bolts but don't tighten them completely at this time.

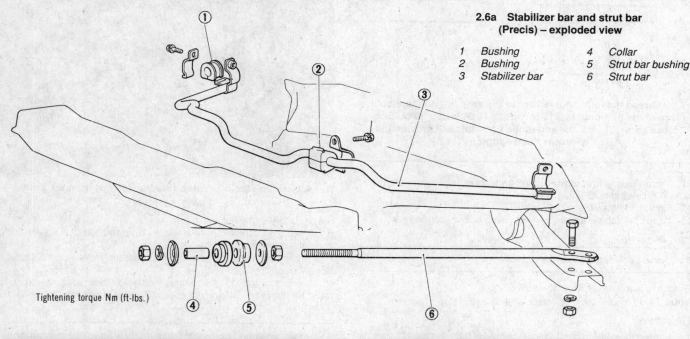

2.6a Stabilizer bar and strut bar (Precis) – exploded view

1 Bushing
2 Bushing
3 Stabilizer bar
4 Collar
5 Strut bar bushing
6 Strut bar

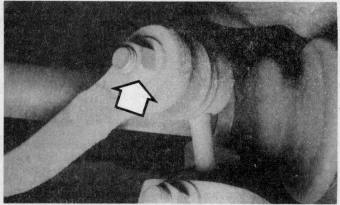

2.6b Remove the bolt (arrow) and separate the stabilizer bar from the stabilizer link (Galant shown)

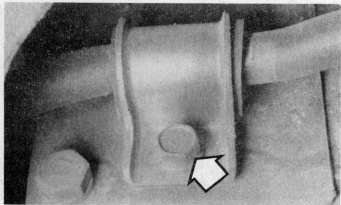

2.7 Remove the bolt (arrow) from each stabilizer bar bracket, swing the brackets down and maneuver the bar out from under the vehicle

Chapter 10 Suspension and steering systems

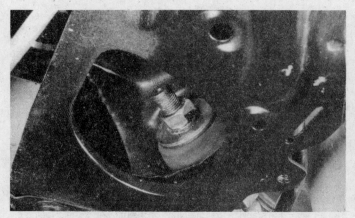

3.3a To remove the strut bar, first unscrew the large nut at the front of the crossmember and remove the washer and bushing, ...

3.3b ... then remove the two bolts and nuts (arrows) that secure the bar to the control arm

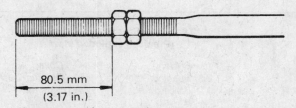

3.6 Thread the inner locknuts onto the strut bar until the outer face of the outer nut is 3.17-inches (3-11/64 inches should be close enough) from the end of the bar – this will provide the proper caster angle

11 Install the stabilizer bar-to-control arm rubber bushings, clamps and bolts. Tighten the bolts finger tight.
12 Tighten the bracket bolts.
13 Lower the vehicle and tighten the clamp bolts securely.

3 Strut bar – removal and installation

Refer to illustrations 3.3a, 3.3b and 3.6
Note: *Some models are not equipped with a strut bar.*

Removal

1 Loosen the wheel lug nuts, raise the front of the vehicle and support it securely on jackstands. Apply the parking brake and position blocks behind the rear wheels. Remove the front wheel.
2 Remove the stabilizer bar-to-strut bar clamp **(see illustration 2.6a)**.
3 Remove the strut bar-to-front crossmember nut and the two bolts and nuts that secure the other end of the bar to the control arm **(see illustrations)**.
4 Remove the large washer and rubber bushing from the front of the bar, then pull the bar straight back out of the crossmember.
5 Check the bar for cracks and distortion. If the bar is bent more than 0.120-inch (3 mm) it must be replaced. Inspect the bushings for wear, cracks, hardness and general deterioration, replacing them if necessary.

Installation

6 Before installing the bar, or if installing a new one, thread the inner locknuts onto the bar the specified distance **(see illustration)**.
7 Assemble the inner washer, rubber bushing and sleeve on the bar and insert it into the hole in the crossmember. Install the two bolts and nuts that secure the bar to the control arm and tighten them to the torque listed in this Chapter's Specifications.
8 Install the outer rubber bushing, washer and nut on the front of the bar,

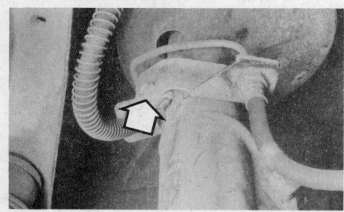

4.2 Unbolt the brake hose bracket (arrow) from the strut

but don't tighten the nut with a tool yet.
9 Install the stabilizer bar-to-strut bar clamp and bolt. Lower the vehicle and tighten the clamp bolt securely.
10 Tighten the outer strut bar nut to the torque listed in this Chapter's Specifications.
11 It would be a good idea to drive the vehicle to an alignment shop to have the front wheel alignment checked and, if necessary, adjusted.

4 Front strut and coil spring assembly – removal, inspection and installation

Refer to illustrations 4.2, 4.3, 4.4 and 4.5

Removal

1 Loosen the wheel lug nuts, raise the front of the vehicle and support it securely on jackstands. Apply the parking brake and position blocks behind the rear wheels. Remove the front wheel.
2 Unbolt the brake hose bracket from the strut **(see illustration)**.
3 Remove the strut-to-knuckle nuts and knock the bolts out with a hammer and punch **(see illustration)**.
4 Separate the strut from the steering knuckle **(see illustration)**. Be careful not to over-extend the inner CV joint. It's a good idea to wire the top of the steering knuckle to the body to prevent this from happening.
5 Support the strut and spring assembly with one hand and remove the strut-to-body nuts **(see illustration)**. Remove the assembly out through the fenderwell.

Inspection

6 Check the strut body for leaking fluid, dents, cracks and other obvious

Chapter 10 Suspension and steering systems

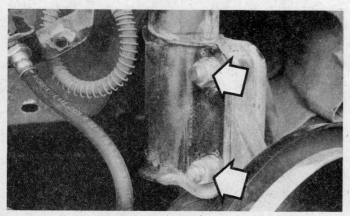

4.3 Remove the two strut-to-knuckle nuts and bolts (arrows) – if the bolts won't pull out easily, knock them out with a hammer and punch

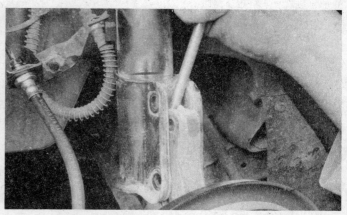

4.4 Pry the steering knuckle out of the strut flanges

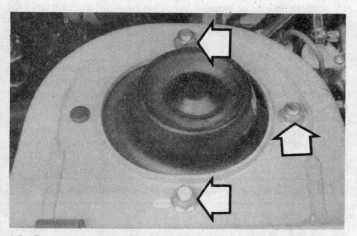

4.5 Be sure to support the strut while removing the strut-to-body nuts (arrows) – the strut will fall if you don't

5.4 Hold the collar on the spring seat with a large pair of pliers while loosening the piston rod nut

damage which would warrant repair or replacement.
7 Check the coil spring for chips or cracks in the spring coating (this will cause premature spring failure due to corrosion).
8 If any undesirable conditions exist, proceed to Section 5 for the strut disassembly procedure.

Installation

9 Guide the strut assembly up into the fenderwell and insert the upper mounting studs through the holes in the body. Once the studs protrude through the body, install the nuts so the strut won't fall back through. This may require an assistant, as the strut is quite heavy and awkward.
10 Slide the steering knuckle into the strut flange and insert the two bolts. Install the nuts and tighten them to the torque listed in this Chapter's Specifications.
11 Position the brake hose bracket on the strut and install the bolt, tightening it securely.
12 Install the wheel, lower the vehicle and tighten the lug nuts to the torque listed in the Chapter 1 Specifications.
13 Tighten the upper mounting nuts to the torque listed in this Chapter's Specifications.

5 Strut cartridge – replacement

Refer to illustrations 5.4, 5.5a, 5.5b, 5.7, 5.9, 5.13, 5.17a, 5.17b and 5.17c
Caution: *Vehicles equipped with the Active Electronic Control Suspension (ECS), require expensive spring compressing tools and alignment tools in order to remove the strut cartridge from the coil spring assembly. Have the job performed by a dealer service department or other repair shop.*

1 If the struts exhibit the telltale signs of wear (leaking fluid, loss of dampening capability) explore all options before beginning any work. The strut cartridges can be replaced. However, rebuilt strut assemblies (some complete with springs) are available on an exchange basis which eliminates much time and work. Whichever route you choose to take, check on the cost and availability of parts before disassembling the vehicle. **Warning:** *Disassembling a strut is a dangerous undertaking and extreme care must be taken during the procedure or serious bodily injury may result. Use only a high quality spring compressor and carefully follow the manufacturer's instructions furnished with the tool. After removing the coil spring from the strut assembly, set it aside in a safe, isolated area (a steel cabinet is preferred).*
2 Remove the strut and spring assembly following the procedure described in Section 4. Mount the strut assembly in a vise with the jaws of the vise cushioned with rags or blocks of wood.
3 Following the tool manufacturers instructions, install the spring compressor (which can be rented at most auto parts stores or equipment yards on a daily basis) on the spring and compress it sufficiently to relieve all pressure from the spring seat. This can be verified by wiggling the spring seat.
4 Unscrew the piston rod nut while using a large pair of pliers on the spring seat to prevent it from turning **(see illustration)**.

Chapter 10 Suspension and steering systems

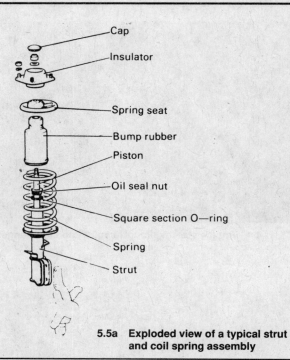

5.5a Exploded view of a typical strut and coil spring assembly

5.5b Remove the insulator and upper spring seat from the piston rod

5   Lift the insulator and upper spring seat off the piston rod **(see illustrations)**. Inspect the bearing in the insulator for smooth operation and replace it if necessary.

6   Carefully remove the compressed spring assembly and set it in a safe place, such as inside a steel cabinet. **Warning:** *Never place your head near the end of the spring!*

7   Slide the bump rubber up off the piston rod. Unscrew the oil seal nut **(see illustration)** and discard it.

8   Remove the strut from the vise and pour the damper fluid into an approved oil container.

9   Place the strut back in the vise and remove the O-ring from the strut body **(see illustration)**. Pull the piston rod out of the strut.

10 Wash the strut tube out with solvent and blow it dry with compressed air, if available. If the old piston rod is to be reinstalled, wash it too.

11 Lubricate the shock absorber cylinder and piston with clean damper fluid. Install the piston rod in the shock absorber cylinder. Assemble the cylinder and piston unit to the strut tube, after lubricating all parts with damper fluid.

12 Fill the strut tube with damper fluid. Do this slowly, taking time to pump the piston rod up and down to expel any air from under the piston.

13 Slide the piston guide over the piston rod, with the flange facing up **(see illustration)**.

14 Install a new O-ring between the piston guide and the strut tube.

15 Coat the oil seal lips and the threads of the oil seal nut with damper fluid and install the nut. Tighten it securely.

16 Stroke the damper shaft up and down a few times to verify proper operation.

17 Fully extend the damper shaft. Assemble the strut beginning with the bump rubber, coil spring and upper spring seat. Make sure the spring is properly positioned and the upper seat is installed so the flat side of the hole in the seat mates with the flat on the piston rod **(see illustrations)**.

18 Install the insulator and a new piston rod nut, tightening it to the torque listed in this Chapter's Specifications. Remove the spring compressor.

19 Install the strut and spring assembly on the vehicle as outlined in Section 4.

6   Control arm – removal, inspection and installation

Refer to illustrations 6.3a, 6.3b and 6.3c

Removal

1   Loosen the wheel lug nuts on the side to be dismantled, raise the front of the vehicle, support it securely on jackstands, apply the parking brake and place blocks behind the rear wheels. Remove the wheel.

2   Remove the strut bar-to-control arm bolts and nuts **(see illustration 3.3b)** and the remaining balljoint-to-control arm bolt and nut (see Section 7).

3   Remove the bolt(s) and nut(s) from the control arm pivot **(see illustrations)**. Pry the arm from the crossmember if it is stuck.

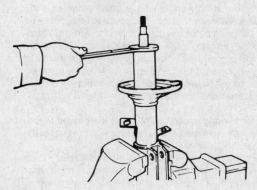

5.7 Unscrew the oil seal nut, then pour out the fluid

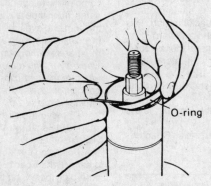

5.9 Remove the square section O-ring from the strut body

5.13 Lubricate the piston rod and install the guide with the flange facing up

Chapter 10 Suspension and steering systems

5.17a The tail on the bottom of the coil spring must be positioned in the seat pocket in the strut

5.17b When installing the upper seat, note that the flat side of the hole in the seat must mate with the flat on the piston rod (arrows)

5.17c The upper spring tail must seat in the recessed area of the upper seat (arrow)

6.3a Remove the control arm nut (arrow) and pivot bolt – if the bolt is difficult to remove, drive it out with a hammer and punch (Precis)

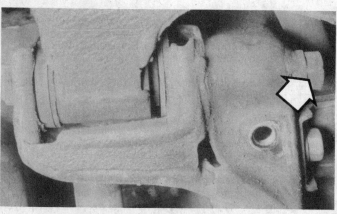

6.3b The front inner control arm pivot is supported with a bolt (arrow) (Galant)

Inspection

4 Check the control arm for distortion and the bushings for wear, damage and deterioration. Replace a damaged or bent control arm with a new one. If the inner pivot bushing is worn, take the control arm to a dealer service department or other repair shop, as special tools are required to replace it.

Installation

5 Place the control arm into the crossmember. Install the bolt and nut, but don't tighten them completely yet.
6 Connect the outer end of the arm to the balljoint and strut bar, tightening the three nuts to the torque listed in this Chapter's Specifications.
7 Place a jack under the balljoint, with a block of wood on the jack head as a cushion. Raise the control arm to simulate normal ride height, then tighten the pivot bolt nut to the torque listed in this Chapter's Specifications.
8 Install the wheel and lug nuts, lower the vehicle and tighten the lug nuts to the torque listed in the Chapter 1 Specifications.

7 Balljoints – check and replacement

Refer to illustrations 7.6 and 7.7

Check

1 Raise the vehicle and support it securely on jackstands.
2 Visually inspect the rubber boot for cuts, tears or leaking grease. If

6.3c The rear inner control arm pivot is supported with a bracket. Remove the bolts and nuts (arrows) to separate the control arm from the chassis (Galant)

any of these conditions are noticed, the balljoint should be replaced.
3 Place a large pry bar under the balljoint and attempt to push the balljoint up. Next, position the pry bar between the steering knuckle and the

10-12 Chapter 10 Suspension and steering systems

7.6 The balljoint is fastened to the control arm by two bolts and nuts (arrows) (Precis only)

7.7 Use a "pickle-fork" type balljoint separator to break the balljoint loose from the steering knuckle – boot damage can be minimized by applying grease to the boot first

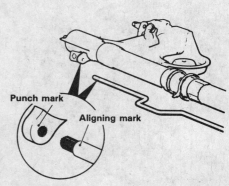

10.2 Before removing the stabilizer bar, make marks on the bar that are aligned with the punch marks on the bar brackets (Precis only)

control arm and apply downward pressure. If any movement is seen or felt during either of these checks, a worn out balljoint is indicated.

4 Have an assistant grasp the tire at the top and bottom and shake the top of the tire in an in-and-out motion. Touch the balljoint stud nut. If any looseness is felt, suspect a worn out balljoint stud or a widened hole in the steering knuckle boss. If the latter problem exists, the steering knuckle should be replaced as well as the balljoint.

Replacement

Note: *The following procedure applies to Precis models only. The balljoint is pressed into the control arm on other models. On these models, if a replacement balljoint is available, remove the control arm (see Section 6) and take it to an automotive machine shop to have the old balljoint pressed out and the new one pressed in. If a replacement balljoint is not available, you'll have to replace the control arm.*

5 Loosen the wheel lug nuts, raise the vehicle and support it securely on jackstands. Remove the wheel.
6 Break the balljoint-to-control arm bolts loose, but don't remove them yet **(see illustration)**.
7 Loosen the balljoint stud nut a couple of turns. Separate the balljoint from the steering knuckle with a balljoint separator **(see illustration)**, then remove the nut.
8 Unscrew the two balljoint-to-control arm bolts and remove the balljoint.
9 To install the balljoint, position it on the control arm and install the two bolts, but don't tighten them yet.
10 Insert the balljoint stud into the steering knuckle and install the nut, tightening it to the torque listed in this Chapter's Specifications.
11 Tighten the balljoint-to-control arm bolts to the torque listed in this Chapter's Specifications.
12 Install the wheel and lug nuts. Lower the vehicle and tighten the lug nuts to the torque listed in the Chapter 1 Specifications.

8 Steering knuckle and hub assembly – removal and installation

Warning: *Dust created by the brake system may contain asbestos, which is harmful to your health. Never blow it out with compressed air and don't inhale any of it. Do not, under any circumstances, use petroleum-based solvents to clean brake parts. Use brake cleaner or denatured alcohol only.*

Removal

1 Remove the wheel cover and unscrew the hub nut (see Chapter 8).
2 Loosen the wheel lug nuts, raise the vehicle and support it securely on jackstands. Remove the wheel. Remove the brake caliper and support it with a piece of wire as described in Chapter 9.
3 Loosen, but do not remove the strut-to-steering knuckle bolts/nuts **(see illustration 4.3)**.
4 Separate the tie-rod from the steering knuckle arm as outlined in Section 19.
5 Separate the balljoint from the steering knuckle (see Section 7). The strut-to-knuckle bolts can now be removed.
6 Remove the steering knuckle and hub assembly from the strut, balljoint and driveaxle. If the driveaxle sticks in the hub splines, push it from the hub as described in Chapter 8. Support the end of the driveaxle with a piece of wire.

Installation

7 Guide the knuckle and hub assembly into position, inserting the driveaxle into the hub.
8 Push the knuckle into the strut flange and install the bolts, but don't tighten them yet.
9 Insert the balljoint stud into the steering knuckle hole and install the nut, but don't tighten it yet.
10 Attach the tie-rod to the steering knuckle arm as described in Section 19. Tighten the strut bolt nuts, the balljoint nut and the tie-rod nut to the torque listed in this Chapter's Specifications.
11 Install the caliper as outlined in Chapter 9.
12 Install the hub nut and tighten it securely (see Chapter 9).
13 Install the wheel and lug nuts.
14 Lower the vehicle and tighten the lug nuts to the torque listed in the Chapter 1 Specifications.
15 Tighten the hub nut to the torque listed in the Chapter 8 Specifications.

9 Front axle hub and bearing – replacement

Due to the need for special tools to replace the hub and bearing assembly in the steering knuckle and to properly adjust the hub bearing preload, the steering knuckle and hub assembly must be taken to a dealer service department or other repair shop equipped with the necessary tools for this procedure. Refer to Section 8 for the steering knuckle removal procedure.

10 Rear stabilizer bar (Precis only) – removal and installation

Refer to illustration 10.2

1 Remove the rear axle assembly as outlined in Section 14.
2 Using a marker or a scribe, place match marks on the stabilizer bar aligned with the punch marks on the stabilizer bar brackets **(see illustra-**

Chapter 10 Suspension and steering systems

11.2 The left end of the lateral rod is secured to the axle with a nut (arrow)

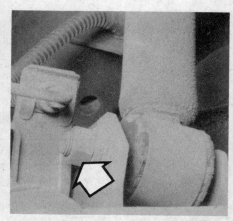

12.2 Remove the shock absorber lower mounting bolt and nut (arrow)

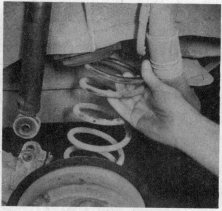

12.3a Once the coil spring is fully extended, it can be removed from the seat – it may be necessary to push down on the suspension arm to obtain adequate clearance for removal

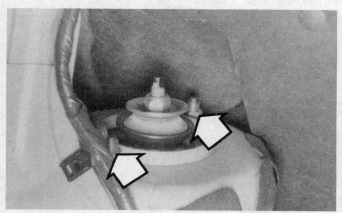

12.3b The rear shock assembly mounting bolts (arrows) are located inside the trunk near the fenderwell.

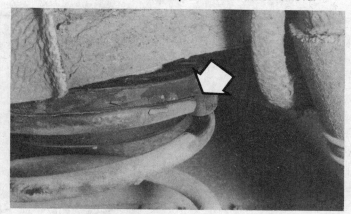

12.7 The upper end of the coil spring must be positioned in the upper seat as shown

tion). This will ensure proper alignment of the stabilizer bar splines during installation.
3   Pull the trailing arms apart far enough to allow the stabilizer bar to slide out of the splines.
4   Check the bar for cracks and distortion. Install a new one if necessary.
5   Installation is the reverse of the removal procedure.
6   Install the rear axle assembly.

11 Lateral rod – removal and installation

Refer to illustration 11.2

1   Raise the rear of the vehicle and support it securely on jackstands.
2   Remove the nut from the left end of the rod and the nut and bolt from the right end of the rod **(see illustration)**. Detach the rod from its brackets and remove it from the vehicle.
3   Installation is the reverse of removal. Tighten the mounting fasteners to the torque listed in this Chapter's Specifications.

12 Rear coil spring – removal and installation

Refer to illustrations 12.2, 12.3a, 12.3b and 12.7
Note: *Precis models are equipped with a shock absorber and separate coil spring; all other models use coil-over shock absorber assemblies.*

Removal

1   Loosen the rear wheel lug nuts, raise the rear of the vehicle and support it securely on jackstands. The jackstands must be placed under the vehicle at the lifting points – not under the rear axle assembly. Remove the wheel.
2   Position a floor jack under the spring pocket area of the suspension arm and raise it just enough to feel resistance from the spring. Remove the shock absorber lower mounting bolt **(see illustration)**.
3   On models equipped with a separate shock absorber and coil spring, slowly lower the jack until the coil spring is fully extended, then guide the spring out of the mounts **(see illustration)**. On models equipped with coil-over shock absorbers, locate the upper mounting bolts inside the trunk of the vehicle **(see illustration)** and remove the bolts to separate the assembly from the body. Follow the procedure in Section 5 to separate the coil spring from the shock absorber.

Installation

4   Before installing the coil spring, check the upper seat for cracks, hardening and general deterioration. Replace it if necessary.
5   Check the bump stop on the suspension arm for cracks and wear and replace it if necessary. It's retained to the arm by one nut on the under side of the arm.
6   Check the coil spring for cracks, chips, excessive rust and deformation. Replace the spring if any of these conditions exist.
7   On models with a separate shock absorber and coil spring, place the coil spring in the seat on the suspension arm. Install the upper seat, making sure it is positioned properly **(see illustration)**. On models equipped

Chapter 10 Suspension and steering systems

with coil-over shock absorbers, install the coil spring to the shock absorber (see Section 5).

8 On models equipped with a separate shock absorber and coil spring, raise the floor jack high enough to enable the lower end of the shock absorber to be reconnected. Tighten the bolt securely. On models equipped with coil-over shocks, install the top mounting bolts located inside the trunk of the vehicle and then the lower bolt and nut.

9 Install the wheel and lower the vehicle. Tighten the lug nuts to the torque listed in the Chapter 1 Specifications

13 Rear shock absorber – removal and installation

Refer to illustration 13.3

1 Loosen the rear wheel lug nuts, raise the rear of the vehicle and support it securely on jackstands. Position blocks in front of the front wheels. Remove the rear wheel(s).

2 Position a floor jack under the rear suspension arm and raise it just enough to take some of the spring pressure off the shock absorber. Remove the shock absorber lower mounting bolt **(see illustration 12.2)**.

3 Remove the upper mounting fasteners **(see accompanying illustration and illustration 12.3b)** and detach the shock absorber.

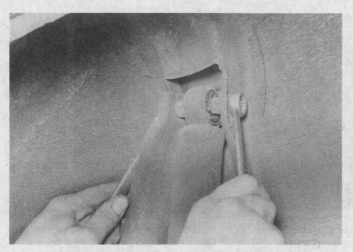

13.3 Two wrenches are needed to remove the upper mounting bolt and nut on Precis models

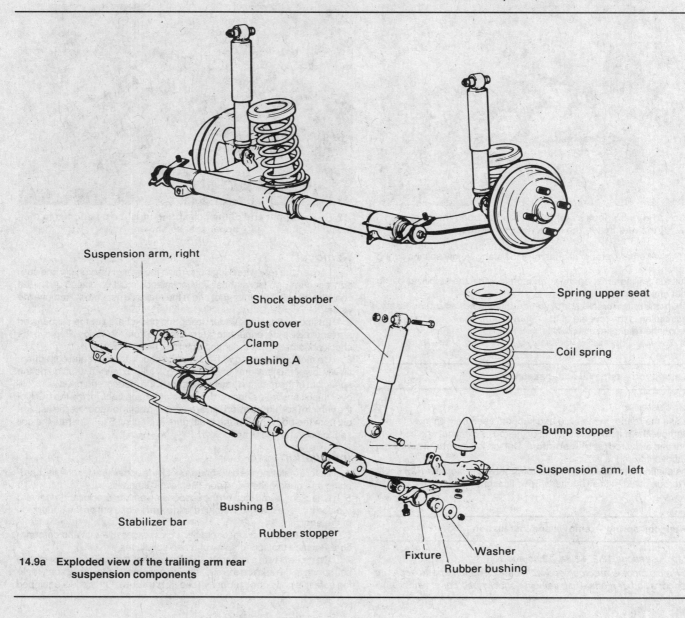

14.9a Exploded view of the trailing arm rear suspension components

Chapter 10 Suspension and steering systems

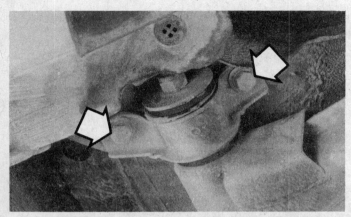

14.9b The rear axle assembly mounting fixtures are retained to the floorpan by two bolts (arrows)

14.13 Exploded view of the torsion axle rear suspension components

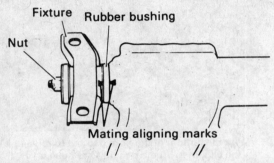

14.15 Make match marks on the fixtures and suspension arms before separating them

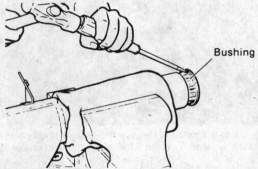

14.19 Tap around the outer edge of the bushing when removing it, so it doesn't become cocked in the tube

4 Installation is the reverse of the removal procedure. Be sure to tighten the fasteners securely.

14 Rear axle assembly – removal, overhaul and installation

Removal
1 Loosen the rear wheel lug nuts (both wheels), raise the rear of the vehicle and support it securely on jackstands placed underneath the lifting points, not under the axle assembly. Position blocks in front of the front wheels.
2 Disconnect the brake hydraulic lines from the rear wheel cylinders or calipers. Refer to Chapter 9 if necessary.
3 Remove the rear brake drums or discs (see Chapter 9).
4 Unbolt the brake line clamps from the undersides of the suspension arms or torsion axle, then disconnect the lines from the flexible hoses at the brackets. After removing the brake lines, plug the ends to keep contaminants out.
5 Detach the brake backing plates from the rear axle by removing the hub assembly (disc brake systems) and bolts (see Chapter 9). Disconnect the parking brake cables from the suspension arms and reposition the brake assemblies out of the way. **Note:** *If the hub assembly (disc brake system) is difficult to remove, use a special wheel and hub puller.*
6 Remove the rear section of the exhaust system (refer to Chapter 4 if necessary).

Trailing arm suspension
Refer to illustrations 14.9a and 14.9b
7 Remove the coil springs (see Section 12).
8 Support the rear axle with a floor jack positioned under the center of the assembly where the two suspension arms meet. If two floor jacks are available, support the assembly on each side, just inside the mounting points.
9 Unbolt the mounting fixtures from the floorpan **(see illustrations)**. Slowly lower the axle assembly, making sure nothing is left connected.

Torsion axle suspension
Refer to illustration 14.13
10 Remove the lateral rod (see Section 11).
11 Remove the shock absorber lower mounting bolt and nut (see Section 12).
12 Support the rear axle with a floor jack positioned under the center of the assembly. If two floor jacks are available, support the assembly on each side, just inside the mounting points.
13 Remove the torsion axle mounting bolts **(see illustration)** and slowly lower the torsion axle assembly.

Overhaul (trailing arm suspension only)
Refer to illustrations 14.15, 14.19, 14.20 and 14.24
14 The rear axle assembly is composed of two separate suspension arms which may be separated for servicing. The arms contain bushings that can be replaced if they're worn out.
15 Using a sharp scribe or paint, apply match marks to the fixtures and the suspension arms **(see illustration)**.
16 Remove the nuts that retain the fixtures to the suspension arms, then pull the fixtures off the arms. Check the rubber bushings on each fixture for wear, cracks and deterioration. If they must be replaced, drive them out with a hammer and drift and drive the new ones in (lubricate them with soapy water first).
17 If the vehicle is equipped with a rear stabilizer bar, mark the relationship of the bar to the bar brackets (see Section 10).
18 Remove the dust cover clamp and pull the two arms apart.
19 Mount the right suspension arm in a vise lined with rags, remove the rubber stopper, then tap the bushing out of the end of the tube **(see illustration)**.

Chapter 10 Suspension and steering systems

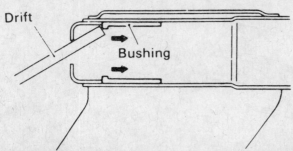

14.20 Use a hammer and drift punch to drive bushing B out of the left suspension arm

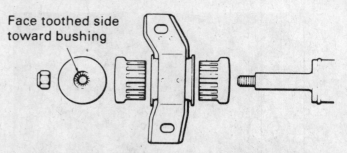

14.24 The fixtures must be positioned as shown here during installation

16.2a On some models, remove the screws that retain the horn pad to the steering column – they are mounted under the horn pad (Galant shown).

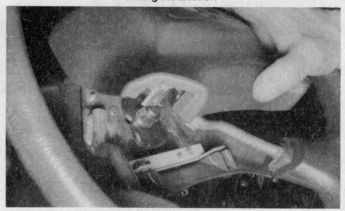

16.2b First lift the bottom of the horn pad out to release it from the clip, then lift the pad up to remove it

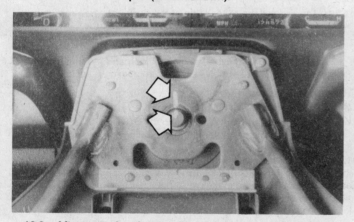

16.3 After removing the steering wheel nut, make alignment marks on the steering column shaft and the steering wheel hub – this will ensure correct steering wheel alignment during installation

16.4 Use a steering wheel puller to separate the steering wheel from the shaft – DO NOT hammer on the shaft in an attempt to remove the wheel!

20 Mount the left suspension arm in the vise and drive out the bushing from inside the tube **(see illustration)**.
21 Lubricate the new bushings and seats with chassis grease and install them with a bushing driver. If a bushing driver isn't available, use a large socket or a piece of pipe.
22 Apply chassis grease to the surface of the right suspension arm. Install the rubber stopper.
23 Push the right and left suspension arms together and install the stabilizer bar, if equipped. Make sure the marks on the stabilizer bar line up with the marks on the brackets.
24 Install the mounting fixtures on the suspension arms (line up the marks made in Step 11). Make sure the fixtures and the washers are installed properly **(see illustration)**. Don't tighten the fixture nuts completely yet.
25 Pack grease into the dust cover and lips and tighten the clamp.

Installation (all models)
26 Install the axle assembly by reversing the removal steps. Be sure to tighten the fixture-to-floorpan bolts to the torque listed in this Chapter's Specifications. Lower the vehicle to the ground and tighten the fixture nuts to the torque listed in this Chapter's Specifications.
27 Bleed the brake system (see Chapter 9).

15 Steering system – general information

All models are equipped with rack-and-pinion steering. The steering gear is bolted to the crossmember at the firewall and operates the steering arms via tie-rods. The inner ends of the tie-rods are protected by rubber boots which should be inspected periodically for secure attachment, tears and leaking lubricant.
The power assist system consists of a belt-driven pump and asso-

Chapter 10 Suspension and steering systems

ciated lines and hoses. The power steering fluid level should be checked periodically (see Chapter 1).

The steering wheel operates the steering shaft, which actuates the steering gear through universal joints. Looseness in the steering can be caused by wear in the steering shaft universal joints, the steering gear, the tie-rod ends and loose retaining bolts.

Note: *Some later Galant models are equipped with a four-wheel steering system. The system consists of a power cylinder with tie-rod ends attached to the rear suspension assembly, a hydraulic control valve, an oil pump mounted on top of the differential case and hydraulic pressure lines and hoses.* **Warning:** *Special tools are required for bleeding the system and measuring oil pump output. In addition, the bleeding procedure and function check are potentially hazardous operations if not performed at a professional repair facility. It is strongly recommended that repairs to or involving this system be performed only by a dealer or other authorized repair shop.*

16 Steering wheel – removal and installation

Refer to illustrations 16.2a, 16.2b, 16.3 and 16.4

1 Disconnect the cable from the negative terminal of the battery.
2 Pull the horn pad from the steering wheel. On some models, the pad is retained by screws **(see illustrations)**.
3 Remove the steering wheel retaining nut then mark the relationship of the steering shaft to the hub to simplify installation and ensure steering wheel alignment **(see illustration)**.
4 Use a steering wheel puller to disconnect the steering wheel from the shaft **(see illustration)**.
5 To install the wheel, align the mark on the steering wheel hub with the mark on the shaft and slip the wheel onto the shaft. Install the hub nut and tighten it to the torque listed in this Chapter's Specifications.
6 Install the horn pad.
7 Connect the negative battery cable.

17 Intermediate shaft – removal and installation

Refer to illustrations 17.2a, 17.2b and 17.4

1 Turn the front wheels to the straight ahead position. Raise the vehicle and support it securely on jackstands.
2 Using white paint, place alignment marks on the upper universal joint, the steering shaft, the lower universal joint and the steering gear input

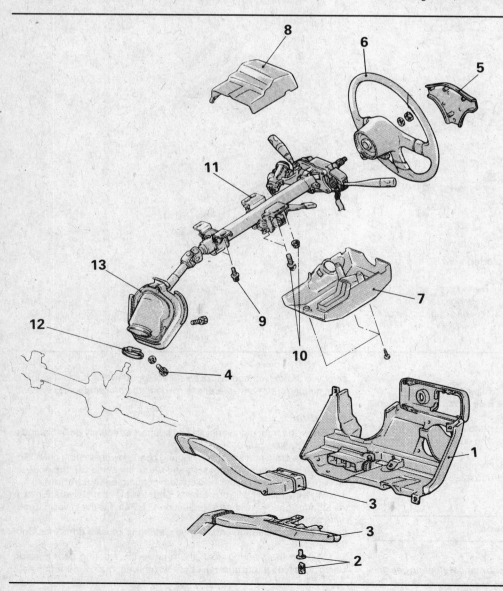

17.2a Typical steering column and components

1 Instrument trim panel
2 Trim clip
3 Air ducts
4 Clamp bolt
5 Horn pad
6 Steering wheel
7 Column cover lower
8 Column cover upper
9 Lower bracket installation bolts
10 Upper bracket installation bolts and nuts
11 Steering column assembly
12 Band
13 Intermediate shaft cover

Chapter 10 Suspension and steering systems

17.2b Make alignment marks on the intermediate shaft and the steering column shaft (do the same to the other end of the intermediate shaft, at the steering gear input shaft) (Precis shown)

17.4 The dust cover is fastened to the inner firewall with three bolts (Precis shown)

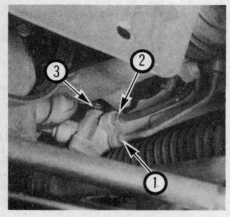

18.2 Remove the power steering pressure (1) and return (2) lines – (3) is the intermediate shaft lower pinch bolt (Precis shown)

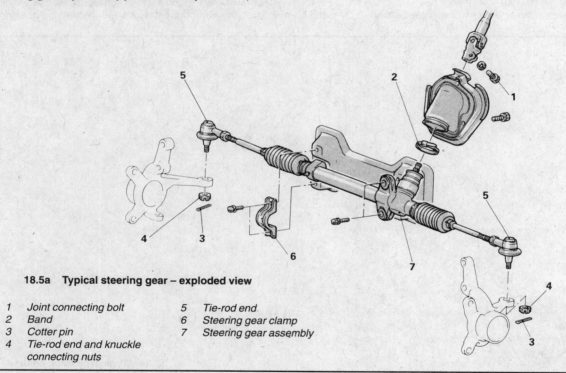

18.5a Typical steering gear – exploded view

1. Joint connecting bolt
2. Band
3. Cotter pin
4. Tie-rod end and knuckle connecting nuts
5. Tie-rod end
6. Steering gear clamp
7. Steering gear assembly

shaft **(see illustrations)**.
3 Remove the upper and lower universal joint pinch bolts.
4 If equipped, remove the dust cover bolts **(see illustration)** and pull the cover into the driver's compartment.
5 Pry the intermediate shaft universal joint off the steering gear input shaft. Separate the shaft from the steering column and pull the assembly into the driver's compartment.
6 Installation is the reverse of the removal procedure. Be sure to align the marks and tighten the pinch bolts to the torque listed in this Chapter's Specifications.

18 Steering gear – removal and installation

Refer to illustrations 18.2, 18.5a and 18.5b
Note: *This procedure applies to both power and manual steering gear as-*

semblies. When working on a vehicle equipped with a manual steering gear, simply ignore any references made to the power steering system.

Removal

1 Raise the front of the vehicle and support it securely on jackstands. Apply the parking brake.
2 Place a drain pan under the steering gear (power steering only). Remove the power steering pressure and return lines and cap the ends to prevent excessive fluid loss and contamination **(see illustration)**.
3 Mark the relationship of the lower intermediate shaft universal joint to the steering gear input shaft (see Section 17). Remove the lower intermediate shaft pinch bolt.
4 Separate the tie-rod ends from the steering knuckle arms (see Section 19).
5 Support the steering gear and remove the steering gear housing clamp bolts **(see illustrations)**. Lower the unit, separate the intermediate

Chapter 10 Suspension and steering systems

18.5b Remove the two bolts from each steering gear housing clamp (Precis shown)

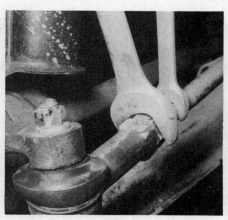

19.2a Loosen the jam nut while holding the tie-rod with a wrench (or a pair of locking pliers) on the flat portion of the rod to prevent it from turning

19.2b Mark the relationship of the tie-rod end to the tie-rod

19.4 A gear puller works well for separating the tie-rod end from the steering knuckle arm – note that the nut hasn't been removed (leaving it in place will prevent the two from separating violently)

shaft from the steering gear input shaft and remove the steering gear from the vehicle.

Installation

6   Raise the steering gear into position and connect the intermediate shaft, aligning the marks.
7   Install the gear housing clamps and bolts and tighten the bolts to the torque listed in this Chapter's Specifications.
8   Connect the tie-rod ends to the steering knuckle arms (see Section 19).
9   Install the intermediate shaft lower pinch bolt and tighten it to the torque listed in this Chapter's Specifications.
10 Connect the power steering pressure and return hoses to the steering gear and fill the power steering pump reservoir with the recommended fluid (Chapter 1).
11 Lower the vehicle and bleed the steering system as outlined in Section 22.

19 Tie-rod ends - removal and installation

Refer to illustrations 19.2a, 19.2b and 19.4

Removal

1   Loosen the wheel lug nuts. Raise the front of the vehicle, support it securely, block the rear wheels and apply the parking brake. Remove the front wheel(s).

2   Hold the tie-rod with a pair of locking pliers or a wrench and loosen the jam nut enough to mark the position of the tie-rod end in relation to the threads **(see illustrations)**.
3   Remove the cotter pin and loosen the nut on the tie-rod end stud.
4   Disconnect the tie-rod from the steering knuckle arm with a puller **(see illustration)**. Remove the nut and separate the tie-rod.
5   Unscrew the tie-rod end from the tie-rod.

Installation

6   Thread the tie-rod end on to the marked position and insert the tie-rod stud into the steering knuckle arm. Tighten the jam nut securely.
7   Install the castellated nut on the stud and tighten it to the torque listed in this Chapter's Specifications. Install a new cotter pin.
8   Install the wheel and lug nuts. Lower the vehicle and tighten the lug nuts to the torque listed in the Chapter 1 Specifications.
9   Have the alignment checked by a dealer service department or an alignment shop.

20 Steering gear boots – replacement

1   Loosen the lug nuts, raise the vehicle and support it securely on jackstands. Remove the front wheel(s).
2   Refer to Section 19 and remove the tie-rod end and jam nut.
3   Remove the steering gear boot clamps and slide the boot off.
4   Before installing the new boot, wrap the threads and serrations on the end of the tie-rod with a layer of tape so the small end of the new boot isn't damaged.
5   Slide the new boot into position on the steering gear until it seats in the groove in the tie-rod. Install new clamps.
6   Remove the tape and install the tie-rod end (Section 19).
7   Install the wheel and lug nuts. Lower the vehicle and tighten the lug nuts to the torque listed in the Chapter 1 Specifications.

21 Power steering pump – removal and installation

Refer to illustration 21.4

1   Disconnect the cable from the negative battery terminal.
2   Loosen the tension on the pump drivebelt (see Chapter 1) and remove the belt.
3   Using a suction gun, suck out as much fluid from the power steering fluid reservoir as possible. Place a drain pan under the vehicle to catch any fluid that may spill out when the hoses are disconnected.

21.4 To detach the power steering pump, remove the pressure line fitting nut (1), the return line hose (2), the pressure line bracket (3), the adjuster bolt (4) and the pivot bolt (not visible in this photo) – have a container ready to catch spilled fluid when removing the pump (Precis shown)

4 Disconnect the pressure and return lines from the pump (see illustration). Unbolt the high pressure line bracket from the pump.
5 Remove the adjuster bolt and pivot bolt from the pump (see illustration 21.4) and detach the pump from the vehicle.
6 Installation is the reverse of the removal procedure. Be sure to adjust the power steering pump drivebelt tension and top up the power steering fluid reservoir (see Chapter 1).
7 Bleed the power steering system as described in Section 22.

22 Power steering system – bleeding

1 Following any operation in which the power steering lines have been disconnected, the power steering system must be bled to remove all air and obtain proper steering performance.
2 With the front wheels in the straight ahead position, check the power steering fluid level and, if low, add fluid (see Chapter 1).
3 Start the engine and allow it to run at fast idle. Recheck the fluid level and add more if necessary.
4 Bleed the system by turning the steering wheel from side-to-side, without hitting the stops. This will work the air out of the system. Keep the reservoir full of fluid as this is done.
5 When the air is out of the system, return the wheels to the straight ahead position and leave the vehicle running for several more minutes before shutting it off.
6 Road test the vehicle to be sure the steering system is functioning normally and noise free.
7 Recheck the fluid level to be sure it's correct. Add fluid if necessary (see Chapter 1).

23 Wheels and tires – general information

Refer to illustration 23.1

All vehicles covered by this manual are equipped with metric-sized fiberglass or steel-belted radial tires (see illustration). Use of other size or type of tires may affect the ride and handling of the vehicle. Don't mix different types of tires, such as radials and bias belted, on the same vehicle as handling may be seriously affected. It's recommended that tires be replaced in pairs on the same axle, but if only one tire is being replaced, be sure it's the same size, structure and tread design as the other.

Because tire pressure has a substantial effect on handling and wear, the pressure on all tires should be checked at least once a month or before any extended trips (see Chapter 1).

Wheels must be replaced if they are bent, dented, leak air, have elongated bolt holes, are heavily rusted, out of vertical symmetry or if the lug nuts won't stay tight. Wheel repairs that use welding or peening are not recommended.

Tire and wheel balance is important in the overall handling, braking and performance of the vehicle. Unbalanced wheels can adversely affect handling and ride characteristics as well as tire life. Whenever a tire is installed on a wheel, the tire and wheel should be balanced by a shop with the proper equipment.

24 Front end alignment – general information

A front end alignment refers to the adjustments made to the front wheels so they are in proper angular relationship to the suspension and the ground. Front wheels that are out of proper alignment not only affect steering control, but also increase tire wear. The only front end adjustment normally required on this vehicle is toe-in. Caster angle is slightly adjustable on Precis models.

Getting the proper front wheel alignment is a very exacting process, one in which complicated and expensive machines are necessary to perform the job properly. Because of this, you should have a technician with the proper equipment perform these tasks. We will, however, use this space to give you a basic idea of what is involved with front end alignment so you can better understand the process and deal intelligently with the shop that does the work.

Toe-in is the turning in of the front wheels. The purpose of a toe specification is to ensure parallel rolling of the front wheels. In a vehicle with zero toe-in, the distance between the front edges of the wheels will be the same as the distance between the rear edges of the wheels. The actual amount of toe-in is normally only a fraction of an inch. Toe-in adjustment is controlled by the tie-rod end position on the tie-rod. Incorrect toe-in will cause the tires to wear improperly by making them scrub against the road surface.

Caster is the tilting of the front steering axis from the vertical. A tilt toward the rear is positive caster and a tilt toward the front is negative caster. Caster is adjusted by changing the position of the adjusting nuts on the strut bar (Precis models only).

Camber angle is not adjustable on these vehicles.

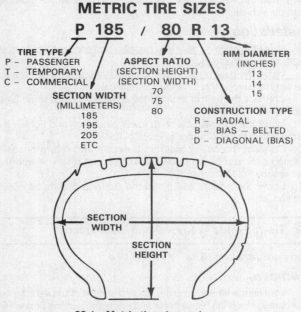

23.1 Metric tire size code

Chapter 11 Body

Contents

Automatic shoulder harnesses – general information	19
Body – maintenance	2
Body repair – major damage	6
Body repair – minor damage	5
Bumpers – removal and installation	10
Door – removal, installation and adjustment	12
Door latch, lock cylinder and handle – removal and installation	15
Door trim panel – removal and installation	11
Door window glass – removal and installation	16
Fixed glass – replacement	8
Front fender inner panels – removal and installation	18
General information	1
Hinges and locks – maintenance	7
Hood – removal, installation and adjustment	9
Liftgate – removal, installation and adjustment	14
Outside mirror – removal and installation	17
Seat belt check	20
Trunk lid – removal, installation and adjustment	13
Upholstery and carpets – maintenance	4
Vinyl trim – maintenance	3

1 General information

These models feature a "unibody" layout, using a floor pan with front and rear frame side rails which support the body components, front and rear suspension systems and other mechanical components.

Certain components are particularly vulnerable to accident damage and can be unbolted and repaired or replaced. Among these parts are the body moldings, bumpers, hood and trunk lids and all glass.

Only general body maintenance practices and body panel repair procedures within the scope of the do-it-yourselfer are included in this Chapter.

2 Body – maintenance

1 The condition of your vehicle's body is very important, because the resale value depends a great deal on it. It's much more difficult to repair a neglected or damaged body than it is to repair mechanical components. The hidden areas of the body, such as the wheel wells, the frame and the engine compartment, are equally important, although they don't require as frequent attention as the rest of the body.

2 Once a year, or every 12,000 miles, it's a good idea to have the underside of the body steam cleaned. All traces of dirt and oil will be removed and the area can then be inspected carefully for rust, damaged brake lines, frayed electrical wires, damaged cables and other problems. The front suspension components should be greased after completion of this job.

3 At the same time, clean the engine and the engine compartment with a steam cleaner or water soluble degreaser.

4 The wheel wells should be given close attention, since undercoating can peel away and stones and dirt thrown up by the tires can cause the paint to chip and flake, allowing rust to set in. If rust is found, clean down to the bare metal and apply an anti-rust paint.

5 The body should be washed about once a week. Wet the vehicle thoroughly to soften the dirt, then wash it down with a soft sponge and plenty of clean soapy water. If the surplus dirt is not washed off very carefully, it can wear down the paint.

6 Spots of tar or asphalt thrown up from the road should be removed with a cloth soaked in solvent.

7 Once every six months, wax the body and chrome trim. If a chrome cleaner is used to remove rust from any of the vehicle's plated parts, remember that the cleaner also removes part of the chrome, so use it sparingly.

These photos illustrate a method of repairing simple dents. They are intended to supplement *Body repair - minor damage* in this Chapter and should not be used as the sole instructions for body repair on these vehicles.

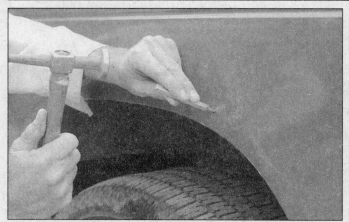

1 If you can't access the backside of the body panel to hammer out the dent, pull it out with a slide-hammer-type dent puller. In the deepest portion of the dent or along the crease line, drill or punch hole(s) at least one inch apart . . .

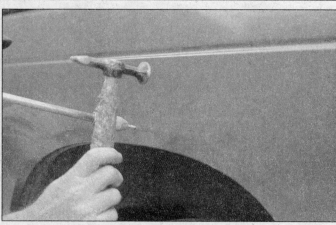

2 . . . then screw the slide-hammer into the hole and operate it. Tap with a hammer near the edge of the dent to help 'pop' the metal back to its original shape. When you're finished, the dent area should be close to its original contour and about 1/8-inch below the surface of the surrounding metal

3 Using coarse-grit sandpaper, remove the paint down to the bare metal. Hand sanding works fine, but the disc sander shown here makes the job faster. Use finer (about 320-grit) sandpaper to feather-edge the paint at least one inch around the dent area

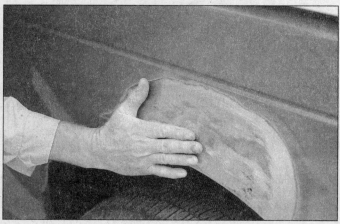

4 When the paint is removed, touch will probably be more helpful than sight for telling if the metal is straight. Hammer down the high spots or raise the low spots as necessary. Clean the repair area with wax/silicone remover

5 Following label instructions, mix up a batch of plastic filler and hardener. The ratio of filler to hardener is critical, and, if you mix it incorrectly, it will either not cure properly or cure too quickly (you won't have time to file and sand it into shape)

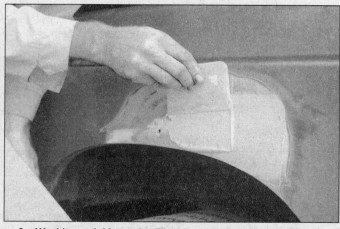

6 Working quickly so the filler doesn't harden, use a plastic applicator to press the body filler firmly into the metal, assuring it bonds completely. Work the filler until it matches the original contour and is slightly above the surrounding metal

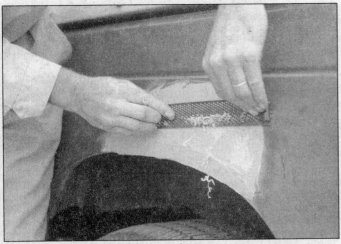

7 Let the filler harden until you can just dent it with your fingernail. Use a body file or Surform tool (shown here) to rough-shape the filler

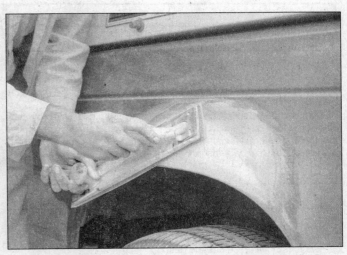

8 Use coarse-grit sandpaper and a sanding board or block to work the filler down until it's smooth and even. Work down to finer grits of sandpaper - always using a board or block - ending up with 360 or 400 grit

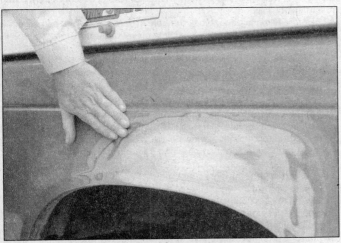

9 You shouldn't be able to feel any ridge at the transition from the filler to the bare metal or from the bare metal to the old paint. As soon as the repair is flat and uniform, remove the dust and mask off the adjacent panels or trim pieces

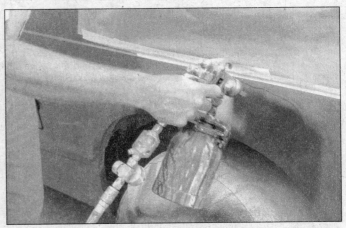

10 Apply several layers of primer to the area. Don't spray the primer on too heavy, so it sags or runs, and make sure each coat is dry before you spray on the next one. A professional-type spray gun is being used here, but aerosol spray primer is available inexpensively from auto parts stores

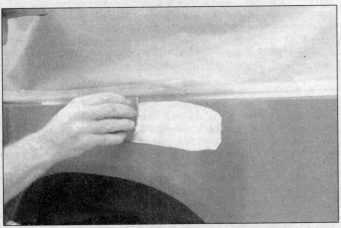

11 The primer will help reveal imperfections or scratches. Fill these with glazing compound. Follow the label instructions and sand it with 360 or 400-grit sandpaper until it's smooth. Repeat the glazing, sanding and respraying until the primer reveals a perfectly smooth surface

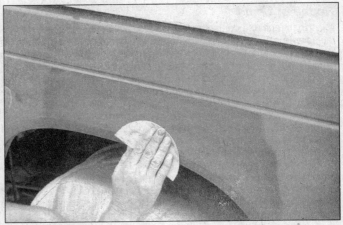

12 Finish sand the primer with very fine sandpaper (400 or 600-grit) to remove the primer overspray. Clean the area with water and allow it to dry. Use a tack rag to remove any dust, then apply the finish coat. Don't attempt to rub out or wax the repair area until the paint has dried completely (at least two weeks)

3 Vinyl trim – maintenance

Don't clean vinyl trim with detergents, caustic soap or petroleum-based cleaners. Plain soap and water works just fine, with a soft brush to clean dirt that may be ingrained. Wash the vinyl as frequently as the rest of the vehicle.

After cleaning, application of a high quality rubber and vinyl protectant will help prevent oxidation and cracks. The protectant can also be applied to weatherstripping, vacuum lines and rubber hoses, which often fail as a result of chemical degradation, and to the tires.

4 Upholstery and carpets – maintenance

Every three months remove the carpets or mats and clean the interior of the vehicle (more frequently if necessary). Vacuum the upholstery and carpets to remove loose dirt and dust.

5 Body repair – minor damage

See photo sequence

Repair of minor scratches

1 If the scratch is superficial and does not penetrate to the metal of the body, repair is very simple. Lightly rub the scratched area with a fine rubbing compound to remove loose paint and built-up wax. Rinse the area with clean water.

2 Apply touch-up paint to the scratch, using a small brush. Continue to apply thin layers of paint until the surface of the paint in the scratch is level with the surrounding paint. Allow the new paint at least two weeks to harden, then blend it into the surrounding paint by rubbing with a very fine rubbing compound. Finally, apply a coat of wax to the scratch area.

3 If the scratch has penetrated the paint and exposed the metal of the body, causing the metal to rust, a different repair technique is required. Remove all loose rust from the bottom of the scratch with a pocket knife, then apply rust inhibiting paint to prevent the formation of rust in the future. Using a rubber or nylon applicator, coat the scratched area with glaze-type filler. If required, the filler can be mixed with thinner to provide a very thin paste, which is ideal for filling narrow scratches. Before the glaze filler in the scratch hardens, wrap a piece of smooth cotton cloth around the tip of a finger. Dip the cloth in thinner and then quickly wipe it along the surface of the scratch. This will ensure that the surface of the filler is slightly hollow. The scratch can now be painted over as described earlier in this Section.

Repair of dents

4 When repairing dents, the first job is to pull the dent out until the affected area is as close as possible to its original shape. There is no point in trying to restore the original shape completely as the metal in the damaged area will have stretched on impact and cannot be restored to its original contours. It is better to bring the level of the dent up to a point which is about 1/8-inch below the level of the surrounding metal. In cases where the dent is very shallow, it is not worth trying to pull it out at all.

5 If the back side of the dent is accessible, it can be hammered out gently from behind using a soft-face hammer. While doing this, hold a block of wood firmly against the opposite side of the metal to absorb the hammer blows and prevent the metal from being stretched.

6 If the dent is in a section of the body which has double layers, or some other factor makes it inaccessible from behind, a different technique is required. Drill several small holes through the metal inside the damaged area, particularly in the deeper sections. Screw long, self-tapping screws into the holes just enough for them to get a good grip in the metal. Now the dent can be pulled out by pulling on the protruding heads of the screws with locking pliers.

7 The next stage of repair is the removal of paint from the damaged area and from an inch or so of the surrounding metal. This is done with a wire brush or sanding disk in a drill motor, although it can be done just as effectively by hand with sandpaper. To complete the preparation for filling, score the surface of the bare metal with a screwdriver or the tang of a file, or drill small holes in the affected area. This will provide a good grip for the filler material. To complete the repair, see the subsection on filling and painting later in this Section.

Repair of rust holes or gashes

8 Remove all paint from the affected area and from an inch or so of the surrounding metal using a sanding disk or wire brush mounted in a drill motor. If these are not available, a few sheets of sandpaper will do the job just as effectively.

9 With the paint removed, you will be able to determine the severity of the corrosion and decide whether to replace the whole panel, if possible, or repair the affected area. New body panels are not as expensive as you may think and it is often quicker to install a new panel than to repair large areas of rust.

10 Remove all trim pieces from the affected area except those which will act as a guide to the original shape of the damaged body, such as headlight shells, etc. Using metal snips or a hacksaw blade, remove all loose metal and any other metal that is badly affected by rust. Hammer the edges of the hole inward to create a slight depression for the filler material.

11 Wire brush the affected area to remove the powdery rust from the surface of the metal. If the back of the rusted area is accessible, treat it with rust inhibiting paint.

12 Before filling is done, block the hole in some way. This can be done with sheet metal riveted or screwed into place, or by stuffing the hole with wire mesh.

13 Once the hole is blocked off, the affected area can be filled and painted. See the following subsection on filling and painting.

Filling and painting

14 Many types of body fillers are available, but generally speaking, body repair kits which contain filler paste and a tube of resin hardener are best for this type of repair work. A wide, flexible plastic or nylon applicator will be necessary for imparting a smooth and contoured finish to the surface of the filler material. Mix up a small amount of filler on a clean piece of wood or cardboard (use the hardener sparingly). Follow the manufacturer's instructions on the package, otherwise the filler will set incorrectly.

15 Using the applicator, apply the filler paste to the prepared area. Draw the applicator across the surface of the filler to achieve the desired contour and to level the filler surface. As soon as a contour that approximates the original one is achieved, stop working the paste. If you continue, the paste will begin to stick to the applicator. Continue to add thin layers of paste at 20-minute intervals until the level of the filler is just above the surrounding metal.

16 Once the filler has hardened, the excess can be removed with a body file. From then on, progressively finer grades of sandpaper should be used, starting with a 180-grit paper and finishing with 600-grit wet-or-dry paper. Always wrap the sandpaper around a flat rubber or wooden block, otherwise the surface of the filler will not be completely flat. During the sanding of the filler surface, the wet-or-dry paper should be periodically rinsed in water. This will ensure that a very smooth finish is produced in the final stage.

17 At this point, the repair area should be surrounded by a ring of bare metal, which in turn should be encircled by the finely feathered edge of good paint. Rinse the repair area with clean water until all of the dust produced by the sanding operation is gone.

18 Spray the entire area with a light coat of primer. This will reveal any imperfections in the surface of the filler. Repair the imperfections with fresh filler paste or glaze filler and once more smooth the surface with sandpaper. Repeat this spray-and-repair procedure until you are satisfied that the surface of the filler and the feathered edge of the paint are perfect. Rinse the area with clean water and allow it to dry completely.

19 The repair area is now ready for painting. Spray painting must be carried out in a warm, dry, windless and dust free atmosphere. These conditions can be created if you have access to a large indoor work area, but if you are forced to work in the open, you will have to pick the day very carefully. If you are working indoors, dousing the floor in the work area with water will help settle the dust which would otherwise be in the air. If the repair

Chapter 11 Body

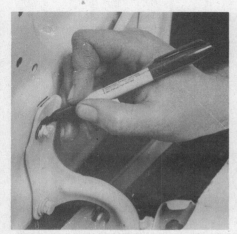

9.2 Use paint or a permanent marker to mark on the hinge plate around the bolts – mark around the entire hinge plate before adjusting the hood (front hinged hood shown)

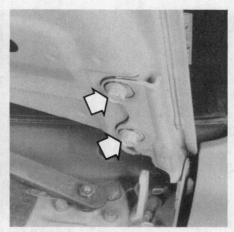

9.7 Loosen the bolts (arrows) and move the hood to adjust its position (rear hinged hood shown)

9.10a To adjust the hood latch, mark lines around the hood latch mounting screws (arrows), loosen the screws, position the latch where required and retighten the screws (front hinged hood shown)

area is confined to one body panel, mask off the surrounding panels. This will help minimize the effects of a slight mismatch in paint color. Trim pieces such as chrome strips, door handles, etc., will also need to be masked off or removed. Use masking tape and several thicknesses of newspaper for the masking operations.

20 Before spraying, shake the paint can thoroughly, then spray a test area until the spray painting technique is mastered. Cover the repair area with a thick coat of primer. The thickness should be built up using several thin layers of primer rather than one thick one. Using 600-grit wet-or-dry sandpaper, rub down the surface of the primer until it is very smooth. While doing this, the work area should be thoroughly rinsed with water and the wet-or-dry sandpaper periodically rinsed as well. Allow the primer to dry before spraying additional coats.

21 Spray on the top coat, again building up the thickness by using several thin layers of paint. Begin spraying in the center of the repair area and then, using a circular motion, work out until the whole repair area and about two inches of the surrounding original paint is covered. Remove all masking material 10 to 15 minutes after spraying the final coat of paint. Allow the new paint at least two weeks to harden, then use a very fine rubbing compound to blend the edges of the new paint into the existing paint. Finally, apply a coat of wax.

6 Body repair – major damage

1 Major damage must be repaired by an auto body shop specifically equipped to perform unibody repairs. These shops have the specialized equipment required to do the job properly.

2 If the damage is extensive, the body must be checked for proper alignment or the vehicle's handling characteristics may be adversely affected and other components may wear at an accelerated rate.

3 Due to the fact that all of the major body components (hood, fenders, etc.) are separate and replaceable units, any seriously damaged components should be replaced rather than repaired. Sometimes the components can be found in a wrecking yard that specializes in used vehicle components, often at considerable savings over the cost of new parts.

7 Hinges and locks – maintenance

Once every 3000 miles, or every three months, the hinges and latch assemblies on the doors, hood and trunk should be given a few drops of light oil or lock lubricant. The door latch strikers should also be lubricated with a thin coat of grease to reduce wear and ensure free movement. Lubricate the door and trunk locks with spray-on graphite lubricant.

8 Fixed glass – replacement

Replacement of the windshield and fixed glass requires the use of special fast-setting adhesive/caulk materials and some specialized tools. It is recommended that these operations be left to a dealer or a shop specializing in glass work.

9 Hood – removal, installation and adjustment

Refer to illustrations 9.2, 9.7, 9.10a, 9.10b and 9.11

Note: *The hood is heavy and somewhat awkward to remove and install – at least two people should perform this procedure.*

Removal and installation

1 Use blankets or pads to cover the cowl area of the body and fenders. This will protect the body and paint as the hood is lifted free.

2 Use a scribe or a permanent marker to make alignment marks around the bolt heads to ensure proper alignment upon installation **(see illustration)**.

3 Disconnect any cables or electrical connectors which will interfere with removal.

4 Have an assistant support the weight of the hood. Remove the hinge-to-hood screws or bolts.

5 Lift off the hood.

6 Installation is the reverse of removal.

Adjustment

7 Front-and-rear and side-to-side adjustment of the hood is made by moving the hinge plate slot after loosening the bolts or nuts **(see illustration)**.

8 Mark a line around the entire hinge plate so you can judge the amount of movement **(see illustration 9.2 or 9.7)**.

9 Loosen the bolts and move the hood into correct alignment. Move the hood only a little at a time. Tighten the hinge bolts and carefully lower the hood to check the position.

10 If necessary after installation, the entire hood latch assembly can be adjusted up-and-down as well as side-to-side on the firewall (front hinged hood) or radiator brace (rear hinged hood) so the hood closes securely, with the hood flush with the fenders. To perform this adjustment, mark

Chapter 11 Body

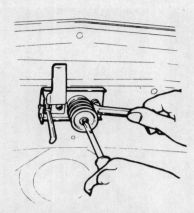

9.10b Use a screwdriver and a wrench to screw the hood latch hook in-or-out (not all models)

9.11 Adjust the hood vertically using the hood bumpers (arrow)

around the hood latch mounting screws to provide a reference point, then loosen them and reposition the latch assembly, as necessary **(see illustration)**. Following adjustment, retighten the mounting bolts. On some models the hood can be further adjusted up-and-down by screwing the hood latch hook assembly in or out **(see illustration)**.

11 Finally, adjust the hood bumpers so that the hood, when closed, is flush with the fenders **(see illustration)**.

12 The hood latch assembly, as well as the hinges, should be periodically lubricated with white lithium base grease to prevent sticking or jamming.

10 Bumpers – removal and installation

Refer to illustrations 10.1a and 10.1b
Note: *This procedure applies to either the front or rear bumper.*

1 Remove the bumper cover-to-body bolts **(see illustrations)**.
2 Disconnect any wiring or other components which would interfere with removal of the bumper.

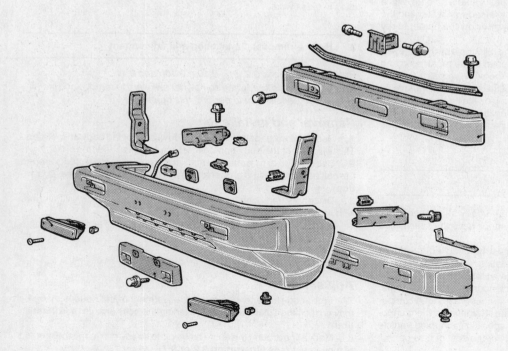

10.1a Typical front bumper details

1 Air dam side extension
2 Air dam center extension
3 Side stay
4 Bumper cover
5 Upper bracket
6 Lower bracket
7 Back beam
8 Upper plate
9 Bumper core
10 License support bracket
11 License plate bracket
12 Bumper stay

Chapter 11 Body

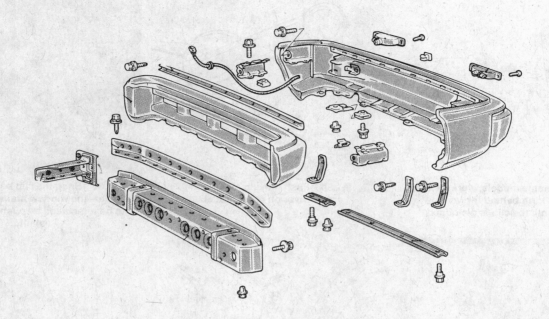

10.1b Typical rear bumper details

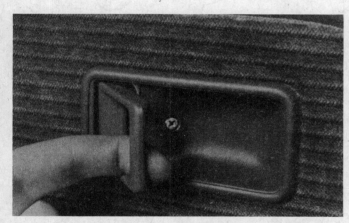

11.2a On some models, the door handle screw can be reached by pulling out the handle

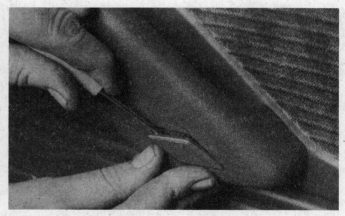

11.2b Door panel screws are often hidden under covers which can be pried off using a small screwdriver

3   Support the bumper's weight with a jack or jackstand. Alternatively, have an assistant support the bumper.
4   Remove the retaining bolts and detach the bumper.
5   Installation is the reverse of removal.
6   Tighten the retaining bolts securely.
7   Install the bumper cover bolts and any other components which were removed.

11 Door trim panel – removal and installation

Refer to illustrations 11.2a, 11.2b, 11.3a, 11.3b, 11.3c and 11.6
1   Disconnect the negative cable from the battery.
2   Remove all door trim panel retaining screws and door/armrest assemblies **(see illustrations)**.

3   On manual window regulator equipped models, remove the crank by working a cloth back and forth behind the handle to dislodge the clip or push it off with a screwdriver **(see illustrations)**. On power regulator models, pry out the control switch assembly and unplug it.
4   Insert a large putty knife or a thin pry bar between the trim panel and the door and disengage the retaining clips, working around the outer edge until the panel is free of the door.
5   Once all of the clips are disengaged, unscrew the lock knob, detach the trim panel, lift it away, unplug any electrical connectors and remove the trim panel from the vehicle.
6   For access to the inner door, peel back the plastic service hole cover, taking care not to tear it **(see illustration)**. To install, place the service hole cover in position and press it in place.
7   Prior to installation of the door panel, make sure to reinstall any clips in the panel which may have come out during the removal procedure and remain in the door itself.

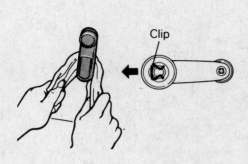
11.3a On some models, work a cloth back and forth behind the window handle until the clip is dislodged

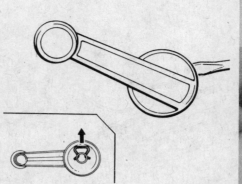
11.3b Push the clip off with a screwdriver on other models

11.3c Snap the clip back into the groove – the window crank is now ready to be reinstalled by pushing it on the shaft

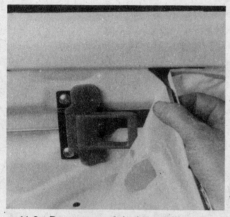
11.6 Be very careful when pulling the plastic service hole cover off – don't tear or distort it

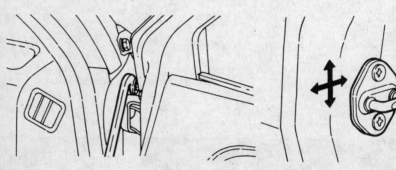
12.5a A cranked wrench such as this one may be required to reach the hinge-to-body bolts when adjusting the doors

12.5b Adjust the door lock striker by loosening the screws and tapping the striker in the desired direction with a plastic mallet

8 Plug in any electrical connectors and place the panel in position in the door. Press the door panel into place until the clips are seated and install any retaining screws and armrest/door pulls. Install the manual regulator window crank.

12 Door – removal, installation and adjustment

Refer to illustrations 12.5a and 12.5b

1 Remove the door trim panel. Disconnect any electrical connectors and push them through the door opening so they won't interfere with door removal.
2 Place a jack or stand under the door or have an assistant on hand to support it when the hinge bolts are removed. **Note:** *If a jack or stand is being used, place a rag between it and the door to protect the door's painted surfaces.*
3 Mark around the hinge-to-door bolts so you can return the door to its proper adjustment when installing it.
4 Remove the hinge-to-door bolts and carefully lift off the door. Installation is the reverse of removal.
5 Following installation of the door, check that it is in proper alignment and adjust it, if necessary, as follows:
   a) Up-and-down and forward-and-backward adjustments are made by loosening the hinge-to-body bolts and moving the door, as necessary. A special cranked tool may be required to reach some of the bolts **(see illustration)**.
   b) The door lock striker can also be adjusted both up-and-down and sideways to provide a positive engagement with the locking mechanism. This is done by loosening the retaining screws and moving the striker, as necessary **(see illustration)**.

13 Trunk lid – removal, installation and adjustment

Refer to illustrations 13.3, 13.7a and 13.7b

1 Open the trunk lid and cover the edges of the trunk compartment with pads or cloths to protect the painted surfaces when the lid is removed.
2 Disconnect any cable or electrical connectors which are attached to the trunk lid and would interfere with removal.
3 Scribe or paint alignment marks around the hinge bolt mounting flanges **(see illustration)**.
4 While an assistant supports the lid, remove the lid-to-hinge bolts on both sides and lift off the lid **(see illustration 13.3)**.
5 Installation is the reverse of removal. **Note:** *When reinstalling the trunk lid, align the lid-to-hinge bolts with the marks made during removal.*
6 After installation, close the lid and check that it is in proper alignment with the surrounding panels. Front and rear and side-to-side adjustments of the lid are controlled by the position of the lid-to-hinge bolts in their slots. To adjust, loosen the lid-to-hinge bolts, reposition the lid and retighten the bolts **(see illustration 13.3, if necessary)**.
7 The height of the lid in relation to the surrounding body panels when

Chapter 11 Body

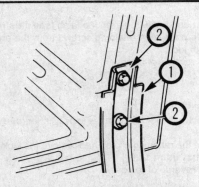

13.3 Scribe or paint around the hinge bolt mounting flange (1) so you can re-install the trunk lid to its proper location – unscrew or loosen the lid-to-hinge bolts (2) to remove or adjust the trunk lid

13.7a After loosening the bolts (arrows), you can move the striker to adjust the trunk lid position

closed can be adjusted by loosening the striker bolts, repositioning the striker and tightening the bolts **(see illustration)**. Some models also are equipped with rubber bumpers which can be screwed in or out for a final adjustment the trunk lid so that it is even with the surrounding body **(see illustration)**.

14 Liftgate – removal, installation and adjustment

Refer to illustration 14.3

1 Open the liftgate and cover the upper body area around the opening with pads or cloths to protect the painted surfaces when the liftgate is removed.
2 Disconnect any cables, hoses or electrical connectors which would interfere with removal of the liftgate. Detach the headliner for access to the tailgate hinge retaining nuts.
3 While an assistant supports the liftgate, detach the support struts **(see illustration)**.

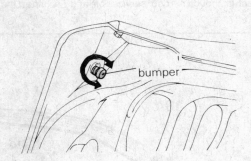

13.7b Some models have adjustable bumpers at each corner of the trunk lid

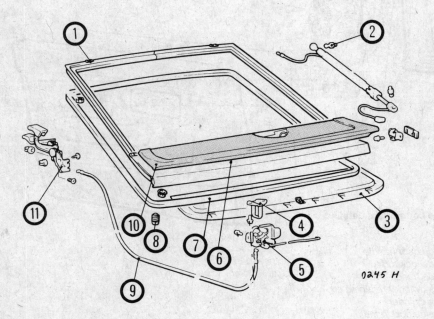

14.3 Typical liftgate details – note the rubber bumpers that can be screwed in or out to raise or lower the rear of the liftgate in relation to the surrounding panels

1 Tailgate hinge
2 Ball joint
3 Tailgate trim
4 Tailgate striker
5 Tailgate latch
6 Tailgate
7 Tailgate weatherstrip
8 Rubber bumper
9 Tailgate lock lelease cable
10 Air spoiler (vehicles with a turbocharger)
11 Release handle

Chapter 11 Body

4 Remove the hinge nuts and detach the liftgate from the vehicle.
5 Installation is the reverse of removal.
6 After installation, close the liftgate and check that the door is in proper alignment with the surrounding panels. Adjustments to the liftgate are made by moving the position of the hinge bolts in their slots. To adjust, loosen the hinge nuts and reposition the hinge either side-to-side or front-to-rear the desired amount and retighten the nuts (**see illustration 14.3, if necessary**).
7 The engagement of the liftgate can be adjusted by loosening the latch mounting bolts and/or the lock striker screws, repositioning the latch and/or striker and tightening the screws.

15 Door latch, lock cylinder and handle – removal and installation

1 Remove the door trim panel and service hole cover (Section 12).
2 Remove the door window glass (Section 16).

Door latch
Refer to illustrations 15.3a and 15.3b
3 Reach in through the door service hole and disconnect the door safety rod and the door inside handle rod from the latch (**see illustrations**).
4 Remove the three door latch retaining screws from the end of the door. Remove the door lock.
5 Installation is the reverse of removal.

Lock cylinder
Refer to illustration 15.6
6 Disconnect the lock cylinder rod from the lock cylinder (**see illustration**).
7 Use pliers to slide the retaining clip off and remove the lock cylinder from the door (**see illustration 15.6**).
8 Installation is the reverse of removal.

Inside handle
Refer to illustrations 15.11a and 15.11b
9 Remove the retaining screw(s) (**see illustrations 15.3a and 15.3b**).
10 Rotate the handle away and detach it from the door.
11 Installation is the reverse of removal. Before tightening the retaining screws on 1988 and 1989 Precis models, adjust the position of the handle assembly so the handle stroke travel is approximately 48-degrees (**see illustration**). On all other models, after installation, the handle should

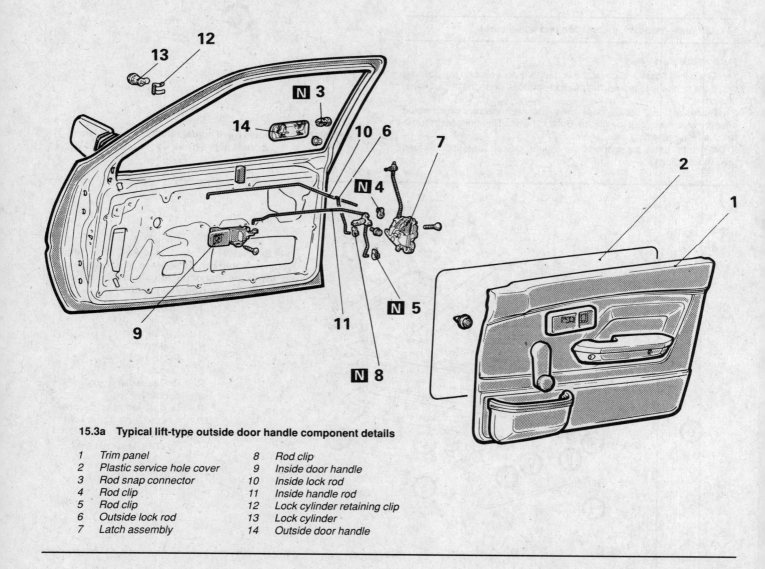

15.3a Typical lift-type outside door handle component details

1	Trim panel	8	Rod clip
2	Plastic service hole cover	9	Inside door handle
3	Rod snap connector	10	Inside lock rod
4	Rod clip	11	Inside handle rod
5	Rod clip	12	Lock cylinder retaining clip
6	Outside lock rod	13	Lock cylinder
7	Latch assembly	14	Outside door handle

Chapter 11 Body

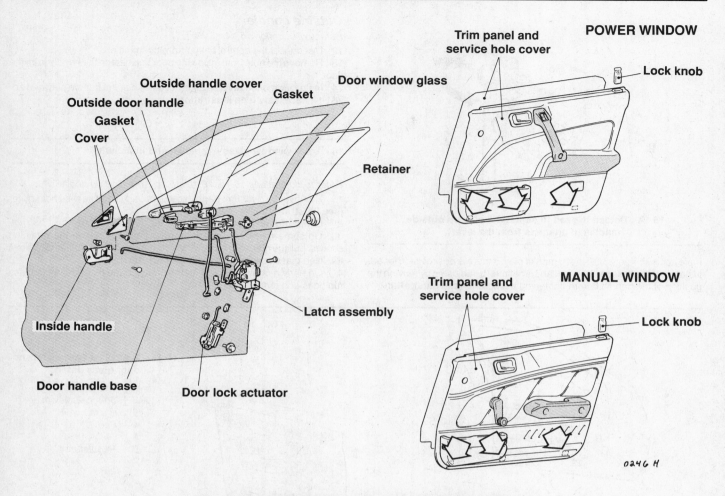

15.3b Typical pull-type outside door handle component details

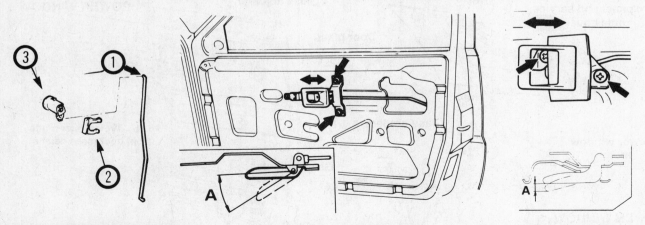

15.6 Disconnect the lock cylinder rod (1) from the lock cylinder, then use pliers to slide the retaining clip (2) off the lock cylinder (3)

15.11a On 1988 and 1989 Precis models, adjust the position of the handle assembly so the handle has a full stroke of 48-degrees (dimension A) before tightening the screws

15.11b Loosen the screws and move the handle, if necessary, to adjust the play

Chapter 11 Body

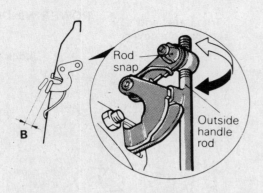

15.14 Detach the rod to adjust the play on outside handles (it unsnaps from the lever)

have approximately 3/32-in (10 mm) of play. If there is no provision for adjustment of the linkage, the play can sometimes be adjusted by moving the position of the handle before tightening the screws **(see illustration)**.

Outside handle

Refer to illustration 15.14

12 Disconnect the control link(s) from the handle.
13 Remove the nuts from inside the door and detach the handle from the door.
14 Installation is the reverse of removal. Check the handle play and adjust it if necessary **(see illustration)**.

16 Door window glass – removal and installation

Refer to illustrations 16.1a, 16.1b, 16.1c, 16.5a, 16.5b and 16.7

1 Remove the door trim panel (see Section 11) and service hole cover **(see illustrations)**.
2 Lower the window glass.
3 On some rear doors, it will be necessary to remove the retaining screws, pull the glass run and division channel out of the door and remove the fixed glass assembly **(see illustration 16.1c)**.
4 On early models, scribe or paint marks on the two window-to-glass run bolts and remove them.

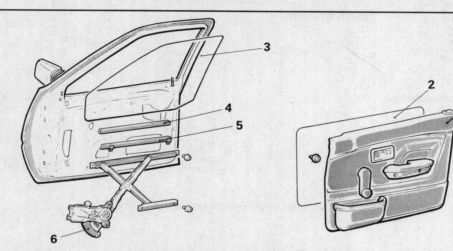

16.1a Typical early model front door glass details

1 Trim panel
2 Service hole cover
3 Glass
4 Pad
5 Glass run
6 Regulator

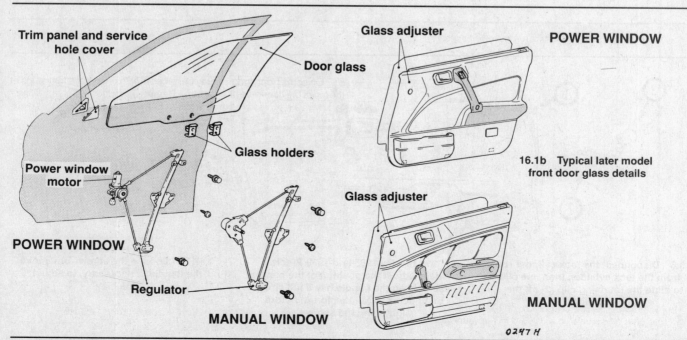

16.1b Typical later model front door glass details

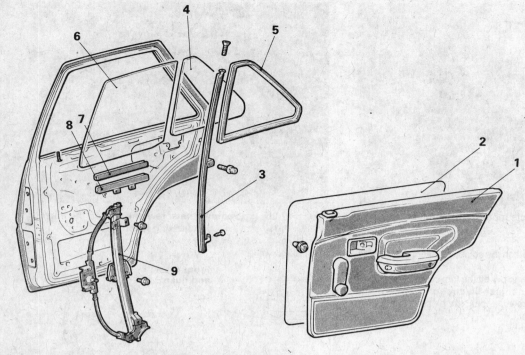

16.1c Typical later model rear door glass details

1 Trim panel
2 Service hole cover
3 Division channel
4 Fixed Glass
5 Weatherstrip
6 Door glass
7 Pad
8 Glass run
9 Regulator

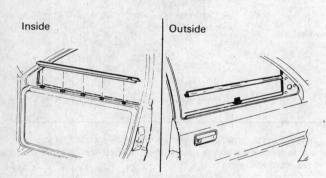

16.5a On some front doors it will be necessary to pry out the trim strips before removing the glass

16.5b Tilt the glass as shown when lifting it out of the opening

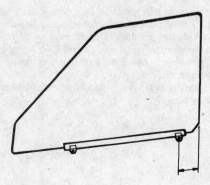

16.7 Make sure to measure the rear edge-to-glass run hole distance before removing the glass from the run so the new glass can be installed in the same position

5   On some front doors, it may be necessary to pry out the trim strips at the top of the door (see illustration). Remove the window glass by tilting it to detach the glass from the glass channel studs and then sliding the glass up and out of the door (see illustration).
6   If necessary, remove the regulator retaining bolts and then slide the regulator out of the opening in the door.
7   Installation is the reverse of removal. If the glass is to be removed from the glass run, make sure to measure from the rear edge of the glass to the run bolt hole so the new glass can be installed in the same position (see illustration).
8   If it is necessary to adjust the glass position, loosen the mounting screws, then move the glass to the desired position, then tighten the screws.

17 Outside mirror – removal and installation

Refer to illustrations 17.2a and 17.2b

1   Detach the battery negative cable (power mirror models).
2   On 1989 and earlier Precis models, pry off the adjusting lever screw

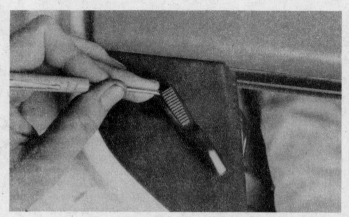

17.2a On 1989 and earlier Precis models, pry off the cover . . .

17.2b . . . remove the screw and detach the mirror adjusting lever

cover, then remove the screw and detach the adjusting lever (see illustrations).
3   On all models, use a screwdriver to pry off the trim cover on the inside of the door, directly opposite the mirror (see illustration).
4   Remove the three retaining screws, unplug the electrical connector (power models) and lift off the mirror.
5   Installation is the reverse of removal.

18 Front fender inner panels – removal and installation

Refer to illustration 18.2
1   Raise the vehicle, support it securely and remove the front wheel(s).
2   Remove the bolts and lower the panel(s) from the fender well (see illustration).
3   Installation is the reverse of removal.

19 Automatic shoulder harnesses – general information

Many late model vehicles are equipped with automatic front seat shoulder harnesses. They are termed automatic because you don't have to buckle them – the shoulder harness automatically positions itself when the door is closed and the key is turned on. An emergency release lever allows the harness to be manually removed for exit in an emergency.
Warning: *Be sure to fasten the manual seatbelt as well. The automatic shoulder harness will not work properly unless the seatbelt is fastened.*
Most systems have a warning light and buzzer that indicate the emergency release lever has been pulled up, releasing the shoulder harness. Make sure the release lever is down and the light/buzzer are off to ensure proper operation of the automatic shoulder harness. Also, if you disconnect any wires or remove any automatic shoulder harness components when performing repair procedures on other vehicle components, be sure to reinstall everything and check the harness for proper operation when the repairs are complete.
Since the automatic shoulder harness is operated by several electrical switches and is computer controlled, diagnosis and repair must be done by a dealer service department. Do not jeopardize the safety of front seat occupants – if the automatic shoulder harness malfunctions, or you have questions regarding the proper use or operation of the system, contact a dealer service department.

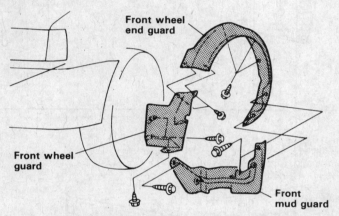

18.2 Typical front fender inner panel details

20 Seat belt check

1   Check the seat belts, buckles, latch plates and guide loops for any obvious damage or signs of wear.
2   Check that the seat belt reminder light comes on when the key is turned to the On or Start positions. A chime should also sound.
3   The seat belts are designed to lock up during a sudden stop or impact, yet allow free movement during normal driving. Check that the retractors return the belt against your chest while driving and rewind the belt fully when the buckle is unlatched.
4   If any of the above checks reveal problems with the seat belt system, replace parts as necessary.

Chapter 12 Chassis electrical system

Contents

Battery check and maintenance	See Chapter 1
Battery – removal and installation	See Chapter 5
Bulb replacement	11
Brake light switch – removal, installation and adjustment	See Chapter 9
Combination switch – removal and installation	8
Cruise control system – description and checking	14
Electrical troubleshooting – general information	2
Fuses – general information	3
Fusible links – general information	4
General information	1
Headlights – adjustment	10
Headlights – removal and installation	9
Ignition switch – removal and installation	7
Instrument cluster – removal and installation	13
Neutral start switch – removal, installation and adjustment	See Chapter 7B
Power door lock system – description and check	16
Power window system – description and checking	15
Relays – general information	5
Speedometer cable – removal and installation	12
Turn signal/hazard flashers – check and replacement	6
Wiring diagrams – general information	17

1 General information

The electrical system is a 12-volt, negative ground type. Power for the lights and all electrical accessories is supplied by a lead/acid-type battery which is charged by the alternator.

This Chapter covers repair and service procedures for the various electrical components not associated with the engine. Information on the battery, alternator, distributor and starter motor can be found in Chapter 5.

It should be noted that when portions of the electrical system are serviced, the cable should be disconnected from the negative battery terminal to prevent electrical shorts and/or fires.

2 Electrical troubleshooting – general information

A typical electrical circuit consists of an electrical component, any switches, relays, motors, fuses, fusible links or circuit breakers related to that component and the wiring and connectors that link the component to both the battery and the chassis. To help you pinpoint an electrical circuit problem, wiring diagrams are included at the end of this book.

Before tackling any troublesome electrical circuit, first study the appropriate wiring diagrams to get a complete understanding of what makes up that individual circuit. Trouble spots, for instance, can often be narrowed down by noting if other components related to the circuit are operating properly. If several components or circuits fail at one time, chances are the problem is in a fuse or ground connection, because several circuits are often routed through the same fuse and ground connections.

Electrical problems usually stem from simple causes, such as loose or corroded connections, a blown fuse, a melted fusible link or a bad relay. Visually inspect the condition of all fuses, wires and connections in a problem circuit before troubleshooting it.

If testing instruments are going to be utilized, use the diagrams to plan ahead of time where you will make the necessary connections in order to accurately pinpoint the trouble spot.

The basic tools needed for electrical troubleshooting include a circuit tester or voltmeter (a 12-volt bulb with a set of test leads can also be used), a continuity tester, which includes a bulb, battery and set of test leads, and a jumper wire, preferably with a circuit breaker incorporated, which can be used to bypass electrical components. Before attempting to locate a problem with test instruments, use the wiring diagram(s) to decide where to make the connections.

Chapter 12 Chassis electrical system

3.1a The fuse block is located at the left side of the dash – some are under a hinged cover, as shown here

3.1b Later models have a fuse block in the engine compartment next to the battery

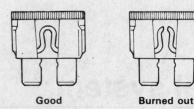

3.3 The fuses used in these models can be easily checked visually to determine if they are blown

Voltage checks

Voltage checks should be performed if a circuit is not functioning properly. Connect one lead of a circuit tester to either the negative battery terminal or a known good ground. Connect the other lead to a connector in the circuit being tested, preferably nearest to the battery or fuse. If the bulb of the tester lights, voltage is present, which means that the part of the circuit between the connector and the battery is problem free. Continue checking the rest of the circuit in the same fashion. When you reach a point at which no voltage is present, the problem lies between that point and the last test point with voltage. Most of the time the problem can be traced to a loose connection. **Note:** *Keep in mind that some circuits receive voltage only when the ignition key is in the Accessory or Run position.*

Finding a short

One method of finding shorts in a circuit is to remove the fuse and connect a test light or voltmeter in its place to the fuse terminals. There should be no voltage present in the circuit. Move the wiring harness from side to side while watching the test light. If the bulb goes on, there is a short to ground somewhere in that area, probably where the insulation has rubbed through. The same test can be performed on each component in the circuit, even a switch.

Ground check

Perform a ground test to check whether a component is properly grounded. Disconnect the battery and connect one lead of a self-powered test light, known as a continuity tester, to a known good ground. Connect the other lead to the wire or ground connection being tested. If the bulb goes on, the ground is good. If the bulb does not go on, the ground is not good.

Continuity check

A continuity check is done to determine if there are any breaks in a circuit – if it is passing electricity properly. With the circuit off (no power in the circuit), a self-powered continuity tester can be used to check the circuit. Connect the test leads to both ends of the circuit (or to the "power" end and a good ground), and if the test light comes on the circuit is passing current properly. If the light doesn't come on, there is a break somewhere in the circuit. The same procedure can be used to test a switch, by connecting the continuity tester to the power in and power out sides of the switch. With the switch turned On, the test light should come on.

Finding an open circuit

When diagnosing for possible open circuits, it is often difficult to locate them by sight because oxidation or terminal misalignment are hidden by the connectors. Merely wiggling a connector on a sensor or in the wiring harness may correct the open circuit condition. Remember this when an open circuit is indicated when troubleshooting a circuit. Intermittent problems may also be caused by oxidized or loose connections.

Electrical troubleshooting is simple if you keep in mind that all electrical circuits are basically electricity running from the battery, through the wires, switches, relays, fuses and fusible links to each electrical component (light bulb, motor, etc.) and to ground, from which it is passed back to the battery. Any electrical problem is an interruption in the flow of electricity to and from the battery.

3 Fuses – general information

Refer to illustrations 3.1a, 3.1b and 3.3

The electrical circuits of the vehicle are protected by a combination of fuses, circuit breakers and fusible links. The fuse block is located on the left side of the dashboard **(see illustration)**. Later models also have a fuse block in the engine compartment, adjacent to the battery **(see illustration)**.

Each of the fuses is designed to protect a specific circuit, and the various circuits are identified on the fuse panel itself.

Miniaturized fuses are employed in the fuse block. These compact fuses, with blade terminal design, allow fingertip removal and replacement. If an electrical component fails, always check the fuse first. A blown fuse is easily identified through the clear plastic body. Visually inspect the element for evidence of damage **(see illustration)**. If a continuity check is called for, the blade terminal tips are exposed in the fuse body.

Be sure to replace blown fuses with the correct type. Fuses of different ratings are physically interchangeable, but only fuses of the proper rating should be used. Replacing a fuse with one of a higher or lower value than specified is not recommended. Each electrical circuit needs a specific amount of protection. The amperage value of each fuse is molded into the fuse body.

If the replacement fuse immediately fails, don't replace it again until the cause of the problem is isolated and corrected. In most cases, this will be a short circuit in the wiring caused by a broken or deteriorated wire.

Chapter 12 Chassis electrical system

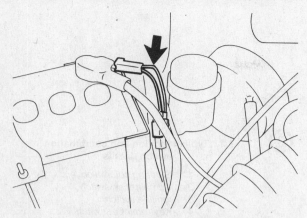

4.3a On earlier models the main fusible link (arrow) is located on the battery cable

4.3b On later models, unplug the fusible link and replace it with a new one – always determine the cause of an overload before replacement

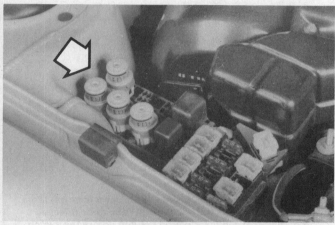

5.2 On most later models, relays for the air conditioner, tail lights and various other components are located next to the fuse block in the engine compartment

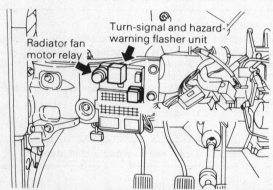

6.1 The turn signal and hazard flasher unit is located in the fuse block on most models

4 Fusible links – general information

Refer to illustrations 4.3a and 4.3b

Some circuits are protected by fusible links. The links are used in circuits which are not ordinarily fused, such as the ignition circuit.

The fusible links on later models are similar to fuses in that they can be visually checked to determine if they are melted. On earlier models, check the fusible links with an ohmmeter to verify that they aren't burned out. **Caution:** *Be sure the battery is disconnected first or the ohmmeter could be damaged.*

To replace a fusible link, first disconnect the negative cable from the battery. On earlier models, unplug the burned out link and replace it with a new one (available from your dealer) **(see illustration)**. On later models the fusible link can be simply unplugged and replaced with a new one much like a fuse **(see illustration)**. Always determine the cause for the overload which melted the fusible link before installing a new one.

5 Relays – general information

Refer to illustration 5.2

Several electrical accessories in the vehicle use relays to transmit the electrical signal to the component. If the relay is defective, the component will not operate properly.

The various relays are grouped together in several locations under the dash, adjacent to the component, or in the engine compartment for convenience in the event of needed replacement **(see illustration)**.

If a faulty relay is suspected, it can be removed and tested by a dealer service department or a repair shop. Defective relays must be replaced as a unit.

6 Turn signal/hazard flashers – check and replacement

Refer to illustration 6.1

1 The turn signal/hazard flasher, a small canister shaped unit located under the dash in the wiring harness or fuse block, flashes the turn signals and hazard flashers **(see illustration)**.

2 When the flasher unit is functioning properly, and audible click can be heard during its operation. If the turn signals fail on one side or the other and the flasher unit does not make its characteristic clicking sound, a faulty turn signal bulb is indicated.

3 If both turn signals fail to blink, the problem may be due to a blown fuse, a faulty flasher unit, a broken switch or a loose or open connection. If a quick check of the fuse box indicates that the turn signal fuse has blown, check the wiring for a short before installing a new fuse.

4 To replace the flasher, simply pull it out of the fuse block or wiring harness.

5 Make sure that the replacement unit is identical to the original. Compare the old one to the new one before installing it.

6 Installation is the reverse of removal.

Chapter 12 Chassis electrical system

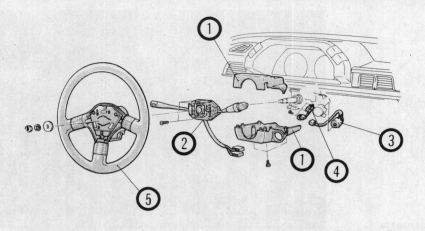

7.2 Typical ignition and combination switch details

1. Steering column covers
2. Combination switch
3. Ignition switch
4. Key reminder switch
5. Steering wheel

7 Ignition switch – removal and installation

Refer to illustration 7.2

1. Disconnect the negative cable from the battery.
2. Remove the steering column covers **(see illustration)**.
3. Detach the electrical harness cable tie-straps.
4. Unplug the switch electrical connector.
5. Remove the screw and lower the ignition switch and harness assembly from the steering column.
6. Installation is the reverse of removal.

8 Combination switch – removal and installation

Refer to illustration 8.4

1. Disconnect the negative cable from the battery.
2. Remove the steering wheel (Chapter 10).
3. Remove the under cover panel and steering column cover.
4. Remove the retaining screws **(see the accompanying illustration and illustration 7.2)**.
5. Trace the wiring harness down the steering column to the connector. Cut the wiring tie-straps (if equipped), unplug the connector and slide the switch up off the column.
6. Installation is the reverse of removal.

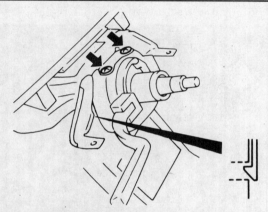

8.4 On earlier Mirage models, the screws (arrows) are located at the top and the clip will have to be released from the recess in the steering column before the switch can be removed

9 Headlights – removal and installation

1. Disconnect the negative cable from the battery.

Sealed-beam type

Refer to illustration 9.3

2. On some models it may be necessary to remove the radiator grille.
3. Unscrew and remove the headlight bezel. Remove the headlight retainer screws, taking care not to disturb the adjustment screws **(see illustration)**.
4. Remove the mounting ring and pull the headlight out far enough to unplug the connector.
5. Remove the headlight.
6. To install the headlight, plug the connector in, place the headlight in position, then install the retainer and screws. Tighten the screws securely.
7. Install the radiator grille, if removed.

Bulb-type

Refer to illustrations 9.8, 9.9 and 9.10
Warning: *The halogen gas filled bulbs used on these models are under pressure and may shatter if the surface is scratched or the bulb is dropped.*

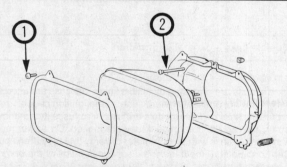

9.3 To remove a sealed beam headlight, unscrew the retainer screws (1) – to adjust the headlight vertically, loosen or tighten the adjustment screw (2) (the horizontal adjustment screw is on the side of the headlight)

Wear eye protection and handle the bulbs carefully, grasping only the base whenever possible. Do not touch the surface of the bulb with your fingers because the oil from your skin could cause it to overheat and fail prematurely. If you do touch the bulb surface, clean it with rubbing alcohol.

8. Open the hood. Unplug the electrical connector **(see illustration)**.
9. Reach behind the headlight assembly, grasp the retainer and turn it counterclockwise to remove it **(see illustration)**.

Chapter 12 Chassis electrical system

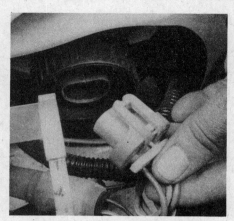

9.8 Pry up on the locking tab and unplug the connector from the bulb assembly

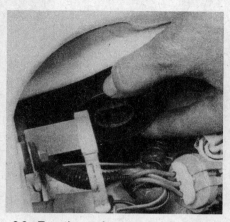

9.9 Turn the retainer counterclockwise to remove it, ...

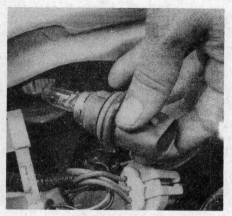

9.10 ... then pull the bulb straight out

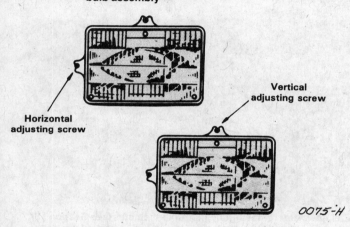

10.6a Typical sealed beam adjusting screw locations

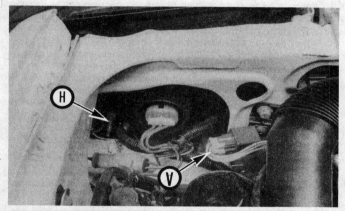

10.6b On 1989 and earlier Precis bulb-type headlights, the upper adjusting screw (H) moves the beam horizontally and lower screw (V) moves it vertically

10 Pull the bulb straight out to remove it **(see illustration)**.
11 Install the new bulb in the headlight assembly.
12 Install the retainer and turn it clockwise to lock it in place.
13 Plug in the electrical connector.

10 Headlights – adjustment

Refer to illustrations 10.6a, 10.6b and 10.6c

Note: *It is very important to have the headlights adjusted correctly. If adjusted incorrectly they could blind the driver of an oncoming vehicle and cause a serious accident or seriously reduce your ability to see the road. The headlights should be checked for proper aim every 12 months and any time a new headlight is installed or front end body work is performed. It should be emphasized that the following procedure is only an interim step which will provide temporary adjustment until the headlights can be adjusted by a properly equipped shop.*

1 Headlights have two adjusting screws, one on the top controlling up and down movement and one on the side controlling left and right movement.
2 There are several methods of adjusting the headlights. The simplest method requires a blank wall 25 feet in front of the vehicle and a level surface in front of the wall.
3 Attach masking tape vertically to the wall in reference to the vehicle centerline and the centerlines of both headlights.
4 Position a horizontal tape line in reference to the centerline of all the

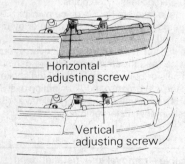

10.6c Typical Mirage bulb-type headlight adjusting screw locations

headlights. **Note:** *It may be easier to position the tape on the wall with the vehicle parked only a few inches away.*

5 Adjustment should be made with the vehicle sitting level, the gas tank half-full and no unusually heavy load in the trunk or cargo area.
6 Starting with the low beam adjustment, position the high intensity zone so it is two inches below the horizontal line and two inches to the right of the headlight vertical line. Adjustment on sealed beam equipped models is made by turning the top adjusting screw clockwise to raise the beam and counterclockwise to lower the beam **(see illustration)**. The adjusting screw on the side should be used in the same manner to move the beam

Chapter 12 Chassis electrical system

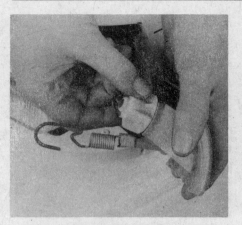

11.3a Some bulbs are accessible after rotating the holder out of the housing (turn it counterclockwise)

11.3b On some bulbs, like the one on this side marker light, simply pull the holder straight out for access to the bulb

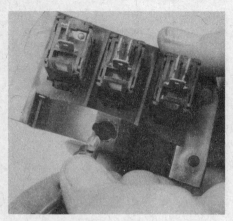

11.4a On this switch housing, turn the holder counterclockwise to remove it for access to the bulb

left or right. On bulb-type headlights, the adjusting screws can be found in a variety of locations **(see illustrations)**. On Cordia/Tredia models, the adjusting screws are located near the upper right and lower left-hand corners of the headlight assembly. On Galant models, the vertical adjusting screw is mounted vertically at the upper right corner, and the horizontal adjusting screw is mounted horizontally at the upper left corner.

7 With the high beams on, the high intensity zone should be vertically centered with the exact center just below the horizontal line. **Note:** *It may not be possible to position the headlight aim exactly for both high and low beams. If a compromise must be made, keep in mind that the low beams are the most used and have the greatest effect on driver safety.*

8 Have the headlights adjusted by a dealer service department or service station at the earliest opportunity.

11 Bulb replacement

Refer to illustrations 11.3a, 11.3b, 11.4a and 11.4b

1 The lenses of many lights are held in place by screws, which makes it a simple procedure to gain access to the bulbs.

2 On some lights, the lenses are held in place by clips. On these, the lenses can either be removed by unsnapping them or by using a small screwdriver to pry them off.

3 Several types of bulbs are used. Some are removed by pushing in and turning counterclockwise **(see illustrations)**. Others can simply be unclipped from the terminals or pulled straight out of the socket.

4 To gain access to the instrument panel illumination or switch housing lights **(see illustrations)**, the cluster or housing will have to be removed. See Section 13 for instrument cluster removal.

12 Speedometer cable – removal and installation

Refer to illustration 12.4

1 Disconnect the negative cable from the battery.
2 Unscrew the collar and disconnect the speedometer cable from the transaxle.
3 Detach the cable from any routing clips in the engine compartment and pull it up if necessary to provide enough slack to allow disconnection from the speedometer.
4 Remove the instrument cluster screws, pull the cluster out, then disconnect the speedometer cable **(see illustration)**.
5 Detach the firewall grommet, pull the cable through into the engine compartment, then remove the it from the vehicle.
6 Prior to installation, lubricate the speedometer end of the cable with spray-on speedometer cable lubricant (available at auto parts stores).

11.4b After removing the instrument cluster (see Section 13), turn the bulb holder, lift them out and then withdraw the bulbs

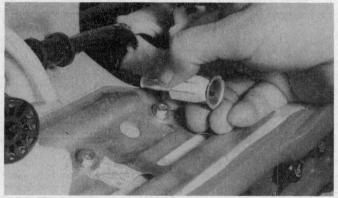

12.4 Squeeze the collar to disconnect the cable from the speedometer (cluster removed for clarity)

7 Installation is the reverse or removal. After installation, make sure the cable has a gentle bend where it enters the firewall grommet, or if there are any markings they are visible on the engine compartment side of the grommet or the cable could be kinked.

13 Instrument cluster – removal and installation

1 Disconnect the negative cable from the battery.
2 Remove the steering wheel (Chapter 10).

Chapter 12 Chassis electrical system

3   Remove the steering column cover, bezel and any other components which would interfere with removal.
4   Remove the retaining screws, then pull out the instrument cluster and disconnect the speedometer cable (see Section 12) and electrical connectors
5   To replace any faulty components within the cluster, simply unbolt or unscrew them and replace them with new units.
6   Installation is the reverse of removal.

14 Cruise control system – description and checking

The cruise control system maintains vehicle speed through a vacuum actuated servo motor located in the engine compartment, which is connected to the throttle linkage by a cable. The system consists of the servo motor, clutch switch, brake light switch, control switches, a relay and associated vacuum hoses.

Because of the complexity of the cruise control system and the special tools and techniques required for diagnosis and repair, it should be left to a dealer service department or properly equipped shop. However, it is possible for the home mechanic to make simple checks of the wiring and vacuum connections for minor faults which can be easily repaired. These include:

a) Inspecting the cruise control actuating switches and wiring for broken wires or loose connections.
b) Checking the cruise control fuse.
c) Checking the hoses in the engine compartment for tight connections, cracked hoses and obvious vacuum leaks. The cruise control system is operated by vacuum so it is critical that all vacuum switches, hoses and connections be secure.

15 Power window system – description and check

The power window system operates the electric motors mounted in the doors, which lower and raise the windows. The system consists of the control switches, the motors (regulators), glass mechanisms and associated wiring.

Because of the complexity of the power window system and the special tools and techniques required for diagnosis and repair, it should be left to a dealer service department or properly equipped shop. However, it is possible for the home mechanic to make simple checks of the wiring connections and motors for minor faults which can be easily repaired. These include:

a) Inspecting the power window actuating switches and wiring for broken wires or loose connections.
b) Checking the power window fuse and/or circuit breaker.
c) Removing the door panel(s) and checking the power window motor wiring connections for looseness and damage, and inspecting the glass mechanisms for damage which could cause binding.

16 Power door lock system – description and check

The power door lock system operates the door lock actuators mounted in each door. The system consists of the switches, actuators and associated wiring. Since special tools and techniques are required to diagnose the system, it should be left to a dealer service department or a repair shop. However, it is possible for the home mechanic to make simple checks of the wiring connections and actuators for minor faults which can be easily repaired. These include:

a) Check the system fuse and/or circuit breaker.
b) Check the switch wires for damage and loose connections. Check the switches for continuity.
c) Remove the door panel(s) and check the actuator wiring connections to see if they're loose or damaged. Inspect the actuator rods (if equipped) to make sure they aren't bent or damaged. Inspect the actuator wiring for damaged or loose connections. The actuator can be checked by applying battery power momentarily. A discernible click indicates that the solenoid is operating properly.

17 Wiring diagrams – general information

Since it isn't possible to include all wiring diagrams for every year covered by this manual, the following diagrams are those that are typical and most commonly needed.

Prior to troubleshooting any circuits, check the fuse and circuit breakers (if equipped) to make sure they are in good condition. Make sure the battery is properly charged and has clean, tight cable connections (Chapter 1).

When checking the wiring system, make sure that all connectors are clean, with no broken or loose pins. When unplugging a connector, do not pull on the wires, only on the connector housings themselves.

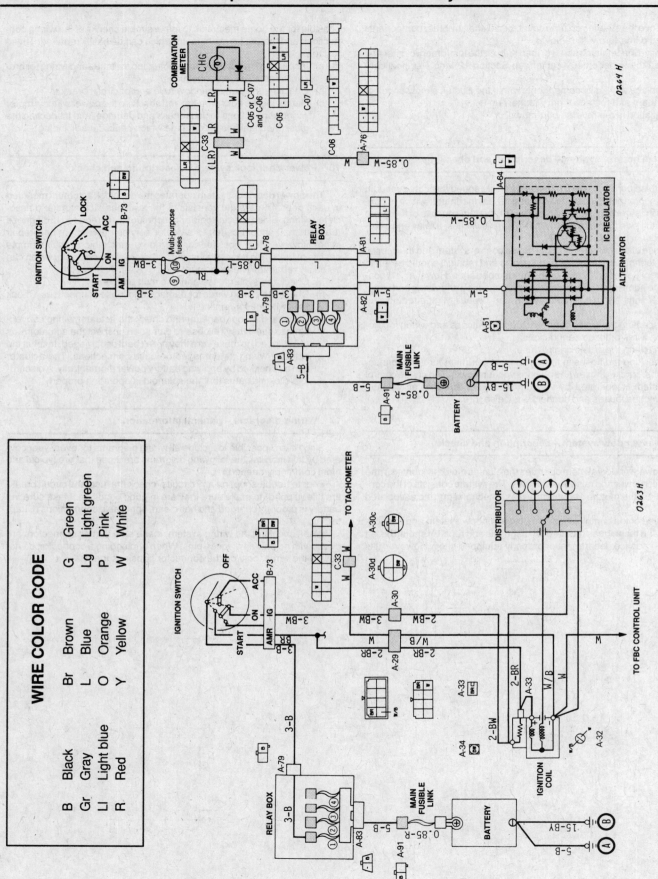

Chapter 12 Chassis electrical system

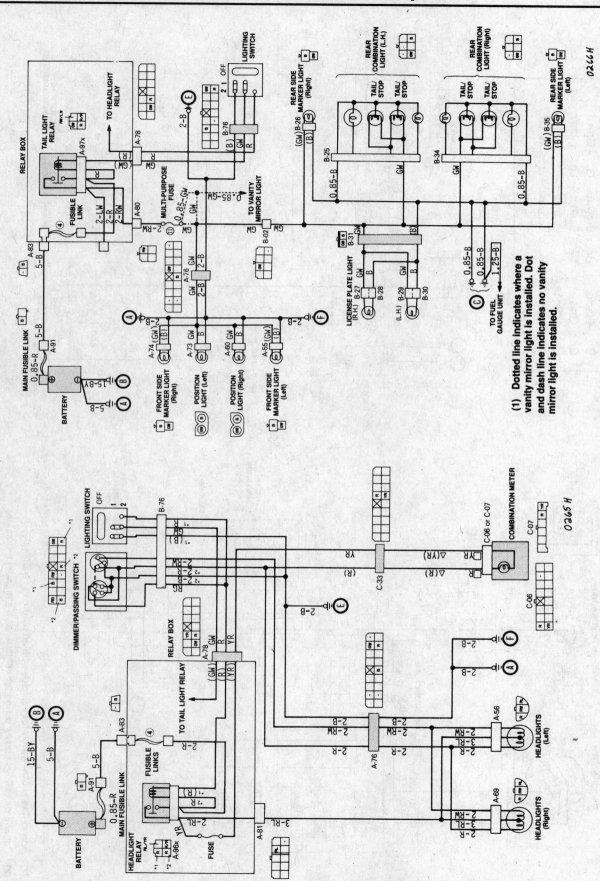

Typical tail light, parking light and license plate light circuit wiring diagram

Typical headlight circuit wiring diagram

12-10 Chapter 12 Chassis electrical system

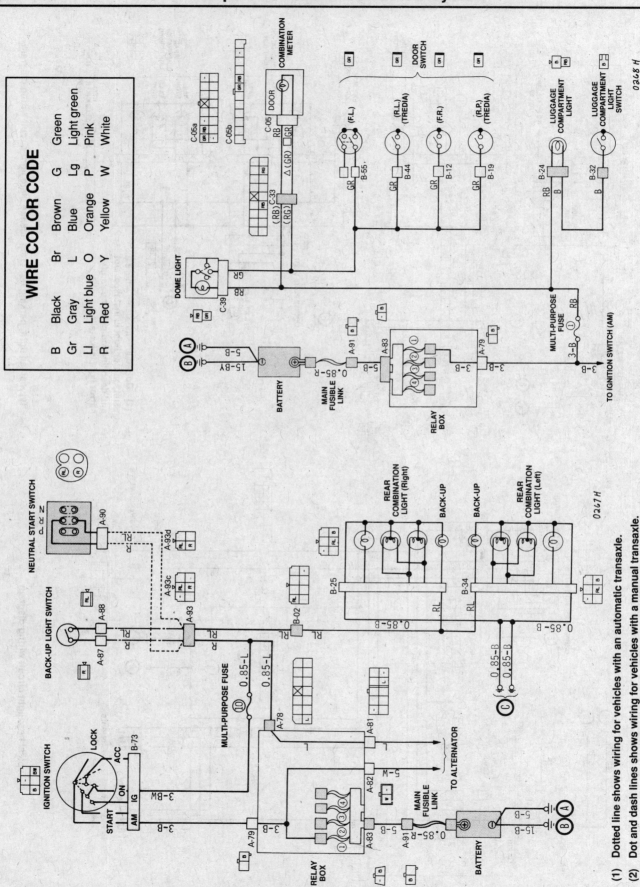

(1) Dotted line shows wiring for vehicles with an automatic transaxle.
(2) Dot and dash lines shows wiring for vehicles with a manual transaxle.

Typical dome and interior light circuit wiring diagram

Typical backup light circuit wiring diagram

Chapter 12 Chassis electrical system

12 – 11

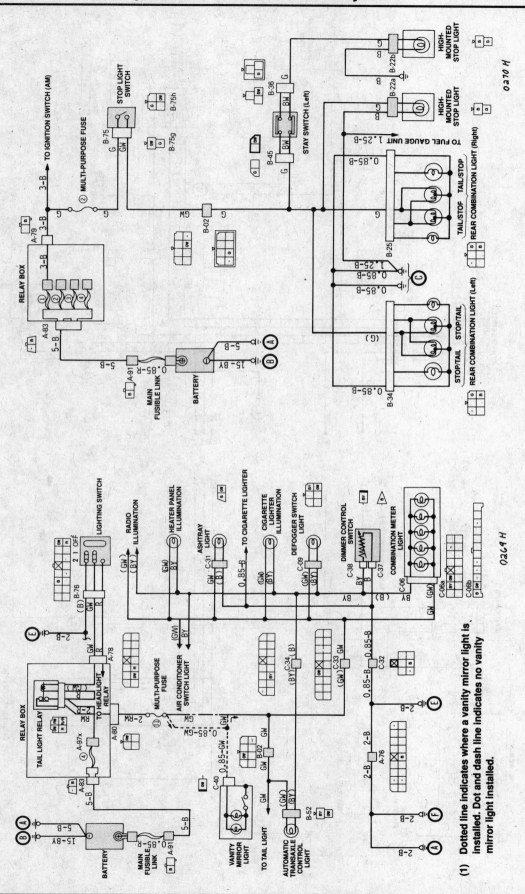

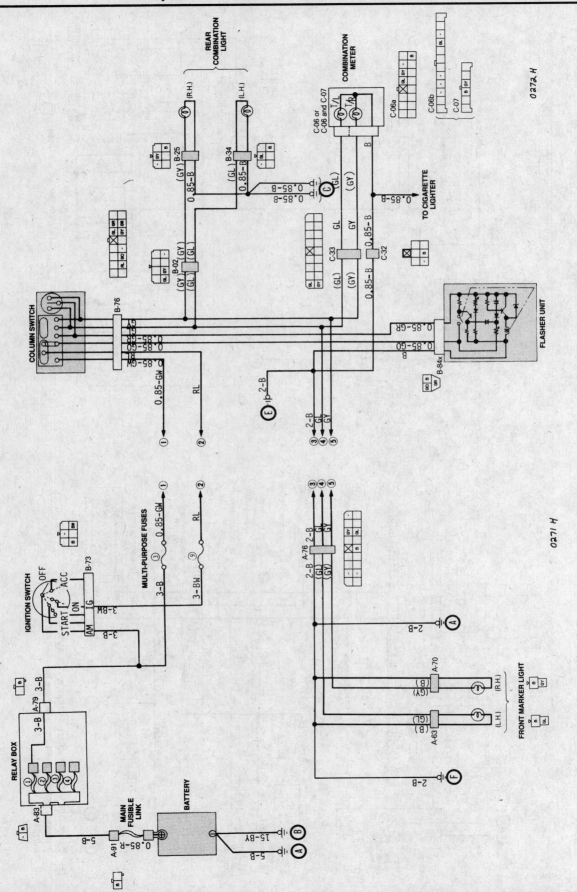

Chapter 12 Chassis electrical system

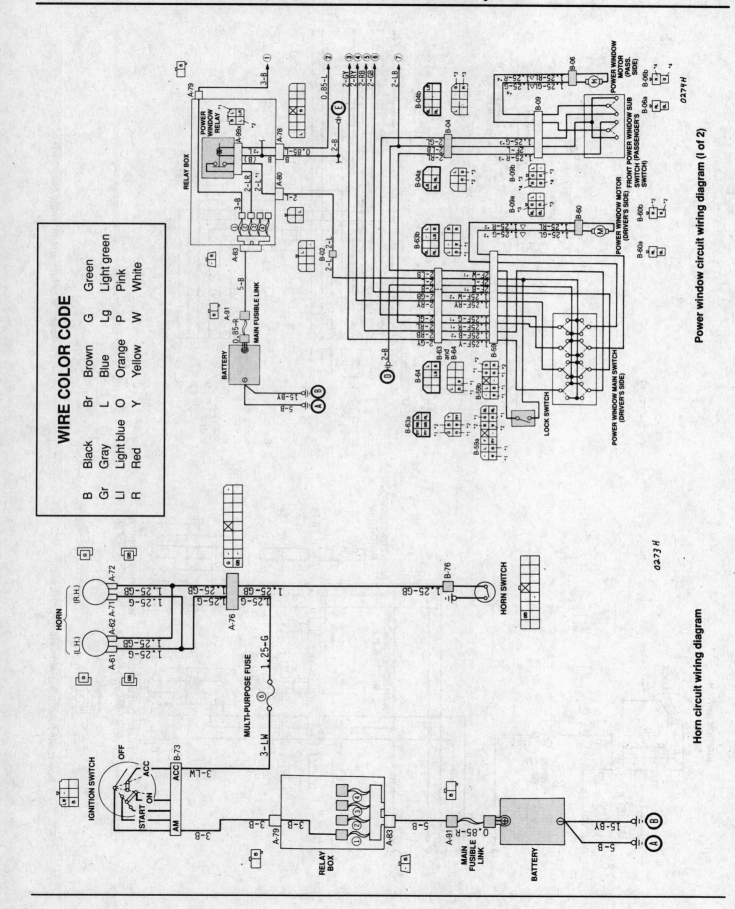

Power window circuit wiring diagram (1 of 2)

Horn circuit wiring diagram

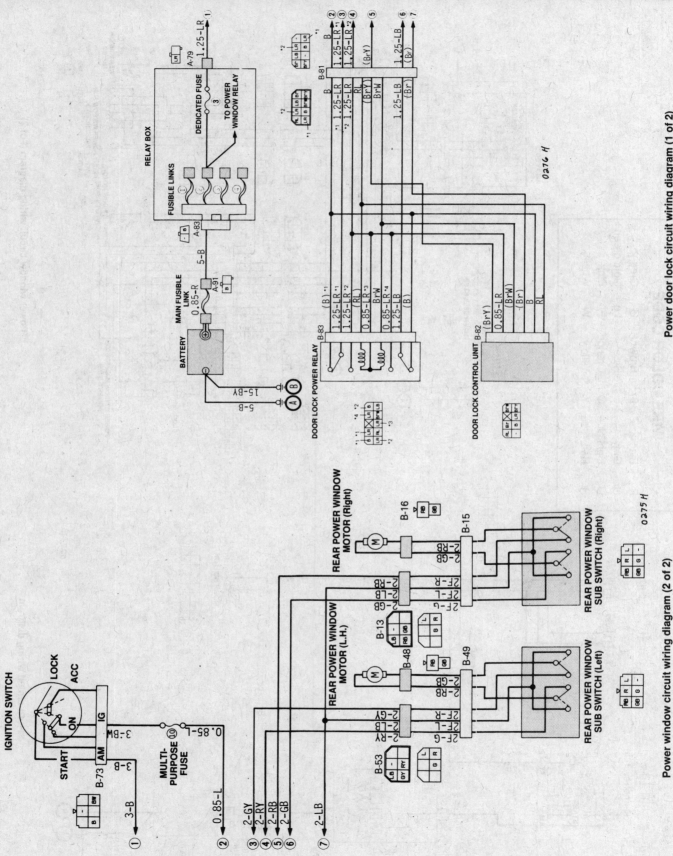

Chapter 12 Chassis electrical system

12-15

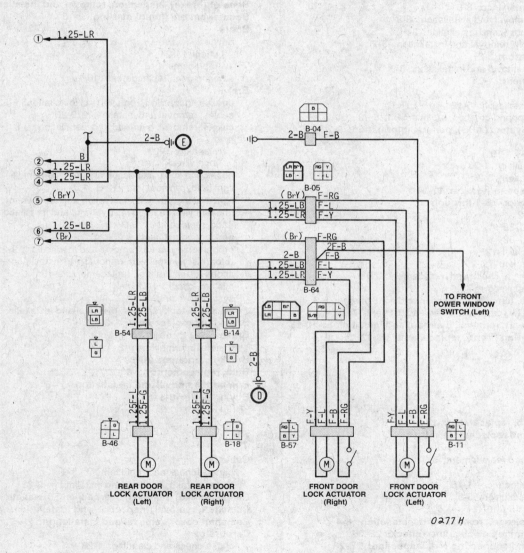

Power door lock circuit wiring diagram (2 of 2)

Index

A

A-arm, removal and installation: 10-10
About this manual: 0-6
ABS (Anti-lock Brake System), general information: 9-2
Air cleaner, removal and installation: 4-7
Air conditioning system: 3-1 through 12
    blower motor, removal and installation: 3-7
    check and maintenance: 3-8
    compressor, removal and installation: 3-9
    condenser, removal and installation: 3-10
    control assembly, removal and installation: 3-7
    general information: 3-2
    receiver/drier, removal and installation: 3-9
Air filter replacement: 1-21
Alternator
    removal and installation: 5-12
    voltage regulator and brushes, replacement: 5-13
Anti-lock Brake System (ABS), general information: 9-2
Antifreeze
    condition check: 1-19
    draining and refilling: 1-30
    general information: 3-2
Automatic shoulder harnesses: 11-14
Automatic transaxle: 7B-1 through 10
    diagnosis: 7B-2
    filter, change: 1-32
    fluid, change: 1-32
    fluid level, check: 1-12
    general information: 7B-2
    neutral start switch, check and replacement: 7B-9
    removal and installation: 7B-9
    selector linkage, removal, installation and adjustment: 7B-4
    Throttle Valve (TV) cable, check and adjustment: 7B-9
Axle assembly (rear), removal, overhaul and installation: 10-15
Axle hub and bearing (front), replacement: 10-12

B

Back-up light bulb, replacement: 12-6
Balljoint, check and replacement: 10-11
Battery
    cables, check and replacement: 5-3
    charging: 1-15
    check and maintenance: 1-15
    removal and installation: 5-3
Battery jump starting: 0-15
Bearing (clutch release), removal and installation: 8-4
Bearing (main, engine), oil clearance check: 2B-26
Bearings, main and connecting rod, inspection: 2B-23
Bleeding
    brake system: 9-20
    clutch: 8-8
    power steering system: 10-20
Block (engine)
    cleaning: 2B-20
    honing: 2B-21
    inspection: 2B-21

Blower motor, heater and air conditioner, removal and installation: 3-7
Body: 11-1 through 14
    general information: 11-1
    maintenance: 11-1
Body repair
    major damage: 11-5
    minor damage: 11-4
Booster (brake), inspection, removal and installation: 9-22
Booster battery (jump) starting: 0-15
Boots
    driveaxle
        check: 1-24
        replacement: 8-10
    steering gear, replacement: 10-19
Brake
    booster, inspection, removal and installation: 9-22
    cables, removal and installation: 9-23
    caliper, removal, overhaul and installation: 9-8
    fluid
        bleeding: 9-20
        level check: 1-8
    hoses and lines, inspection and replacement: 9-20
    light bulb, replacement: 12-6
    light switch, removal, installation and adjustment: 9-24
    master cylinder, removal, overhaul and installation: 9-18
    pads, replacement: 9-2
    parking, adjustment: 9-23
    proportioning valve, removal and installation: 9-20
    rotor (disc), inspection, removal and installation: 9-11
    shoes (drum brakes), replacement: 9-13
    system, check: 1-19
    system bleeding: 9-20
    wheel cylinder, removal, overhaul and installation: 9-16
Brake fluid replacement: 1-32
Brakes: 9-1 through 24
    general information: 9-2
Break-in (engine): 2B-29
Bulb, replacement: 12-6
Bumpers, removal and installation: 11-6
Buying parts: 0-8

C

Cable
    battery, check and replacement: 5-3
    parking brake, removal and installation: 9-23
Caliper (brake), removal, overhaul and installation: 9-8
Camshaft, removal, inspection and installation: 2A-20
Camshaft cover, removal and installation: 2A-5
Carburetor
    choke check and cleaning: 1-28
    diagnosis and overhaul: 4-8
    on-vehicle checks and adjustments: 4-10
    removal and installation: 4-7
Carpets, maintenance: 11-4
Cartridge (strut), replacement: 10-9
Catalytic converter: 6-4
CEC (Computerized Engine Controls): 6-7
Centrifugal advance mechanism, check and replacement: 5-7

Index

Charging system
   check: 5-11
   general information and precautions: 5-11
Chassis electrical system: 12-1 through 15
Chemicals and lubricants: 0-17
Choke, check: 1-28
Clutch
   bleeding: 8-8
   cable, removal, installation and adjustment: 8-6
   components, removal and installation: 8-3
   description and check: 8-2
   fluid, level check: 1-8
   general information: 8-1
   master cylinder, removal, overhaul and installation: 8-7
   release bearing and lever, removal and installation: 8-4
   release cylinder, removal, overhaul and installation: 8-8
Clutch and drivetrain: 8-1 through 14
Clutch pedal freeplay check and adjustment: 1-15
Coil (ignition), check and replacement: 5-4
Coil spring (rear), removal and installation: 10-13
Combination switch, removal and installation: 12-4
Compression check: 2B-8
Compressor (air conditioning), removal and installation: 3-9
Computerized Engine Controls (CEC): 6-7
Condenser (air conditioning), removal and installation: 3-10
Connecting rods
   bearing inspection: 2B-23
   inspection: 2B-22
   installation and oil clearance check: 2B-28
   removal: 2B-19
Constant velocity joint, overhaul: 8-10
Control arm, removal and installation: 10-10
Converter (catalytic): 6-4
Coolant
   general information: 3-2
   level check: 1-8
   reservoir, removal and installation: 3-5
   temperature sending unit, check and replacement: 3-4
Cooling fan, check and replacement: 3-3
Cooling system: 3-1 through 12
   coolant, general information: 3-2
   general information: 3-2
   servicing: 1-30
Cooling system check: 1-19
Core (heater), removal and installation: 3-7
Crankshaft
   front oil seal, replacement: 2A-20
   inspection: 2B-23
   installation and main bearing oil clearance check: 2B-26
   rear oil seal, replacement: 2A-32
   removal: 2B-19
Cruise control, description and check: 12-7
CV (Constant Velocity) joint, overhaul: 8-10
Cylinder block
   cleaning: 2B-20
   honing: 2B-21
   inspection: 2B-21
Cylinder head
   cleaning and inspection: 2B-16
   disassembly: 2B-15
   reassembly: 2B-17
   removal and installation: 2A-25
   valves, servicing: 2B-17
Cylinder honing: 2B-21

D

Deceleration devices: 6-4

Dent repair
   major damage: 11-5
   minor damage: 11-4
Diagnosis: 0-20 through 26
Disc (brake), inspection, removal and installation: 9-11
Disc brake
   caliper, removal, overhaul and installation: 9-8
   pads, replacement: 9-2
Distributor
   cap, check and replacement: 1-26
   centrifugal advance mechanism, check and replacement: 5-7
   pickup air gap, check and adjustment: 5-11
   removal and installation: 5-5
   vacuum advance mechanism, check and replacement: 5-6
Distributor rotor, check and replacement: 1-26
Door
   latch, lock cylinder and handle, removal and installation: 11-10
   lock system (power), description and check: 12-7
   removal, installation and adjustment: 11-8
   trim panel, removal and installation: 11-7
   window glass, removal and installation: 11-12
Driveaxle, boot
   check: 1-24
   replacement: 8-10
Driveaxles: 8-1 through 14
   general information: 8-1, 9
   removal and installation: 8-9
Drivebelt check, adjustment and replacement: 1-17
Driveplate, removal and installation: 2A-31
Drum brake, shoes, replacement: 9-13
Drum brake wheel cylinder, removal, overhaul and installation: 9-16

E

EECS (Evaporative Emission Control System): 6-2
EGR (Exhaust Gas Recirculation) system: 6-4
Electric
   door lock system, description and check: 12-7
   windows, description and check: 12-7
Electrical
   circuit diagrams: 12-7
   system, general information: 12-1
   troubleshooting, general information: 12-1
Electrical system (chassis): 12-1 through 15
Electrical systems (engine): 5-1 through 16
Emergency battery jump starting: 0-15
Emissions control systems: 6-1 through 10
   general information: 6-1
Engine
   block
      cleaning: 2B-20
      inspection: 2B-21
   compression check: 2B-8
   coolant
      general information: 3-2
      level check: 1-8
   cooling fan, check and replacement: 3-3
   cylinder head, removal and installation: 2A-25
   drivebelt check, adjustment and replacement: 1-17
   general information: 2A-5
   identification: 2A-5
   idle speed check and adjustment: 1-23
   in-vehicle repairs: 2A-1 through 34
   mount, check and replacement: 2A-33
   oil, level check: 1-8
   oil seal, replacement: 2A-20

Index

overhaul: 2B-1 through 30
    disassembly sequence: 2B-11
    general information: 2B-7
    reassembly sequence: 2B-24
overhaul alternatives: 2B-11
rebuilding, alternatives: 2B-11
removal, methods and precautions: 2B-9
removal and installation: 2B-10
repair operations possible with engine in vehicle: 2A-5
tune-up: 1-1 through 36
Engine electrical systems: 5-1 through 16
    general information: 5-3
Engine emissions control systems: 6-1 through 10
Engine fuel and exhaust systems: 4-1 through 22
Engine oil and filter change: 1-12
EVAP system: 6-2
    check: 1-33
Evaporative Emission Control System: 6-2
Exhaust, manifold, removal and installation: 2A-24
Exhaust Gas Recirculation (EGR) system, general information: 6-4
Exhaust system
    check: 1-24
    general information: 4-2
    servicing, general information: 4-21

F

Fan, engine cooling, check and replacement: 3-3
Fan belt check, adjustment and replacement: 1-17
Fault-finding: 0-20 through 26
Filter
    automatic transaxle, replacement: 1-32
    fuel, general information: 4-16
    fuel, replacement: 1-25
    oil, replacement: 1-12
Fixed glass, replacement: 11-5
Fluid
    automatic transaxle
        change: 1-32
        checking: 1-12
    level checks: 1-8
    manual transaxle
        change: 1-33
        checking: 1-24
    power steering, checking: 1-11
Flywheel, removal and installation: 2A-31
Front axle hub and bearing, replacement: 10-12
Front end alignment, general information: 10-20
Front fender inner panels, removal and installation: 11-14
Front stabilizer bar, removal and installation: 10-2
Front strut and spring assembly, removal, inspection and installation: 10-8
Fuel
    filter
        general information: 4-16
        replacement: 1-25
    injection system, check: 4-15
    lines, inspection and replacement: 4-4
    pressure
        relief procedure: 4-3
        testing: 4-3
    pump
        removal and installation: 4-5
        testing: 4-3
    tank
        cleaning and repair: 4-7
        removal and installation: 4-6

Fuel system: 4-1 through 22
    general information: 4-2
Fuses, general information: 12-2
Fusible links, general information: 12-3

G

Gauge cluster, removal and installation: 12-6
Gear (steering), removal and installation: 10-18
General information
    automatic transaxle: 7B-2
    body: 11-1
    brakes: 9-2
    clutch and driveaxles: 8-1
    cooling, heating and air conditioning systems: 3-2
    driveaxles: 8-9
    electrical system: 12-1
    emissions control systems: 6-1
    engine electrical systems: 5-3
    engine overhaul: 2B-7
    exhaust system: 4-2
    fuel system: 4-2
    front-end alignment: 10-20
    ignition system: 5-3
    manual transaxle: 7A-2
    steering and suspension systems: 10-2
    steering system: 10-16
    tune-up: 1-3
    wheels and tires: 10-20
Generator (alternator), removal and installation: 5-12
Glass (fixed), replacement: 11-5

H

Hand brake, adjustment: 9-23
Hazard, flashers, check and replacement: 12-3
Head (cylinder)
    cleaning and inspection: 2B-16
    disassembly: 2B-15
    reassembly: 2B-17
    removal and installation: 2A-25
Headlight
    adjustment: 12-5
    removal and installation: 12-4
Heated Air Intake (HAI) system: 6-2
    air cleaner check: 1-28
Heater
    blower motor, removal and installation: 3-7
    control assembly, removal and installation: 3-7
    core, removal and installation: 3-7
Heating system: 3-1 through 12
    general information: 3-2
High altitude compensation system: 6-6
Hinges, maintenance: 11-5
Honing, cylinder: 2B-21
Hood
    adjustment: 11-5
    removal and installation: 11-5
Hoses
    brake, inspection and replacement: 9-20
    check and replacement: 1-18
Hub (front), removal and installation: 10-12
Hydraulic clutch system, bleeding: 8-8
Hydraulic valve adjusters, removal, inspection and installation: 2A-9

Index

I

Idle speed check and adjustment: 1-23
Idle-up system: 6-5
Igniter, replacement: 5-6
Ignition
    distributor, removal and installation: 5-5
    igniter, replacement: 5-6
    pickup coil, check and replacement: 5-6
    system
        check: 5-4
        general information: 5-3
Ignition coil, check and replacement: 5-4
Ignition switch, removal and installation: 12-4
Ignition timing, check and adjustment: 1-34
In-vehicle engine repairs: 2A-1 through 34
Initial start-up and break-in after overhaul: 2B-29
Instrument cluster, removal and installation: 12-6
Intake manifold, removal and installation: 2A-23
Intermediate shaft (steering), removal and installation: 10-17
Introduction to the vehicles covered by this manual: 0-6

J

Jacking: 0-16
Jet air system: 6-4
Jump starting: 0-15

K

Knuckle and hub, removal and installation: 10-12

L

Lateral rod, removal and installation: 10-13
Liftgate, removal, installation and adjustment: 11-9
Linings (brake), check: 1-19
Locks
    maintenance: 11-5
    power door, description and check: 12-7
Lubricants: 0-17

M

Main and connecting rod bearings, inspection: 2B-23
Main bearing, oil clearance, check: 2B-26
Main bearing oil seal, installation: 2B-27
Maintenance
    introduction: 1-3
    techniques: 0-8
Maintenance schedule: 1-7
Manifold
    exhaust, removal and installation: 2A-24
    intake, removal and installation: 2A-23
Manual transaxle
    general information: 7A-2
    lubricant
        change: 1-33
        check: 1-24
    overhaul, general information: 7A-5
    removal and installation: 7A-4
    shift assembly: 7A-2
Manual transmission: 7A-1 through 6
Master cylinder, removal, overhaul and installation: 9-18
Mixture control valves: 6-4
Mount, engine, check and replacement: 2A-33
Mount (transaxle), check and replacement: 7B-3
Muffler, check: 1-24
Multi-Point Injection (MPI) system, component removal and installation: 4-17

N

Neutral start switch, check, adjustment and replacement: 7B-9

O

O_2 sensor, replacement: 1-35
Oil
    change: 1-12
    level check: 1-8
    pan, removal and installation: 2A-27
    pump, removal and installation: 2A-28
    seal
        engine: 2A-20, 32
        transaxle: 7B-3
Oil filter, change: 1-12
Oil seal, main bearing, installation: 2B-27
Outside mirror, removal and installation: 11-13
Overhaul
    engine: 2B-1 through 30
        general information: 2B-7
    manual transaxle, general information: 7A-5
Overhaul (engine)
    alternatives: 2B-11
    disassembly sequence: 2B-11
    initial start-up and break-in: 2B-29
    reassembly sequence: 2B-24
Owner maintenance: 1-1 through 36
Oxygen sensor, replacement: 1-35

P

Pads (brake), replacement: 9-2
Pan, oil, removal and installation: 2A-27
Pan (oil), removal and installation: 2A-27
Parking brake
    adjustment: 9-23
    cables, removal and installation: 9-23
Parking lamp bulb, replacement: 12-6
Parts, replacement: 0-8
PCV (Positive Crankcase Ventilation) system: 6-2
PCV system
    filter replacement: 1-21
    valve, check and replacement: 1-35
Pickup coil, check and replacement: 5-6
Piston
    installation: 2B-28
    rings, installation: 2B-24
Pistons and connecting rods
    inspection: 2B-22
    installation: 2B-28
    removal: 2B-19

Index

Positive Crankcase Ventilation (PCV) system: 6-2
    filter replacement: 1-21
    valve, check and replacement: 1-35
Power
    brake booster, inspection, removal and installation: 9-22
    door lock system, description and check: 12-7
    window system, description and check: 12-7
Power steering
    bleeding: 10-20
    fluid level, check: 1-11
    pump, removal and installation: 10-19
Pressure (tire), checking: 1-10
Proportioning valve, removal and installation: 9-20
Pump
    fuel
        removal and installation: 4-5
        testing: 4-3
    oil, removal and installation: 2A-28
    power steering, removal and installation: 10-19
    water
        check: 3-5
        removal and installation: 3-5

R

Radiator
    check: 1-19
    coolant, general information: 3-2
    draining, flushing and refilling: 1-30
    removal and installation: 3-3
Rear
    axle assembly, removal, overhaul and installation: 10-15
    main oil seal, installation: 2B-27
    shock absorber, removal and installation: 10-14
Rear coil spring, removal and installation: 10-13
Rear main oil seal (engine), replacement: 2A-32
Rear stabilizer bar, removal and installation: 10-12
Rear wheel bearing, check and repack: 1-30
Receiver/drier, removal and installation: 3-9
Relays, general information: 12-3
Relieving fuel system pressure: 4-3
Release bearing (clutch), removal and installation: 8-4
Reservoir (coolant), removal and installation: 3-5
Rings (piston), installation: 2B-24
Rocker arm, removal, inspection and installation: 2A-6
Rocker arm cover, removal and installation: 2A-5
Rods (connecting)
    installation: 2B-28
    removal: 2B-19, 22
Rotation (tire): 1-19
Rotor, check and replacement: 1-26
Rotor (brake), inspection, removal and installation: 9-11
Routine maintenance: 1-1 through 36

S

Safe automotive repair practices: 0-18
Safety: 0-18
Seat belts: 11-14
Secondary air supply system: 6-4
Selector linkage (automatic transaxle), removal, installation and adjustment: 7B-4
Sending unit, cooling system temperature, check and replacement: 3-4
Shift assembly (manual transaxle): 7A-2

Shock absorber (rear), removal and installation: 10-14
Shoes (brake), check: 1-19
Smog control systems: 6-1 through 10
Solenoid (starter), removal and installation: 5-16
Spark plug
    replacement: 1-25
    wires: 1-26
Speed control system, description and check: 12-7
Speedometer cable, replacement: 12-6
Stabilizer bar
    front, removal and installation: 10-2
    rear, removal and installation: 10-12
Starter
    motor
        removal and installation: 5-16
        testing in vehicle: 5-16
    solenoid, removal and installation: 5-16
Starting system, general information and precautions: 5-15
Steering
    gear
        boots, replacement: 10-19
        removal and installation: 10-18
    knuckle and hub, removal and installation: 10-12
    pump (power), removal and installation: 10-19
    system
        check: 1-23
        general information: 10-16
    wheel, removal and installation: 10-17
Steering and suspension systems, general information: 10-2
Steering system: 10-1 through 20
    intermediate shaft, removal and installation: 10-17
Strut and spring assembly, removal, inspection and installation: 10-8
Strut bar, removal and installation: 10-8
Strut cartridge, replacement: 10-9
Suspension
    balljoint, check and replacement: 10-11
    system, check: 1-23
Suspension system: 10-1 through 20
Switch, brake lights, removal, installation and adjustment: 9-24

T

Taillight bulb, replacement: 12-6
Tank (fuel)
    cleaning and repair: 4-7
    removal and installation: 4-6
TDC (top dead center), locating: 2B-9
Temperature sending unit, check and replacement: 3-4
Thermostat, check and replacement: 3-2
Thermostatic air cleaner check: 1-28
Throttle Body Injection (TBI) assembly, removal, overhaul and installation: 4-15
Throttle cable, removal, installation and adjustment: 4-18
Throttle position sensor, check: 1-21
Throttle Valve (TV) cable (automatic transaxle), check and adjustment: 7B-9
Tie-rod ends, removal and installation: 10-19
Timing, ignition: 1-34
Timing belt and sprockets, removal and installation: 2A-11
Tire
    checking: 1-10
    rotation: 1-19
Tires and wheels, general information: 10-20
Tools: 0-8
Top Dead Center (TDC), locating: 2B-9
Towing: 0-16

Transaxle
   mount, check and replacement: 7B-3
   oil seal replacement: 7B-3
Transaxle (automatic): 7B-1 through 10
   fluid level, check: 1-12
   general information: 7B-2
   neutral start switch, check and replacement: 7B-9
   removal and installation: 7B-9
   selector linkage, removal, installation and adjustment: 7B-4
   Throttle Valve (TV) cable, check and adjustment: 7B-9
Transaxle (manual)
   general information: 7A-2
   overhaul, general information: 7A-5
   removal and installation: 7A-4
Transmission (manual): 7A-1 through 6
Troubleshooting: 0-20 through 26
   electrical, general information: 12-1
Trunk lid, removal, installation and adjustment: 11-8
Tune-up
   general information: 1-3
   introduction: 1-3
Tune-up and routine maintenance: 1-1 through 36
Turbocharger
   general information: 4-21
   inspection: 4-22
   removal and installation: 4-22
Turn signal
   bulb, replacement: 12-6
   flasher, check and replacement: 12-3
TV (Throttle Valve) cable (automatic transaxle), check and adjustment: 7B-9

U

Underhood hose check and replacement: 1-18

Upholstery, maintenance: 11-4
Using this manual: 0-6

V

Vacuum advance unit, check and replacement: 5-6
Valve
   adjustment: 1-22
   clearance check and adjustment: 1-22
   lash adjustment: 1-22
   springs, retainers and seals, replacement: 2A-8
Valve cover, removal and installation: 2A-5
Valve job: 2B-17
Valves, servicing: 2B-17
Vehicle identification numbers: 0-7
Vinyl trim, maintenance: 11-4
Voltage regulator and brushes, replacement: 5-13

W

Water pump
   check: 3-5
   removal and installation: 3-5
Wheel
   alignment, general information: 10-20
   cylinder, removal, overhaul and installation: 9-16
   steering, removal and installation: 10-17
Wheel bearing, check and repack: 1-30
Wheels and tires, general information: 10-20
Window, power, description and check: 12-7
Windshield, washer fluid, level check: 1-8
Windshield wiper blade inspection and replacement: 1-13
Wiper blade (windshield) inspection and replacement: 1-13
Wiring diagrams, general information: 12-8
Working facilities: 0-8

Haynes Automotive Manuals

NOTE: If you do not see a listing for your vehicle, consult your local Haynes dealer for the latest product information.

ACURA
- 12020 Integra '86 thru '89 & Legend '86 thru '90
- 12021 Integra '90 thru '93 & Legend '91 thru '95
- 12050 Acura TL all models '99 thru '08

AMC
- Jeep CJ - see JEEP (50020)
- 14020 Concord/Hornet/Gremlin/Spirit '70 thru '83
- 14025 (Renault) Alliance & Encore '83 thru '87

AUDI
- 15020 4000 all models '80 thru '87
- 15025 5000 all models '77 thru '83
- 15026 5000 all models '84 thru '88
- 15030 Audi A4 '02 thru '08

AUSTIN
- Healey Sprite - see MG Midget (66015)

BMW
- 18020 3/5 Series '82 thru '92
- 18021 3 Series including Z3 models '92 thru '98
- 18022 3-Series '99 thru '05, including Z4 models
- 18025 320i all 4 cyl models '75 thru '83
- 18050 1500 thru 2002 except Turbo '59 thru '77

BUICK
- 19010 Buick Century '97 thru '05
- Century (front-wheel drive) - see GM (38005)
- 19020 Buick, Oldsmobile & Pontiac Full-size (Front wheel drive) '85 thru '05
- 19025 Buick, Oldsmobile & Pontiac Full-size (Rear wheel drive) '70 thru '90
- 19030 Mid-size Regal & Century '74 thru '87
- Regal - see GENERAL MOTORS (38010)
- Skyhawk - see GM (38030)
- Skylark - see GM (38020, 38025)
- Somerset - see GENERAL MOTORS (38025)

CADILLAC
- 21030 Cadillac Rear Wheel Drive '70 thru '93
- Cimarron, Eldorado & Seville - see GM (38015, 38030, 38031)

CHEVROLET
- 10305 Chevrolet Engine Overhaul Manual
- 24010 Astro & GMC Safari Mini-vans '85 thru '05
- 24015 Camaro V8 all models '70 thru '81
- 24016 Camaro all models '82 thru '92, Cavalier - see GM (38015), Celebrity - see GM (38005)
- 24017 Camaro & Firebird '93 thru '02
- 24020 Chevelle, Malibu, El Camino '69 thru '87
- 24024 Chevette & Pontiac T1000 '76 thru '87
- Citation - see GENERAL MOTORS (38020)
- 24027 Colorado & GMC Canyon '04 thru '08
- 24032 Corsica/Beretta all models '87 thru '96
- 24040 Corvette V8 models '68 thru '82
- 24041 Corvette all models '84 thru '96
- 24045 Full-size Sedans Caprice, Impala, Biscayne, Bel Air & Wagons '69 thru '90
- 24046 Impala SS & Caprice and Buick Roadmaster '91 thru '96
- Lumina '90 thru '94 - see GM (38010)
- 24047 Impala & Monte Carlo all models '06 thru '08
- 24048 Lumina & Monte Carlo '95 thru '05
- Lumina APV - see GM (38035)
- 24050 Luv Pick-up all 2WD & 4WD '72 thru '82
- Malibu - see GM (38026)
- 24055 Monte Carlo all models '70 thru '88
- Monte Carlo '95 thru '01 - see LUMINA
- 24059 Nova all V8 models '69 thru '79
- 24060 Nova/Geo Prizm '85 thru '92
- 24064 Pick-ups '67 thru '87 - Chevrolet & GMC, all V8 & in-line 6 cyl, 2WD & 4WD '67 thru '87; Suburbans, Blazers & Jimmys '67 thru '91
- 24065 Pick-ups '88 thru '98 - Chevrolet & GMC, all full-size models '88 thru '98; C/K Classic '99 & '00; Blazer & Jimmy '92 thru '94; Suburban '92 thru '99; Tahoe & Yukon '95 thru '99
- 24066 Pick-ups '99 thru '06 - Chevrolet Silverado & GMC Sierra '99 thru '06; Suburban/Tahoe/Yukon/Yukon XL/Avalanche '00 thru '06
- 24067 Chevrolet Silverado & GMC Sierra '07 thru '09
- 24070 S-10 & GMC S-15 Pick-ups '82 thru '93
- 24071 S-10, Sonoma & Jimmy '94 thru '04
- 24072 Chevrolet TrailBlazer, GMC Envoy & Oldsmobile Bravada '02 thru '09
- 24075 Sprint '85 thru '88, Geo Metro '89 thru '01
- 24080 Vans - Chevrolet & GMC '68 thru '96
- 24081 Chevrolet Express & GMC Savana Full-size Vans '96 thru '07

CHRYSLER
- 10310 Chrysler Engine Overhaul Manual
- 25015 Chrysler Cirrus, Dodge Stratus, Plymouth Breeze, '95 thru '00
- 25020 Full-size Front-Wheel Drive '88 thru '93
- K-Cars - see DODGE Aries (30008)
- Laser - see DODGE Daytona (30030)
- 25025 Chrysler LHS, Concorde & New Yorker, Dodge Intrepid, Eagle Vision, '93 thru '97
- 25026 Chrysler LHS, Concorde, 300M, Dodge Intrepid '98 thru '04
- 25027 Chrysler 300, Dodge Charger & Magnum '05 thru '09
- 25030 Chrysler/Plym. Mid-size '82 thru '95
- Rear-wheel Drive - see DODGE (30050)
- 25035 PT Cruiser all models '01 thru '09
- 25040 Chrysler Sebring/Dodge Avenger '95 thru '05, Dodge Stratus '01 thru '05

DATSUN
- 28005 200SX all models '80 thru '83
- 28007 B-210 all models '73 thru '78
- 28009 210 all models '78 thru '82
- 28012 240Z, 260Z & 280Z Coupe '70 thru '78
- 28014 280ZX Coupe & 2+2 '79 thru '83
- 300ZX - see NISSAN (72010)
- 28018 510 & PL521 Pick-up '68 thru '73
- 28020 510 all models '78 thru '81
- 28022 620 Series Pick-up all models '73 thru '79
- 720 Series Pick-up - NISSAN (72030)
- 28025 810/Maxima all models, '77 thru '84

DODGE
- 400 & 600 - see CHRYSLER (25030)
- 30008 Aries & Plymouth Reliant '81 thru '89
- 30010 Caravan & Ply. Voyager '84 thru '95
- 30011 Caravan & Ply. Voyager '96 thru '02
- 30012 Challenger/Plymouth Saporro '78 thru '83
- 30013 Challenger/'67-'76 - see DART (30025)
- 30013 Caravan, Chrysler Voyager, Town & Country '03 thru '07
- 30016 Colt/Plymouth Champ '78 thru '87
- 30020 Dakota Pick-ups all models '87 thru '96
- 30021 Durango '98 & '99, Dakota '97 thru '99
- 30022 Durango '00 thru '03, Dakota '00 thru '04
- 30023 Durango '04 thru '06, Dakota '05 and '06
- 30025 Dart, Challenger/Plymouth Barracuda & Valiant 6 cyl models '67 thru '76
- 30030 Daytona & Chrysler Laser '84 thru '89
- Intrepid - see Chrysler (25025, 25026)
- 30034 Dodge & Plymouth Neon '95 thru '99
- 30035 Omni & Plymouth Horizon '78 thru '90
- 30036 Dodge and Plymouth Neon '00 thru '05
- 30040 Pick-ups all full-size models '74 thru '93
- 30041 Pick-ups all full-size models '94 thru '01
- 30042 Pick-ups full-size models '02 thru '08
- 30045 Ram 50/D50 Pick-ups & Raider and Plymouth Arrow Pick-ups '79 thru '93
- 30050 Dodge/Ply./Chrysler RWD '71 thru '89
- 30055 Shadow/Plymouth Sundance '87 thru '94
- 30060 Spirit & Plymouth Acclaim '89 thru '95
- 30065 Vans - Dodge & Plymouth '71 thru '03

EAGLE
- Talon - see MITSUBISHI (68030, 68031)
- Vision - see CHRYSLER (25025)

FIAT
- 34010 124 Sport Coupe & Spider '68 thru '78
- 34025 X1/9 all models '74 thru '80

FORD
- 10320 Ford Engine Overhaul Manual
- 10355 Ford Automatic Transmission Overhaul
- 36004 Aerostar Mini-vans '86 thru '97
- Aspire - see FORD Festiva (36030)
- 36006 Contour/Mercury Mystique '95 thru '00
- 36008 Courier Pick-up all models '72 thru '82
- 36012 Crown Victoria & Mercury Grand Marquis '88 thru '10
- 36016 Escort/Mercury Lynx '81 thru '90
- 36020 Escort/Mercury Tracer '91 thru '02
- Expedition - see FORD Pick-up (36059)
- 36022 Escape & Mazda Tribute '01 thru '07
- 36024 Explorer & Mazda Navajo '91 thru '01
- 36025 Explorer/Mercury Mountaineer '02 thru '10
- 36028 Fairmont & Mercury Zephyr '78 thru '83
- 36030 Festiva & Aspire '88 thru '97
- 36032 Fiesta all models '77 thru '80
- 36034 Focus all models '00 thru '07
- 36036 Ford & Mercury Full-size '75 thru '87
- 36044 Ford & Mercury Mid-size '75 thru '86
- 36048 Mustang V8 all models '64-1/2 thru '73
- 36049 Mustang II 4 cyl, V6 & V8 '74 thru '78
- 36050 Mustang & Mercury Capri '79 thru '86
- 36051 Mustang all models '94 thru '04
- 36052 Mustang '05 thru '07
- 36054 Pick-ups and Bronco '73 thru '79
- 36058 Pick-ups and Bronco '80 thru '96
- 36059 F-150 & Expedition '97 thru '09, F-250 '97 thru '99 & Lincoln Navigator '98 thru '09
- 36060 Super Duty Pick-ups, Excursion '99 thru '10
- 36061 F-150 full-size '04 thru '09
- 36062 Pinto & Mercury Bobcat '75 thru '80
- 36066 Probe all models '89 thru '92
- 36070 Ranger/Bronco II gas models '83 thru '92
- 36071 Ford Ranger '93 thru '10, Mazda Pick-ups '94 thru '09
- 36074 Taurus & Mercury Sable '86 thru '95
- 36075 Taurus & Mercury Sable '96 thru '01
- 36078 Tempo & Mercury Topaz '84 thru '94
- 36082 Thunderbird/Mercury Cougar '83 thru '88
- 36086 Thunderbird/Mercury Cougar '89 thru '97
- 36090 Vans all V8 Econoline models '69 thru '91
- 36094 Vans full size '92 thru '05
- 36097 Windstar Mini-van '95 thru '07

GENERAL MOTORS
- 10360 GM Automatic Transmission Overhaul
- 38005 Buick Century, Chevrolet Celebrity, Olds Cutlass Ciera & Pontiac 6000 '82 thru '96
- 38010 Buick Regal, Chevrolet Lumina, Oldsmobile Cutlass Supreme & Pontiac Grand Prix front wheel drive '88 thru '07
- 38015 Buick Skyhawk, Cadillac Cimarron, Chevrolet Cavalier, Oldsmobile Firenza Pontiac J-2000 & Sunbird '82 thru '94
- 38016 Chevrolet Cavalier/Pontiac Sunfire '95 thru '05
- 38017 Chevrolet Cobalt & Pontiac G5 '05 thru '09
- 38020 Buick Skylark, Chevrolet Citation, Olds Omega, Pontiac Phoenix '80 thru '85
- 38025 Buick Skylark & Somerset, Olds Achieva, Calais & Pontiac Grand Am '85 thru '98
- 38026 Chevrolet Malibu, Olds Alero & Cutlass, Pontiac Grand Am '97 thru '03
- 38027 Chevrolet Malibu '04 thru '07
- 38030 Cadillac Eldorado & Oldsmobile Toronado '71 thru '85, Seville '80 thru '85, Buick Riviera '79 thru '85
- 38031 Cadillac Eldorado & Seville '86 thru '91, DeVille & Buick Riviera '86 thru '93, Fleetwood & Olds Toronado '86 thru '92
- 38032 DeVille '94 thru '05, Seville '92 thru '04
- 38033 Cadillac DTS '06 thru '10
- 38035 Chevrolet Lumina APV, Oldsmobile Silhouette & Pontiac Trans Sport '90 thru '96
- 38036 Chevrolet Venture, Olds Silhouette, Pontiac Trans Sport & Montana '97 thru '05
- General Motors Full-size Rear-wheel Drive - see BUICK (19025)
- 38040 Chevrolet Equinox '05 thru '09, Pontiac Torrent '06 thru '09

GEO
- Metro - see CHEVROLET Sprint (24075)
- Prizm - see CHEVROLET (24060) or TOYOTA (92036)
- 40030 Storm all models '90 thru '93
- Tracker - see SUZUKI Samurai (90010)

GMC
- Vans & Pick-ups - see CHEVROLET

HONDA
- 42010 Accord CVCC all models '76 thru '83
- 42011 Accord all models '84 thru '89
- 42012 Accord all models '90 thru '93
- 42013 Accord all models '94 thru '97
- 42014 Accord all models '98 thru '02
- 42015 Accord models '03 thru '07
- 42020 Civic 1200 all models '73 thru '79
- 42021 Civic 1300 & 1500 CVCC '80 thru '83
- 42022 Civic 1500 CVCC all models '75 thru '79
- 42023 Civic all models '84 thru '91
- 42024 Civic & del Sol '92 thru '95
- 42025 Civic '96 thru '00, CR-V '97 thru '01, Acura Integra '94 thru '00
- Passport - see ISUZU Rodeo (47017)
- 42026 Civic '01 thru '10, CR-V '02 thru '09
- 42035 Odyssey models '99 thru '04
- 42037 Honda Pilot '03 thru '07, Acura MDX '01 thru '07
- 42040 Prelude CVCC all models '79 thru '89

HYUNDAI
- 43010 Elantra all models '96 thru '06
- 43015 Excel & Accent all models '86 thru '09
- 43050 Santa Fe all models '01 thru '06
- 43055 Sonata all models '99 thru '08

ISUZU
- Hombre - see CHEVROLET S-10 (24071)
- 47017 Rodeo, Amigo & Honda Passport '89 thru '02
- 47020 Trooper '84 thru '91, Pick-up '81 thru '93

JAGUAR
- 49010 XJ6 all 6 cyl models '68 thru '86
- 49011 XJ6 all models '88 thru '94
- 49015 XJ12 & XJS all 12 cyl models '72 thru '85

JEEP
- 50010 Cherokee, Comanche & Wagoneer Limited all models '84 thru '01
- 50020 CJ all models '49 thru '86
- 50025 Grand Cherokee all models '93 thru '04
- 50026 Grand Cherokee '05 thru '09
- 50029 Grand Wagoneer & Pick-up '72 thru '91
- 50030 Wrangler all models '87 thru '08
- 50035 Liberty '02 thru '07

KIA
- 54070 Sephia '94 thru '01, Spectra '00 thru '09

LEXUS
- ES 300 - see TOYOTA Camry (92007)

LINCOLN
- Navigator - see FORD Pick-up (36059)
- 59010 Rear Wheel Drive all models '70 thru '10

MAZDA
- 61010 GLC (rear wheel drive) '77 thru '83
- 61011 GLC (front wheel drive) '81 thru '85
- 61015 323 & Protegé '90 thru '00
- 61016 MX-5 Miata '90 thru '09
- 61020 MPV all models '89 thru '98
- Navajo - see FORD Explorer (36024)
- 61030 Pick-ups '72 thru '93
- Pick-ups '94 on - see Ford (36071)
- 61035 RX-7 all models '79 thru '85
- 61036 RX-7 all models '86 thru '91
- 61040 626 (rear wheel drive) '79 thru '82
- 61041 626 & MX-6 (front wheel drive) '83 thru '92
- 61042 626 '93 thru '01, MX-6/Ford Probe '93 thru '01

MERCEDES-BENZ
- 63012 123 Series Diesel '76 thru '85
- 63015 190 Series 4-cyl gas models, '84 thru '88
- 63020 230, 250 & 280 6 cyl sohc '68 thru '72
- 63025 280 123 Series gas models '77 thru '81
- 63030 350 & 450 all models '71 thru '80
- 63040 C-Class: C230/C240/C280/C320/C350 '01 thru '07

MERCURY
- 64200 Villager & Nissan Quest '93 thru '01
- All other titles, see FORD listing.

MG
- 66010 MGB Roadster & GT Coupe '62 thru '80
- 66015 MG Midget & Austin Healey Sprite Roadster '58 thru '80

MITSUBISHI
- 68020 Cordia, Tredia, Galant, Precis & Mirage '83 thru '93
- 68030 Eclipse, Eagle Talon & Plymouth Laser '90 thru '94
- 68031 Eclipse '95 thru '05, Eagle Talon '95 thru '98
- 68035 Galant '94 thru '03
- 68040 Pick-up '83 thru '96, Montero '83 thru '93

NISSAN
- 72010 300ZX all models incl. Turbo '84 thru '89
- 72011 350Z & Infiniti G35 all models '03 thru '08
- 72015 Altima all models '93 thru '06
- 72020 Maxima all models '85 thru '92
- 72021 Maxima all models '93 thru '01
- 72030 Pick-ups '80 thru '97, Pathfinder '87 thru '95
- 72031 Frontier Pick-up, Xterra, Pathfinder '96 thru '04
- 72032 Frontier & Xterra '05 thru '08
- 72040 Pulsar all models '83 thru '86
- 72050 Sentra all models '82 thru '94
- 72051 Sentra & 200SX all models '95 thru '06
- 72060 Stanza all models '82 thru '90
- 72070 Titan pick-ups '04 thru '09, Armada '05 thru '10

OLDSMOBILE
- 73015 Cutlass '74 thru '88
- For other OLDSMOBILE titles, see BUICK, CHEVROLET or GM listings.

PLYMOUTH
- For PLYMOUTH titles, see DODGE.

PONTIAC
- 79008 Fiero all models '84 thru '88
- 79018 Firebird V8 models except Turbo '70 thru '81
- 79019 Firebird all models '82 thru '92
- 79025 G6 all models '05 thru '09
- 79040 Mid-size Rear-wheel Drive '70 thru '87
- For other PONTIAC titles, see BUICK, CHEVROLET or GM listings.

PORSCHE
- 80020 911 Coupe & Targa models '65 thru '89
- 80025 914 all 4 cyl models '69 thru '76
- 80030 924 all models incl. Turbo '76 thru '82
- 80035 944 all models incl. Turbo '83 thru '89

RENAULT
- Alliance, Encore - see AMC (14020)

SAAB
- 84010 900 including Turbo '79 thru '88

SATURN
- 87010 Saturn all S-series models '91 thru '02
- 87011 Saturn Ion '03 thru '07
- 87020 Saturn all L-series models '00 thru '04
- 87040 Saturn VUE '02 thru '07

SUBARU
- 89002 1100, 1300, 1400 & 1600 '71 thru '79
- 89003 1600 & 1800 2WD & 4WD '80 thru '94
- 89100 Legacy models '90 thru '99
- 89101 Legacy & Forester '00 thru '06

SUZUKI
- 90010 Samurai/Sidekick/Geo Tracker '86 thru '01

TOYOTA
- 92005 Camry all models '83 thru '91
- 92006 Camry all models '92 thru '96
- 92007 Camry/Avalon/Solara/Lexus ES 300 '97 thru '01
- 92008 Toyota Camry, Avalon and Solara & Lexus ES 300/330 all models '02 thru '06
- 92015 Celica Rear Wheel Drive '71 thru '85
- 92020 Celica Front Wheel Drive '86 thru '99
- 92025 Celica Supra all models '79 thru '92
- 92030 Corolla all models '75 thru '79
- 92032 Corolla rear wheel drive models '80 thru '87
- 92035 Corolla front wheel drive models '84 thru '92
- 92036 Corolla & Geo Prizm '93 thru '02
- 92037 Corolla models '03 thru '08
- 92040 Corolla Tercel all models '80 thru '82
- 92045 Corona all models '74 thru '82
- 92050 Cressida all models '78 thru '82
- 92055 Land Cruiser FJ40/45/55 '68 thru '82
- 92056 Land Cruiser FJ60/62/80/FZJ80 '80 thru '96
- 92060 Matrix & Pontiac Vibe '03 thru '08
- 92065 MR2 all models '85 thru '87
- 92070 Pick-up all models '69 thru '78
- 92075 Pick-up all models '79 thru '95
- 92076 Tacoma, 4Runner & T100 '93 thru '04
- 92077 Tacoma all models '05 thru '09
- 92078 Tundra '00 thru '06, Sequoia '01 thru '07
- 92079 4Runner all models '03 thru '09
- 92080 Previa all models '91 thru '95
- 92081 Prius '01 thru '08
- 92082 RAV4 all models '96 thru '05
- 92085 Tercel all models '87 thru '94
- 92090 Sienna all models '98 thru '09
- 92095 Highlander & Lexus RX-330 '99 thru '06

TRIUMPH
- 94007 Spitfire all models '62 thru '81
- 94010 TR7 all models '75 thru '81

VW
- 96008 Beetle & Karmann Ghia '54 thru '79
- 96009 New Beetle '98 thru '05
- 96016 Rabbit, Jetta, Scirocco, & Pick-up gas models '75 thru '92 & Convertible '80 thru '92
- 96017 Golf, GTI & Jetta '93 thru '98, Cabrio '95 thru '02
- 96018 Golf, GTI & Jetta '98 thru '05
- 96020 Rabbit, Jetta, Pick-up diesel '77 thru '84
- 96023 Passat '98 thru '05, Audi A4 '96 thru '01
- 96030 Transporter 1600 all models '68 thru '79
- 96035 Transporter 1700, 1800, 2000 '72 thru '79
- 96040 Type 3 1500 & 1600 '63 thru '73
- 96045 Vanagon air-cooled models '80 thru '83

VOLVO
- 97010 120, 130 Series & 1800 Sports '61 thru '73
- 97015 140 Series all models '66 thru '74
- 97020 240 Series all models '76 thru '93
- 97040 740 & 760 Series all models '82 thru '88

TECHBOOK MANUALS
- 10205 Automotive Computer Codes
- 10206 OBD-II & Electronic Engine Management
- 10210 Automotive Emissions Control Manual
- 10215 Fuel Injection Manual, 1978 thru 1985
- 10220 Fuel Injection Manual, 1986 thru 1999
- 10225 Holley Carburetor Manual
- 10230 Rochester Carburetor Manual
- 10240 Weber/Zenith/Stromberg/SU Carburetor
- 10305 Chevrolet Engine Overhaul Manual
- 10310 Chrysler Engine Overhaul Manual
- 10320 Ford Engine Overhaul Manual
- 10330 GM and Ford Diesel Engine Repair
- 10333 Engine Performance Manual
- 10340 Small Engine Repair Manual
- 10345 Suspension, Steering & Driveline
- 10355 Ford Automatic Transmission Overhaul
- 10360 GM Automatic Transmission Overhaul
- 10405 Automotive Body Repair & Painting
- 10410 Automotive Brake Manual
- 10415 Automotive Detailing Manual
- 10420 Automotive Electrical Manual
- 10425 Automotive Heating & Air Conditioning
- 10430 Automotive Reference Dictionary
- 10435 Automotive Tools Manual
- 10440 Used Car Buying Guide
- 10445 Welding Manual
- 10450 ATV Basics
- 10452 Scooters 50cc to 250cc

SPANISH MANUALS
- 98903 Reparación de Carrocería & Pintura
- 98904 Carburadores para los modelos Holley & Rochester
- 98905 Códigos Automotrices de la Computadora
- 98910 Frenos Automotriz
- 98913 Electricidad Automotriz
- 98915 Inyección de Combustible 1986 al 1999
- 99040 Chevrolet & GMC Camionetas '67 al '87
- 99041 Chevrolet & GMC Camionetas '88 al '98
- 99042 Chevrolet & GMC Camionetas Cerradas '68 al '95
- 99043 Chevrolet/GMC Camionetas '94 thru '04
- 99055 Dodge Caravan/Ply. Voyager '84 al '95
- 99075 Ford Camionetas y Bronco '80 al '94
- 99077 Ford Camionetas Cerradas '69 al '91
- 99088 Ford Modelos de Tamaño Mediano '75 al '86
- 99091 Ford Taurus & Mercury Sable '86 al '95
- 99095 GM Modelos de Tamaño Grande '70 al '90
- 99100 GM Modelos de Tamaño Mediano '70 al '88
- 99106 Jeep Cherokee, Wagoneer & Comanche '84 al '00
- 99110 Nissan Camionetas & Pathfinder '80 al '96
- 99118 Nissan Sentra '82 al '94
- 99125 Toyota Camionetas y 4-Runner '79 al '95

Over 100 Haynes motorcycle manuals also available

Haynes North America, Inc., 861 Lawrence Drive, Newbury Park, CA 91320 • (805) 498-6703 • http://www.haynes.com